Lecture Notes in Computer Science 7419

Commenced Publication in 1973
Founding and Former Series Editors:
Gerhard Goos, Juris Hartmanis, and Jan van Leeuwen

Zhifeng Bao   Yunjun Gao   Yu Gu
Longjiang Guo   Yingshu Li   Jiaheng Lu
Zujie Ren   Chaokun Wang   Xiao Zhang (Eds.)

Web-Age
Information Management

WAIM 2012 International Workshops:
GDMM, IWSN, MDSP, USDM, and XMLDM
Harbin, China, August 18-20, 2012
Proceedings

 Springer

Volume Editors

Zhifeng Bao
National University of Singapore, Singapore, E-mail: baozhife@comp.nus.edu.sg

Yunjun Gao
Zhejiang University, Hangzhou, China, E-mail: gaoyj@zju.edu.cn

Yu Gu
Northeastern University, Shenyang, China, E-mail: guyu@ise.neu.edu.cn

Longjiang Guo
Heilongjiang University, Harbin, China, E-mail: longjiangguo@gmail.com

Yingshu Li
Georgia State University, Atlanta, GA, USA, E-mail: yli@cs.gsu.edu

Jiaheng Lu
Renmin University of China, Beijing, China, E-mail: jiahenglu@ruc.edu.cn

Zujie Ren
Hangzhou Dianzi University, Hangzhou, China, E-mail: renzju@gmail.com

Chaokun Wang
Tsinghua University, Beijing, China, E-mail: chaokun@tsinghua.edu.cn

Xiao Zhang
Renmin University of China, Beijing, China, E-mail: zhangxiao@ruc.edu.cn

ISSN 0302-9743   e-ISSN 1611-3349
ISBN 978-3-642-33049-0   e-ISBN 978-3-642-33050-6
DOI 10.1007/978-3-642-33050-6

Springer Heidelberg Dordrecht London New York

Library of Congress Control Number: 2012945067

CR Subject Classification (1998): H.3, H.4, H.2.8, H.2.4, C.2.1, E.1, F.2.2

LNCS Sublibrary: SL 3 – Information Systems and Application, incl. Internet/Web and HCI

Typesetting: Camera-ready by author, data conversion by Scientific Publishing Services, Chennai, India

Printed on acid-free paper

Springer is part of Springer Science+Business Media (www.springer.com)

WAIM 2012 Workshop Chairs' Message

Web-Age Information Management (WAIM) is an annual international conference for researchers, practitioners, developers, and users to share and exchange their cutting-edge ideas, results, experiences, techniques, and tools in connection with all aspects of Web data management. The conference invites original research papers on the theory, design, and implementation of Web-based information systems. WAIM 2012, the 13th in the series, was held in Harbin during August 18–20, 2012.

Along with the main conference, WAIM workshops intend to provide the international group of researchers with a forum for the discussion and exchange of research results contributing to the main themes of the WAIM conference. This WAIM 2012 workshop volume contains the papers accepted for the following five workshops that were held in conjunction with WAIM 2012. These five workshops were selected after a public call-for-proposals process, each of which focuses on a specific area that contributes to the main themes of the WAIM conference. The five workshops were:

- The First International Workshop on Graph Data Management and Mining (GDMM 2012)
- The Second International Wireless Sensor Networks Workshop (IWSN 2012)
- The First International Workshop on Massive Data Storage and Processing (MDSP 2012)
- The Third International Workshop on Unstructured Data Management (USDM 2012) •
- The 4th International Workshop on XML Data Management (XMLDM 2012)

All the organizers of the previous WAIM workshops and conferences have made WAIM a valuable trademark and we are proud to pursue their work. We would like to express our thanks and acknowledgement to all the workshop organizers and Program Committee members who contributed to making the workshop program such a success. They put a tremendous amount of effort into soliciting and selecting research papers with a balance of high quality, novelty, and applications. They also followed a vigorous review process. A total of 34 papers were accepted. We are very grateful to the main conference organizers and the local Organizing Committee for their great support and wonderful arrangements.

August 2012

Xiaochun Yang
Hongzhi Wang

Preface of the 4th International Workshop on XML Data Management

It is our great pleasure to welcome you to the proceedings of the 4th International Workshop on XML Data Management (XMLDM 2012).

XML has gained much attention from database and Web researchers who are actively working in one or more of the emerging XML areas. XML data are self-describing and provide a platform-independent means to describe data and, therefore, can transport data from one platform to another. XML documents can be mapped to one more of the existing data models such as relational and object-relational models, and XML views can be produced dynamically from the pre-existing data models. XML queries can be mapped to the queries of the underlying models and can use their optimization features. XML data integration is useful for E-commerce applications such as comparison-shopping, which requires further study in the domain of data-, schema- and query-based integration. XML change management is another important area that has attracted attention in the context of Web warehouses. XML has been in use in upcoming areas such Web services, sensors and biological data management. The Third International Workshop on XML Data Management focused on the convergence of database technology with XML technology, and brought together academics, practitioners, users, and vendors to discuss the use and synergy between these technologies.

XMLDM attracted 16 submissions from Asia, Europe, and Singapore. The Program Committee accepted seven full papers. These papers cover a variety of topics, including XML keyword search, XML concurrency control protocols, indexing, XPath, uncertain XML dataset, classification and so on. We hope that they will serve as a valuable starting point for much brilliant thinking in XML data management.

The paper "Effective Keyword Search with Synonym Rules over XML Document" introduces a novel XML keyword search that can find the semantic information behind user input queries. The authors use synonyms, acronyms, and abbreviations that define the equality between strings. Finally, they have devised a transformation matching-based IL algorithm (TMIL) with synonym rules to improve the effectiveness of SLCA-based keyword search over XML documents.

In the paper "XML Concurrency Control Protocols: A Survey," Shan et al. present an overview of some of the most important XML concurrency control protocols, such as locking-based, timestamp-based, and optimistic XML concurrency control protocols. In addition, a summary and comparison are given for each protocol. While indexing XML documents for research purposes can be a complex task especially when we consider content and structure, the paper "Using Conceptual Scaling for Indexing XML Native Databases" proposes using conceptual scaling-based formal concept analysis for indexing both content and structure. It aims to provide a combined structure while assuring hierarchical levels of data content and structure representation.

The paper "Indexing Compressed XML Documents" consists of studies and analyzes some suitable compressors to improve the indexing compressed XML documents process in order to exploit the compressed data for querying and information retrieval. The authors propose a new indexing process which leads to compressed XML data by re-indexing compressed XML data under an XMill compressor.

In the paper "Path-Based XML Stream Compression with XPath Query Support," Qian et al. present a compression for XML stream technology which divides XML streams into structure and context, and then encodes them respectively. They also present experimental results that demonstrate the effectiveness and efficiency of the methods proposed.

The paper "Uncertain XML Functional Dependencies Based on Tree Tuple Models" studies the functional dependencies and their applications in uncertain XML datasets. In this paper, Lv et al. propose three new kinds of functional dependencies based on tree tuple models for uncertain XML datasets. Finally, they also provide a sound and complete set of inference rules as well as two applications.

In "XML Document Classification Using Closed Frequent Subtree," Wang et al. propose an efficient SVM- and SLVM-based classification approach for XML documents that combines the content with the structure of XML documents to compute the similarity between the categories and documents. The experimental results show that this approach performs better than any other competitor's approach on XML classification.

Making XMLDM 2012 possible was a team effort. First of all, we would like to thank the authors and panelists for providing the content of the program. We would like to express our gratitude to the Program Committee and external reviewers, who worked very hard in reviewing papers and providing suggestions for their improvement. In particular we extend our special thanks to Linlin Zhang for maintaining the XMLDM 2012 website and for his effort in organizing the workshop.

We hope that you will find these proceedings interesting and thought-provoking.

Zhifeng Bao
Jiaheng Lu
Talel Abdessalem

XMLDM 2012 Workshop Organization

Workshop General Chairs

Tok Wang Ling   National University of Singpaore, Singapore
Ge Yu   North-East University, China

Workshop Co-chairs

Talel Abdessalem   Telecom Paristech Institute, France
Zhifeng Bao   National University of Singapore, Singapore
Jiaheng Lu   Renmin University of China, China

Program Committee

Stephane Bressan   National University of Singapore, Singapore
Mongli Lee   National University of Singapore, Singapore
Guoliang Li   Tsinghua University, China
Jianxin Li   Swinburne University of Technology, Australia
Zhanhuai Li   Northwestern Polytechnical University, China
Chengfei Liu   Swiburne University of Technology, Australia
Lu Qin   Chinese University of Hong Kong, Singapore
Hongzhi Wang   Harbin Institute of Technology, China
Junhu Wang   Griffith University, Australia
Wei Wang   University of New South Wales, Australia
Huayu Wu   Institute for Infocomm Research, Singapore
Xiaochun Yang   North-East University, China
Junfeng Zhou   Yanshan University, China
Rui Zhou   Swinburne University of Technology, Australia

It Is Time to Exploit Unstructured Data
Preface to the Third International Workshop on USDM

The management of unstructured data has become a hot topic in academia, industry, and government. Most data today, such as Web data, media data, sensor data, are generated by a mass with free will, or automatically by software or hardware, and lack explicit, predefined schema. In contrast to the existing relational data, these data are called unstructured data. The large volume, high change ratio, implicit and heterogeneous structures of these unstructured data pose great challenges to database researchers and engineers. So far, we are encountering the challenges of universal data models, highly flexible storage organization, metadata management, content understanding and so on. Fortunately, there are some initial and enlightened ideas.

The Third International Workshop on Unstructured Data Management (USDM 2012) aimed at bringing together researchers, developers, and users to discuss and present current technical developments in this area. The first Workshop on Unstructured Data Management (USDM 2010) was held with APWeb 2010 (April 6, 2010, Busan, Korea) and the Second Workshop on Unstructured Data Management (USDM 2011) was held with APWeb 2011 (April 20, 2011, Beijing, China), which provided a successful international forum for the dissemination of research achievements in unstructured data management. This year, we received 24 submissions on diverse topics of interest, and selected nine of them through a rigorous review progress and extensive discussions. These accepted papers handle issues on unstructured data storing, querying, retrieval, analysis, mining, and applications. The Program Committee composed a diverse and exciting program for USDM 2012.

The workshop was a forum for both presenting new research results and discussing practical experiences, which can help shape and solve critical problems in unstructured data management. We believe it provided participants with a chance to gain more knowledge in the field. We had two accepted papers on the storage and indexing of unstructured data, three on image data retrieval and processing, two on text search, and two on data mining of unstructured data.

This workshop was partially supported by the Unstructured Data Management System Projects in the HGJ program of China. We would like to thank all the people for their help in making the workshop successful. We thank the Steering Committee Chairs (Xiaoyong Du, Jianmin Wang, and Tengjiao Wang) and the Steering Committee members (Dianfu Ma and Yueting Zhuang) for their suggestions and important instructions. We would like to thank all PC members, especially Zhenying He, Jinchuan Chen, and YueguoChen. Finally, we would like to thank all the speakers and presenters at the workshop, and all the participants at the workshop, for their engaged and fruitful contributions.

Xiao Zhang
Chaokun Wang
Jun Gao

Organization

Program Committee Co-chairs

Xiao Zhang	Renmin University of China
Chaokun Wang	Tsinghua University
Jun Gao	Peking University

Program Committee Members

Jinchuan Chen	Renmin University of China
Yueguo Chen	Renmin University of China
Zhenying He	Fudan University
Jun Gao	Peking University
Chaokun Wang	Tsinghua University
Xiao Zhang	Renmin University of China

Steering Committee Co-chairs

Xiaoyong Du	Renmin University of China
Jianmin Wang	Tsinghua University
Tengjiao Wang	Peking University

Steering Committee Members

Xiaoyong Du	Renmin University of China
Dianfu Ma	Beihang University
Jianmin Wang	Tsinghua University
Tengjiao Wang	Peking University
Yueting Zhuang	Zhejiang University

Preface to the 1st International Workshop on MDSP

On behalf of the Program Chairs for MDSP 2012, consisting of two General Co-chairs and two Program Co-chairs, we are pleased to present you with this volume. It contains the papers accepted for presentation in the workshop program of the 13th International Conference on Web-Age Information Management held in Harbin, China, during August 18–20, 2012.

This is the First International Conference on Massive Data Storage and Processing (MDSP). Twenty papers were submitted to the MDSP program, from which eight were accepted for presentation and inclusion in the conference proceedings. An acceptance rate of 40% makes MDSP one of the most selective workshops of WAIM 2012.

We would like to thank all the authors of submitted papers for choosing MDSP 2012 for the presentation of their research results. Owing to the high quality of the submitted papers, selecting the eight papers for the main conference was a very difficult task. We are deeply indebted to the four Program Chairs and 16 Program Committee members for their conscientious and impartial judgment and for the time and effort they contributed in preparation of this year's conference. All Area Chairs and reviewers are listed on the following pages.

The organizers of the conference are very happy with the response to our call for papers, noticing the interest of the data storage and processing community in this field. The workshop is composed of eight papers selected for presentation, covering a wide range of topics and showing interesting experiences. A brief summary of all the contributions, classified in three main areas, is presented below.

- PTL: Partitioned Logging for Database Storage on Flash Solid State Drives by Jun Yang and Qiong Luo from Hong Kong University of Science and Technology. The authors describe a storage scheme for databases on flash solid state drives.
- Adaptation Mechanism of iSCSI Protocol for NAS Storage Solution in Wireless Environments by Shamim Ripon and Sung Park from East West University. This paper presents an architecture to adapt iSCSI protocols with traditional network attached-storage cluster systems with error recovery methods.
- Band Selection for Hyperspectral Imagery with PCA-MIG by Kitti Koonsanit and Chuleerat Jaruskulchai from Kasetart University, Thailand. In this paper, an integrated PCA, maxima–minima functional method and information gain is proposed for hyperspectral band selection.
- NestedCube: Toward Online Analytical Processing on Information-Enhanced Multidimensional Networks by Jing Zhang, Xiaoguang Hong, and Qingzhong Li from Shandong University. This paper presents Nested Cube, a new

data warehousing model, which can support OLAP queries on information-enhanced multidimensional networks.

- MRFM:An Efficient Approach to Spatial Join Aggregates by Yi Liu, Luo Chen, Ning Jing, and Wei Xiong from University of Defense Technology. In this paper, the authors study the problem of answering spatial join aggregate queries under the MapReduce framework.
- A Distributed Inverted Indexing Scheme for Large-scale RDF Data by Xu Li and Xin Wang from. This paper presents a distributed inverted indexing scheme for large-scale RDF data. A scalable inverted index is built using the underlying data structure of Cassandra, which is a distributed key-value storage system.
- MSMapper: An Adaptive Split Assignment Scheme for MapReduce by Wei Pan, Zhanhuai Li, Qun Chen, Shanglian Peng, Suo Bo, and Jiang Xu from Northwestern Polytechnical University. This paper introduces the MSMapper (Multi-Split Mapper), a modified self-tuning mapper in which multiple splits can be assigned to one mapper.
- Driving Environment Reconstruction and Analysis Systems on Multi-sensor Networks by Chunyu Zhang, Yong Su, Jiyang Chen, and Wen Wang from the Research Institute of Highway. The authors construct a driving environment reconstruction and analysis system based on multi-sensors network onboard and some functional subsystems.

We would like to thank everyone who helped us. We greatly appreciate the advice and support by the WAIM 2012 General Co-chairs, Jianzhong Li (Harbin Institute of Technology, China) and Qing Li (City University of Hong Kong, China), Program Co-chairs, Hong Gao (Harbin Institute of Technology, China) and Local Organization Chair, Jizhou Luo (Harbin Institute of Technology, China), Workshops Chairs, Xiaochun Yang (Northeast University, China) and Hongzhi Wang (Harbin Institute of Technology, China).

Weisong Shi
Yunjun Gao
Weiping Wan
Zujie Ren

Organization

General Co-chairs

Weisong Shi   Wayne State University, USA
Yunjun Gao   Zhejiang University, China

Program Co-chairs

Weiping Wang   Chinese Academy of Sciences, China
Zujie Ren   Hangzhou Dianzi University, China

Program Committee

Guoray Cai	Pennsylvania State University, USA
Yong Woo LEE	University of Seoul, Korea
Hung Keng Pung	National University of Singapore
Xiaofei Liao	Huazhong University of Science and Technology, China
Yijun Bei	Zhejiang University, China
Yi Zhuang	Zhejiang GongShang University, China
Tao Jiang	Jiaxing University, China
Weiwei Sun	Fudan University, China
Xiaokui Xiao	Nanyang Technological University, Singapore
Yimin Lin	Singapore Management University, Singapore
Bin Yao	Shanghai Jiaotong University, China
Shaojie Qiao	Southwest Jiaotong University, China
Congfeng Jiang	Hangzhou Dianzi University, China
Jilin Zhang	Hangzhou Dianzi University, China
Qi Qiang	Alibaba Corp., China
Yongjian Ren	Infocore Corp., China

Preface to the 2nd International Workshop on IWSN

Over the past decade, signficant advances in wireless communication and computing technologies have led to the proliferation of reliable and ubiquitous infrastructure and infrastructureless wireless sensor networks all over the world, as well as a diverse range of new applications, such as the surveillance and protection of critical infrastructures and environment monitoring. Wireless sensor networks collect sensing measurements or detect special events, perform node-level processing, and export the combined data from their sensing nodes to the outside world. Sensing, processing, and communication are three key elements whose combination in one small device is instrumental to pervasive computing and gives rise to countless applications. These applications have raised new challenges ranging from the theoretical foundations of these systems, algorithms and protocol design, security and privacy to rigorous and systematic design and evaluation methodologies and new architectures for next-generation wireless sensor networks.

The International Wireless Sensor Networks Workshop 2012 (IWSN 2012) provided a forum for researchers and practitioners worldwide to exchange ideas, share new findings, and discuss challenging issues for the current and next-generation wireless sensor networks.

IWSN 2012, co-located with the 13th International Conference on Web-Age Information Management (WAIM 2012), took place in Harbin during August 18, 2012. Each submission was reviewed by at least three Program Committee members. Following a rigorous review process, a total of six papers were selected for presentations at the workshop.

We thank all the authors for submitting their papers to the conference. Finally, many other people contributed to the success of IWSN 2012 directly and indirectly. Even though their names cannot be listed here because of space limitation, we owe them our gratitude.

August 2012

Yingshu Li
Jinbao Li
Longjiang Guo

Organization

Workshop Committee on International Wireless Sensor Networks Workshop (IWSN)

General Co-chairs

Yingshu Li   Georgia State University, USA
Jinbao Li   Heilongjiang University, China

Program Co-chair

Longjiang Guo   Heilongjiang University, China

Program Committee Members

Arif Selcuk Uluagac   Georgia Institute of Technology, USA
Donghyun Kim   North Carolina Central University, USA
Xiaoming Wang   Shaanxi Normal University, China
Fei Li   George Mason University, USA
Wenwei Li   Hunan University, China
Feng Wang   Arizona State University, USA
Kai Xing   University of Science and Technology of China,
                            China
Yoora Kim   Ohio State University, USA
Chuanhe Huang   Wuhan University, China
Yan Yang   Heilongjiang University, China
Zenghua zhao   Tianjing, University, China
Chunyu Ai   Troy University, USA
Juan Luo   Hunan University, China
Jinghua Zhu   Heilongjiang University, China
Guilin Li   Xiamen University, China
Lei Yu   Clemson University, USA
Hongzi Zhu   Shanghai Jiao Tong University, China
Jing He   Kennesaw State University, USA
Meirui Ren   Heilongjiang University, China
Shouling Ji   Georgia State University, USA
Desheng Zhang   University of Minnesota, USA

GDMM 2012 Workshop Organizers' Message

Graph data have become a powerful tool for representing and understanding objects and their relationships. With the rapid growth of emerging applications like social network analysis, semantic Web analysis, bio-information network analysis and so on, there is an urgent need to support high-performance query processing and mining ability for various graph data structures. Current database researchers have been actively contributing to pressing problems on graph data management including storage for graph data, graph query processing, similarity measure and search, graph analysis and mining, graph query languages proposals, distributed graph data management, compressing large graph data, prototype systems for managing graph data, graph visualization, and browsing.

The First International Workshop on Graph Data Management and Mining (GDMM 2012) was held in August 2012 in Harbin, China, in conjunction with the 13th International Conference on Web-Age Information Management (WAIM 2012). The overall goal of the GDMM workshop is to bring researchers from different fields together, to exchange research ideas and results, share insights about how to provide efficient graph data management and mining techniques, and to understand the research challenges and solutions of this area.

The workshop attracted eight submissions from China and Japan, covering a broad range of interesting topics in graph data management. All submissions were peer reviewed by at least three Program Committee members to ensure that high-quality papers were selected. The Program Committee selected four papers for inclusion in the workshop proceedings (acceptance rate 50%). The accepted papers span exciting topics from graph extraction to graph data compressing, and query processing.

The Program Committee of the workshop consisted of 11 experienced researchers and experts in the area of data management. The workshop would not been successful without the help of many people and organizations. Firstly, we would like to acknowledge the valuable contribution of all the Program Committee members during the peer-review process. Secondly, we would also like to thank the WAIM 2012 workshop chairs for their great support in ensuring the success of GDMM 2012.

July 2012

Yu Gu
Sai Wu
Dawei Jiang

International Workshop on Graph Data Management and Mining (GDMM 2012)

Workshop Co-chairs

Yu Gu   Northeastern University, China
Sai Wu   Zhejiang University, China
Dawei Jiang   National University of Singapore, Singapore

Program Committee

Guoliang Li   Tsinghua University, China
Dongxiang Zhang   National University of Singapore, Singapore
Ke Chen   Zhejiang University, China
Zhenjie Zhang   ADSC, UIUC, Singapore
Hongzhi Wang   Harbin Institute of Technology, China
Quanqing Xu   National ICT Australia (NICTA), Australia
Wei Wu   Institute for Infocomm Research, Singapore
Yuting Lin   Google, USA
Jeffrey Xu Yu   The Chinese University of Hong Kong,
                             Hong Kong, SAR China
TieZheng Nie   Northeastern University, China
Yueguo Chen   Renmin University of China, China

Table of Contents

The First International Workshop on Massive Data Storage and Processing (MSDP 2012)

The Third International Workshop on Unstructured Data Management (USDM 2012)

The Forth International Workshop on XML Data Management (XMLDM 2012)

Algebra for Parallel XQuery Processing*

Haixu Miao, Tiezheng Nie, Dejun Yue, Tiancheng Zhang, and Jinshen Liu

School of Information Science and Engineering, Northeastern University
Shenyang 110819, P.R. China
nietiezheng@ise.neu.edu.cn

Abstract. As XML becomes the standard of data presentation and information exchange, how to efficiently query information from XML documents becomes a hot topic. However, for larger XML documents and complicated XQueries, the performance of query processing which executes in a single node can seldom meet the needs of users. In this paper, algebra PPXA (Pure Parallel XQuery Algebra) is proposed to support parallel processing for XQuery statements. Based on the Algebra, a strategy for query plan decomposition is proposed for complex path queries and Twig queries. Then, we propose three optimization algorithms based on PPXA. The logical parallel execution plan is optimized by rules on operators, which reduce the local query execution costs. We implement the algebra and the query decomposition strategy in a native XML database system PureXBase. The experimental results show that it supports the XQuery parallel query processing effectively, and can significantly improve the efficiency of query processing.

Keywords: XML, algebra, parallel, XQuery, query processing.

1 Introduction

With the rapid development of Web technology and its applications, as an important standard of data presentation and exchange, XML has been widely used in many applications. In the vast amounts of XML data, from which efficient retrieval information to users' interest becomes a hot topic [1, 2]. Currently existing XQuery engines are mainly designed for a single processing node environment. For query processing on large XML documents, it will be limited by memory and performance constraints. To improve the performance, processing XQuery with parallel methods is a very efficient way.

The challenges of parallel XQuery processing include decompose query plans, transforming query into single-path queries, and combining the single-path results. In this paper, based on characteristics of XQuery queries, we design algebra PPXA (Pure Parallel XQuery Algebra) for pure XML database system. PPXA supports the

---

* Project supported by the National Natural Science Foundation of China (Grant No. 61003060)
The Fundamental Research Funds for the Central Universities (Grant No. N110404010).
The 863 High Technology Foundation of China under Grant No. 2012AA010704.

Z. Bao et al. (Eds.): WAIM 2012 Workshops, LNCS 7419, pp. 1–10, 2012.

decomposition of XQuery execution plan and integrating distributed query results. The main contribution of this paper is as follows:

(1) Pattern structure and Constructor structure is proposed to represent the execution plan of decomposition.
(2) XML algebra PPXA is designed for native XML database system, and it consists of 9 operators, including *Position* operator, *Selection* operator, *And* operator etc.
(3) We propose the optimization rules for the logical parallel execution plan of XQuery.
(4) We implements the algebra system in PureXBase, and experimentally verify that PPXA can effectively address the description of query plan decomposition, moreover, its optimization methods can significantly improve the efficiency of parallel XQuery processing.

The rest of this paper is organized as follows: Section2 will introduce the related works. Section3 presents the steps of parallel query processing of XQuery. In Section 4, we present XML algebra PPXA. In Section 5, we propose the optimization rules for logical execution plan expressed by PPXA. Section 6 shows the experimental results. Section 7 concludes the work of this paper.

2   Related Work

XQuery is an XML query language which exacts data from XML documents. It originates in XML query language Quilt[3], and takes XPath2.0 as its subset. XQuery is a form of functional query language, each query contain one or more query expressions, which could be freely combined with other expressions to generate new expressions, among which FLWOR expressions[4] is most commonly used and most effective. XQuery has become an industry standard, and has been supported and universally recognized by famous firms like IBM, Oracle, Microsoft etc.

Many researchers focus on the study of XML query algebra. TAX[5] introduced the notion of pattern tree, which gives pattern information of queries, and witness tree, which gives the instance sections satisfied the model structure. The core structure of Xtasy[6] is path operator and return operator. Firstly, it use path operator to filter the document, and then control the returned results using return operator. Operators defined in XAL[7] are based on relational algebra, including 3 operators: extraction operation; meta operator and constructor operator. QrientXA[8] proposed the notion of node binding and sequence binding, strong binding and weak binding. It leverages tree model structure to filter documents, and use constructor operator to generate results. The above XML algebras are designed for single-node environment, and their efficiency may be limited by memory or other aspects limitations.

In recent years, high-performance parallel query processing technologies for XML data are widely concerned. How to efficient query information from massive XML data gradually becomes one of the hotspots in databases. Despite some XML data parallel query processing technology has been proposed, to my knowledge, parallel processing algorithms for representative complex path query and Twig query are seldom studied so far.

3   Steps for Parallel XQuery Execution

The execution of Parallel query processing for XQuery usually includes three steps. The first step is the partition of XML documents, which makes XML documents distributed stored in multiple nodes. Then, the query statement should be send to each storage node by a master node to be executed independently. At last, the results of each node will be merged together by the master node.

Due to the partition of XML documents, the query statement may involve multiple paths. Moreover, the corresponding information in the document may be distributed in different storage nodes. So, if the storage nodes directly execute the original query, the matched results would not be accessed. Therefore, we need proper decompose the origin XQuery into single-path queries and generate execution plans for each storage node. And we also need merge all results in each single-path query. In this paper, we use the XML document partition strategy proposed in [9], which divides the nodes of XML documents into three types: duplicate node DN, intermediary node IN and trivial node TN. Their properties are shown in following:

(1) The father node of DN is DN or NULL, the father node of IN is DN or NULL, and the father node of TN is IN or TN;

(2) DN will be stored in all storage nodes, while IN and TN will be stored in a single storage node;

(3) $\{IN\} \cap \{DN\} = \Phi, \{IN\} \cap \{TN\} = \Phi$ and $\{TN\} \cap \{DN\} = \Phi$.

IN set determines the partition results in an XML document. For example, an XML document tree is shown in Fig.1. According to the above method, XML document tree is divided into three sub-fragments. The results are shown in Fig.2, intermediary nodes are {b1, c2, c3, d2, b3, c5, d4, b5}.

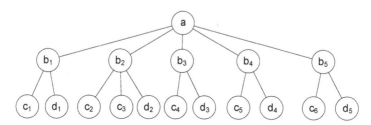

Fig. 1. XML document tree before partition

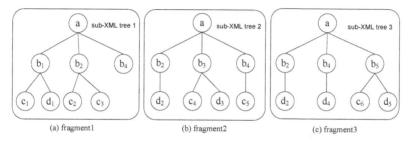

Fig. 2. Sub-fragments of the partition

4   Design of PPXA

XQuery statements are specifically for queries in XML documents. Compared with relational database, XML document has no fixed pattern and distinct table structure. The algebra system designed in this paper, taking XQuery statements as the key point and referencing the existing query algebras. We design the Pattern structure and Constructor structure to show the new query plan of decomposition. Pattern structure is used to filter nodes, while Constructor structure is used to describe how to return results. We also design operators based on these structures in this section.

4.1   Structures of Query Plan

4.1.1   Pattern Structure

XML queries are only interested in nodes which satisfied the user-specified conditions, we use Pattern structure to represent these conditions. Each bind variable is equal to a table, and in the queries after decomposition, each bind variable will be corresponding to a Pattern structure. In XQuery, the filter conditions for each bind variable are mainly some path information. Analyzing the characteristics of XML documents after partition, we can know that single-path can be executed directly in storage node. Therefore, when parsing the query statements, we could decompose the original XQuery according single path. The pattern structure is composed of many single paths, which are connected by And operator or Or operator and make up the entire filter conditions for a bind variable. For example, for the following query statement:

FOR $i in collection("auction.dbxml") /site/closed_auctions/closed_auction
WHERE ($i/price >= 40 and $i/type = Regular) or $i/buyer/@person = "person0"
RETURN $i

The Pattern structure of bind variable i is shown in Fig.3, which contains all restrictions in the WHERE statements for i.

In pattern constructor, every leaf node separately represents a single-path, which is equal to selection operation in relational database. When decomposing queries, each single-path will be extracted out and will be combined with limitations of the same bind variable through *And* or *Or* operator to form a complete filter conditions.

In addition, in XQuery statements, for binding binds variable to one node, while Let binding binds variable to a node sequence, that is, all nodes satisfied one condition will be returned as a result. Different binding ways will have a significant impact on the results. Therefore, in mark of each variable, we should pay attention to distinguish node binding and sequence binding. After the qualification of Let binding, the nodes satisfied the same conditions need to be clustered.

4.1.2   Constructor Structure

In XQuery statements, the results are required to be outputted according to a particular way. In order to present the construction operation for results, we introduce Constructor structure, which mainly responds to the RETURN part in XQuery statements. Constructor structure contains two parts: one is data, and the other is the output format for data. For example, in the RETURN part of an XQuery statement:

RETURN <item person="{$p/name/text ()}"> {count ($a)} </item>

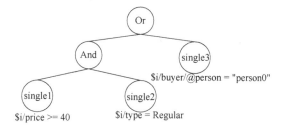

Fig. 3. Pattern structure of binding variable *i*

The data part of outcome, consist of two contents: $p/name/text () and count ($a). Data part is usually the descendant content of bind variable. Output format part is a new tag, constructed by node with an attribute. The attribute value is the content of the first data item, and the text part of this node is the content of the second data item.

(1) Data part
The data part is usually the content of bind variable itself or its descendant. In parallel environment, we usually take a way of set-at-a-time. When decomposing the queries, all data part in the outcome will be extracted, and then processed centrally.

(2) Output format part
The output format part is the structure information of node. The results are usually outputted in XML format, so the output format contains there parts:

(1) Node name, i.e., the tag name of the node. If the node name is null, we could output the text content of this node as itself;
(2) Node attributes. Node attributes contain attribute name and attribute content, which is represent by its serial number in result sets. When outputting the results, we may need use aggregate function on result content;
(3) Node content. Node content could be directly represented by its serial number in result sets. Sometime, it could be the cluster of results or the output construct of another nest, through which we could represent the nested structure of XML.

4.2   PPXA Operators

According to the study about query decomposition before and referencing the existed relational algebra, we design the operator of our query algebra, PPXA. Our system takes node level granularity in storage, so the input and output of operator both are node sets in documents. Each node is represented by its Dewey code in the system. There are limitations between bind variables, which are represented by foreign keys.

Let N denote the universal set of all nodes in document, Ph (n) denotes node n meeting the requirements of path Ph, Ph (n_s, n) denotes node ns and node n meeting path relation Ph, FK (n) denotes the foreign key of node n, FK_j (n) denotes value of bind variable j in the foreign keys of node n. FK(n_s, n) denotes that the foreign keys of node ns contain the corresponding bind variables of node n, $L(n)$ denotes the nodes whose location relation with node n is L, in which $ns \in N$ and $n \in N$.

(1) Position operator

$$L_{ph}(N_s) = \{n \mid ns \in N_s \wedge n \in N \wedge Ph(ns, n), N_s \subseteq N\} \tag{1}$$

N_s denotes the candidate nodes set, and Ph denotes the path condition. The output result is all nodes which meet the path condition between nodes in N and N_s.

(2) Selection operator

$$\sigma_{ph}(N_s) = \{n \mid n \in N_s \wedge Ph(n), N_s \subseteq N\} \tag{2}$$

N_s denotes the candidate nodes set, and Ph denotes the path condition. The output result is the subset of nodes which meet the path condition in N_s.

(3) Or operator

$$\cup(N_1, N_2) = \{n \mid n \in N_1 \vee n \in N_2, N_1 \subseteq N, N_2 \subseteq N\} \tag{3}$$

N_1 and N_2 denote candidate nodes sets. The output result is the union of two sets.

(4) And operator

$$\cap(N_1, N_2) = \{n \mid n \in N_1 \wedge n \in N_2, N_1 \subseteq N, N_2 \subseteq N\} \tag{4}$$

N_1 and N_2 denote the candidate nodes sets. The output result is the join of two sets.

(5) Branch connection operator

$$TJ_{L_1, L_2}(N_1, N_2) = \{n_2 \mid n_1 \in N_1 \wedge n_2 \in N_2 \wedge L_1(n_1) = L_2(n_2), N_1 \subseteq N, N_2 \subseteq N\} \tag{5}$$

N_1 and N_2 denote the candidate nodes sets, and L_1 and L_2 denote the relative position between nodes. The output result is the subset of N_2; moreover, nodes in this subset and nodes in N_1 satisfy that they are the same node in some particular ancestor or descendant positions.

(6) Semi-join operator

$$\propto_{Ph_1 \theta Ph_2} = \{n_2 \mid n_1 \in N_1 \wedge n_1' \in N \wedge Ph_1(n_1, n_1') \wedge n_2 \in N_2 \wedge n_2 \in N \wedge Ph_2(n_2, n_2')$$
$$\wedge(n_1' \theta n_2'), N_1 \subseteq N, N_2 \subseteq N\} \tag{6}$$

N_1 and N_2 denote the candidate nodes sets, and Ph_1 and Ph_2 denote path conditions. The output result is the subset of N_2; moreover, nodes in this subset meeting path Ph_2 and nodes in N_1 meeting path Ph_1 satisfy θ.

(7) Clustering operator

$$CL(N_s) = \{N_c \mid N_c \subseteq N_s \wedge (\forall n_1, n_2 \in N_c, FK(N_1) = FK(N_2)), N_s \subseteq N\} \tag{7}$$

N_s denote the candidate nodes set. The output result is the result set after clustering nodes in N_s according to foreign key conditions.

(8) Match operator

$$\times(N_1, N_2, ..., N_K) = \{(n_1, n_2, ..., n_k) \mid n_1 \in N_1 \wedge n_2 \in N_2 \wedge ... n_k \in N_k \wedge (\forall 1 \leq i, j \leq k, FK(n_i, n_j) \Rightarrow (FK_j(n_i) = n_j))\} \tag{8}$$

$N_1, N_2, ..., N_k$ denote candidate nodes sets. The output result is connecting different bind variables according to the foreign key conditions between bind variables.

(9) Construct operator

$$CP(N_s) = \{P(n_1, n_2, ..., n_k) \mid (n_1, n_2, ..., n_k) \in N_s\} \tag{9}$$

P is the output format, and N_s is the result after node matching. N_s will be outputted in P format as the final result.

We take a typical XQuery statement as example to illustrate the query algebra structure. The corresponding algebra tree is shown in Fig.4.

Example 1: Q1
FOR p in collection("people.dbxml")/site/people/person
LET a := collection("people.dbxml")/site/closed_auctions/closed_auction
WHERE $a/buyer/@person = $p/@id
WHERE $p/name = "John"
RETURN <item person="{$p/name/text()}">{count($a)}</item>

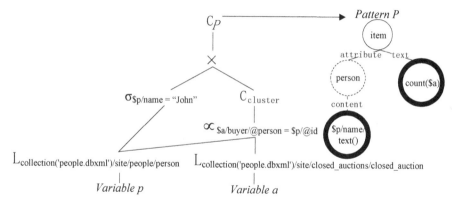

Fig. 4. Query algebra tree for Q1

5   Optimization Rules for PPXA

5.1   Move Down Selection Operator

On optimization, referencing the selection predicates move down strategy in relational database, filter conditions of For bind variables can also be moved down, i.e., the selection operation can be executed ahead of schedule. In this way, we can filter some useless nodes so as to improve the efficiency of query processing. For example, we optimize query Q2 according to this method, moving down the selection condition of variable p and taking the results after filtering as the input of semi-join operation. The results are shown in Fig. 5.

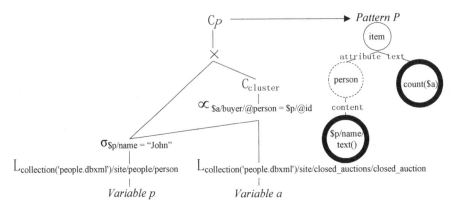

Fig. 5. Optimized Query algebra tree for Q1

5.2   Simplify Operator

Some operators in query plan can be omitted, or can be merged together, which can reduce the generation of intermediate results, dropping the communication cost between processors and improving the efficiency of query execution. For example, when taking the same dataset in two different paths as the input parameters of match operator, and one of the paths as direct input end, the input end can be omitted, e.g., we optimize query Q2 through simplify operator, results are shown in Fig. 6.

Another situation for simplify operator: if there is only one constraint for bind variable, and in the whole returned results, only need return one descendent of the bind variable, moreover, two paths represent same path information, we could transfer the constraint of bind variable to the returned result and therefore reducing the intermediate results and obtaining the output directly.

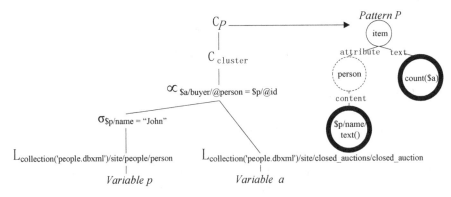

Fig. 6. Simplified Query algebra tree for Q1

5.3   Series Executive Intersection

We perform conversions to And constraints of a bind variable, turning them into operations iteratively. The transformational rule is as follows:

$$\sigma_{Ph_1 \wedge Ph_2}(N_s) = \sigma_{Ph_1}(\sigma_{Ph_2}(N_s)) = \sigma_{Ph_2}(\sigma_{Ph_1}(N_s)) \quad (10)$$

In parallel environment, two branch paths are stored in different nodes. Thus for filter paths in the same sets, we need judge them separately; moreover, we need merge the results. All these operations should be executed in server host. Leveraging (10), we can convert the manner of judgments, reduce the intermediate results and filter some useless nodes; therefore dropping the communication cost and improving the query efficiency in storage nodes, at the mean time, saving the cost of join operation for two result sets in server host.

6   Experimental Result

6.1   Experimental Setting

In the experiments, we used 14 PCs with Window XP of operating system, 1.86GHz of basic frequency, 2GB of memory and 160GB of disk space. The 50M dataset used in our system is generated by XMark generator using factor 1.0. The XQuery statements used in test meet the standard set by the W3C organization. We use two queries in our experiments. Q1 is shown as follow and Q2 is shown in *Example 1*:

FOR $i in collection("auction.dbxml")/site/closed_auctions/closed_auction
WHERE ($i/price >= 40 and $i/type = Regular) or $i/buyer/@person = "person0"
RETURN $i

6.2   Experimental Results

Q1 is a simple query, only concern one bind variable i. Documents in the first dataset take 0.8 as selectivity for the first selection condition. The abscissa denotes the number of processors, and the ordinate denotes the query execution time. The query execution time in single-processor environment and parallel environment are shown in Fig 7(a). We change the selectivity for the first selection condition to 0.2, the execution time are shown in Fig 7(b). With the increase of processor numbers, the times for merging results take more proportion. Even if increasing the storage nodes, query times change very little.

Q2 is a relative complicate query, involving two bind variables p and a. The query results need to be constructed before returning. Documents in the first dataset take 0.8 as selectivity for selection condition of bind variable p. The query execution time in single-processor environment and parallel environment are shown in Fig 7 (c). We change the selectivity for the selection condition of p to 0.2, the execution time are shown in Fig 7(d).

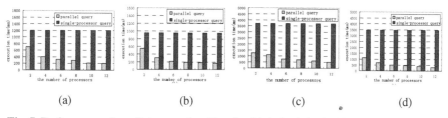

(a)   (b)   (c)   (d)

Fig. 7. Performance of parallel query algorithm for Q1 (selectivity is 0.8 for a and 0.2 for b)

7   Conclusions

In this paper, we firstly introduced the steps for parallel XQuery processing. Then, we proposed Pattern structure and Constructor structure for the decomposition of query plan. Further, we proposed algebra PPXA that consist of a set of operators such as *Selection, Position, And, Match* etc. We also define a set of optimization rules for improving performance of local query processing. Finally, we implement the algebra in a native XML database and verify the effectiveness and efficiency of PPXA.

References

1. Meng, X.: XML Data Management Concepts and Techniques. Tsinghua University Press, Beijing (2009)
2. Feng, J., Qian, Q., Liao, Y., et al.: Survey of Research on Native XML Databases. Application Research of Computers 6, 1–7 (2006)
3. Chamberlin, D., Robie, J., Florescu, D.: Quilt: An XML Query Language for Heterogeneous Data Sources. In: Suciu, D., Vossen, G. (eds.) WebDB 2000. LNCS, vol. 1997, pp. 1–25. Springer, Heidelberg (2001)
4. Wan, C., Liu, X.: XML Database Technology, 2nd edn. Tsinghua University Press, Beijing (2008)
5. Jagadish, H.V., Lakshmanan, L.V.S., Srivastava, D., Thompson, K.: TAX: A Tree Algebra for XML. In: Ghelli, G., Grahne, G. (eds.) DBPL 2001. LNCS, vol. 2397, pp. 149–164. Springer, Heidelberg (2002)
6. Sartiani, C., Albano, A.: Yet Another Query Algebra For XML Data. In: Proc. of IEEE International Database Engineering and Applications Symposium (IDEAS 2002), pp. 106–115 (2002)
7. Frasincar, F., Houben, G., Pau, C.: XAL: An Algebra for XML Query Optimization. In: Proc. of the 13th Australasian Database Conference (ADC 2002), pp. 49–56 (2002)
8. Luo, D., Jiang, Y., Meng, X.: OrientXA: An Effective XQuery Algebra. Journal of Software 15(11), 1648–1660 (2004)
9. Wang, G., Tang, N., Yu, Y., et al.: A Data Placement Strategy for Parallel XML Databases. Journal of Software 17(4), 70–78 (2006)

Graph-Structured Data Compression Based on Frequent Subgraph Contraction[*]

Cong Wang and Hongzhi Wang

Harbin Institute of Technology
wangconghuizi@yeah.net, whongzhi@gmail.com

Abstract. There is much redundant information in graph-structured which has self-describing characteristic, so how to compress graph-structured so as to improve the efficiency of management of data is becoming more and more significant. This paper studies storage oriented graph-structured compression techniques. For the graph given, many subgraph will be generated. A based on graph traversal, the frequently patterns(fp) can be found. With join the fp patterns, a new fp pattern is produced. Followed by this loop until the threshold, the it result in compression results. Analysis and experiments show that the algorithms have high performance.

Keyword: graph-structured data compression, high performance.

1 Introduction

Graph-structured that has self-describing characteristic is similar with XML data. So it has more powerful ability of representing data. For an example social networks, communication networks, web graph.

Query language on tree-structured can also be extended to graph-structured. But one of a disadvantage of Graph-structured is that it has redundancy, as tree-structured. Compression is a practical strategy to solve the problem of redundancy. Compression techniques can save not only storage space but also disk I/O during query processing when graph document is stored on the disk.

This paper focuses on the compression of graph-structure which has self-describing characteristic. An observation of graph is that some patterns exist multiple times in a graph. Such frequently existing patterns are redundancies. Based on this observation, a compression strategy for graph-structured data is presented. The basic idea of this compression strategy is to find frequent existing patterns in graph-structured data and store these patterns separately. In original graph, sub-graphs matching frequent pattern are contracted to stub nodes. Query processing technology on graph-structured XML data can be adapted on compressed data.

---

[*] This paper was partially supported by NGFR 973 grant 2012CB316200 and NSFC grant. 61003046, 6111113089. Doctoral Fund of Ministry of Education of China (No. 20102302120054). the Fundamental Research Funds for the Central Universities (No. HIT.NSRIF.2013064).

Z. Bao et al. (Eds.): WAIM 2012 Workshops, LNCS 7419, pp. 11–18, 2012.

The structure of this paper is as following. Section 2 shows compression algorithm of graph-structured XML data are presented, respectively. Experimental results are presented in Section 3. Section 4 discusses related work and Section 5 draws conclusions.

2   Compression

In this section, compression strategy of graph-structured data is presented. The basic idea of the compression strategy is to find frequent patterns in graph, store these patterns separately and contract subgraphs matching frequent patterns in original graph.

2.1   An Overview of the Compression Strategy

In this subsection, compression strategy is presented. The steps of compression can be summarized as following steps.

1. Frequent patterns(FP for brief) are found in graph. Each FP is given a uniform number. Each node in each FP is given a uniform internal number. The details of FP finding will be discussed in Section 2.2.
2. In graph, each subgraph matching some FP is contracted to a stub node. A stub node n corresponding to subgraph G_n contains following information: the number of corresponding FP, the pointers to the nodes pointed by the outcoming paths of G_n. These pointers are called outcoming pointer. With each outcoming pointer, the interval number of its start node is assigned. For each node u not in G_n but having outcoming edge e pointing to some v node in G_n, the interval number of v in G_n is added to e. The goal of attaching the internal number is to distinguish nodes in the contracted subgraph.
3. Values are classified based on data type and compressed separately. In the corresponding element e with a value v, a pointer to the position of the value is stored. If e is in some contracted node, the internal number of e is assigned to the pointer to value.

2.2   Frequent Pattern Finding Algorithm

In our compression strategy, the major part is to find FPs and the subgraphs matching fps. In this subsection, we will discuss this problem in detail. At first, we give the formal description of FP finding problem in Section. Then in Section 2.2 a naive brute-force algorithm is presented as the base algorithm. In order to accelerate FP finding algorithm, a heuristic algorithm is presented based on the brute-force algorithm. The Definition of FP Finding Problem In this subsection, the formal definition of FP finding problem for graph compression is presented. The object of the problem is to find frequent patterns to make the compression ratio maximal. Here we focus on the compression ratio of structure part. The compression ratio of an graph G can be presented by following formula.

$$ratio = \frac{\sum_{p \in P}((|V_p| \cdot \alpha + |E_p| \cdot \beta) \cdot frequent_p - (|V_p| \cdot \alpha + |E_p| \cdot \beta) - \alpha \cdot frequent_p - \sum_{g \in M_p}}{size_G}$$

$$\log(|V_g|) \cdot (in_g + out_g)$$

Where $P = \{p_1, p_2, cdots, p_k\}$ is the set of selected pattern; V_p and E_p are the sets of vertexes and edges of pattern p respectively; α and β are the size of store a node and an edge respectively; $frequent_p$ is the frequency of pattern p in G; M_p is the set of subgraphs in G matching pattern p; in_g and out_g are the number of incoming edges and outcoming edges of subgraph g; $size_G$ is the storage size of G.

The numerator is the storage size saved by the compression strategy. Intuitively, saved space is the contracted subgraphs. But compression results in additional storage cost, including pattern, stub node and the internal number attached to incoming and outcoming of stub nodes.

The first item $(|V_p| \cdot \alpha + |E_p| \cdot \beta) \cdot frequent_p$ is the size of contracted subgraphs matching pattern p. The second item $|V_p| \cdot \alpha + |E_p| \cdot \beta$ is the space of storing pattern p. Item $\alpha \cdot frequent_p$ is the storage size of stub nodes matching pattern $p \cdot \log(|V_g|) \cdot (in_g + out_g)$is the size of internal numbers attached to incoming edges and outcoming edges of stub node of g.

Since $size_G$ is a constant, maximum compression ratio is to maximum the numerator of Formula 1. Therefore, the problem can be expressed as following optimization problem.

Problem1:

$$\max \sum_{p \in P} ((|V_p| \cdot \alpha + |E_p| \cdot \beta) \cdot frequent_p - (|V_p| \cdot \alpha + |E_p| \cdot \beta) - \alpha \cdot frequent_p - \sum_{g \in M_p} \log(|V_g|) \cdot$$

$$(in_g + out_g))$$

subject to:

$$\forall g_1, g_2 \in \bigcup_{p \in (P)} M_p, g_1 \cap g_2 = \phi$$

$$\forall g \in M_p, g \subseteq G \text{ and } g \text{ matching } p$$

Constraint (1) shows that any two different subgraphs to contract should be overlapping. This point comes from our compression strategy. Constraint(2) shows that each subgraph g in M_p must matching pattern p.

This problem is different from traditional frequent pattern mining problem, because traditional frequent pattern mining problem is only to find frequent pattern without the constraint that any two subgraph belong to frequent patterns should not be overlapping.

Lemma 1. Rroblem1 is not an NP problem.
Theorem 1. Problem1 has no polynomial algorithm.

Brute-Force Algorithm. In this subsection, we present a naive brute-force algorithm for problem1. This algorithm has two phases, the first phase is to generate subgraphs and patterns and the second phase is to determine which pattern is frequent and find subgraphs matching frequent patterns.

The basic idea is to use edges as basic subgraphs and add edges to basic subgraphs iteratively in the first step. Such that all possible subgraphs and patterns will be generated. Then based on these subgraphs and patterns, frequent pattern and corresponding subgraphs are chosen. We will discuss this algorithm in detail.

The process of the first phase is shown in Algorithm .

Algorithm 1. FindFrequentPattern(G)

$T_{edge} = generateEdgeTable(G)$
$T_{current} = T_{edge}$
for $i = 2$ **To** m **do**
  $Filter(T_{current})$
  $T_{temp} = Join(T_{current}, T_{edge})$
  $Classify(T_{temp})$
  $T_{result}[i] = T_{temp}$
  $T_{current} = T_{temp}$

In Algorithm 1, function generateEdgeTable is to generate the edge table containing all edges of G. Function Filter is to filter the subgraphs whose pattern exists only once. Function Join is to join two tables to generate the new subgraphs. In iteration with $i = k$, all subgraphs with k edges are generated. Function Classify is to classify the subgraphs into patterns.

In order to support the process, we use relation with schema($tag_{n1}, id_{n1}, tag_{n2}, id_{n2}, \ldots\ldots,$ $tag_{nk}, id_{nk}, num_{ni}, num_{out}$) to represent subgraphs with k nodes, where tag_{ni} and num_{ni} are the tag and identify of node n_i in graph. Note that a result set $T_{result}[i]$ may contain subgraphs represented multiple schemas because $T_{result}[i]$ contains subgraphs with i edges and such subgraphs may have various number of nodes.

The implementation of function Join is to sort subgraphs with each node and join it with T_{edge}. Here $T_{current}$ is supposed to contain subgraphs with just one schema In order to support the precess, we use relation with schema ($tag_{n1}, id_{n1}, tag_{n2}, id_{n2}, \ldots\ldots, tag_{nk}, id_{nk}$ num_{in}, num_{out}).

In graph, tags and ids of nodes have uniform order. In each tuple, tags in the schema is in this uniform order and nodes with same tag in the order of the uniform order of ids. Keeping such order is to avoid generating isomorphic subgraphs. If it has multiple schemas, with considering the schema with the most number of nodes such schema and the value of other subgraph without node incorresponding position NULL, the algorithm is also feasible. In this algorithm.T_{edge} is suppose to have schema ($tag_{n1}, id_{n1},$ $tag_{n2}, id_{n2}, \ldots\ldots, tag_{nk}, id_{nk}, num_{in}, num_{out}$).

Algorithm 2. $Join(T_{current}, T_{edge})$

  $T_{result} = \emptyset$
  for $i = 1$ **to** k **do**
    $sort\ (T_{current},\ tag_{ni},\ id_{ni})$
    $sort\ (T_{edge},\ tag_1,\ id_{ni})$
    $T_{temp} = MergeJoin\ (\ T_{current},\ T_{edge},\ id_{ni} = id_1)$
    $T_{result} = T_{result}\ \bigcup\ T_{temp}$
    $sort\ (T_{edge},\ tag_1,\ id_1)$
    $T_{temp} = MergeJoin\ (T_{current},\ T_{edge},\ id_{ni} = id_2)$
    $T_{result} = T_{result}\ \bigcup\ T_{temp}$
  $SortResult\ (\ T_{result})$

where SortResult is to sort each tuple in T_{result} based on tag and ids of nodes; MergeJoin(T_A; T_B; p) is to merge join two sorted tables T_A and T_B with prediction p; sort(T; tag; id) is to sort tuples of a table T with *tag* and *id*. Note that in order to accelerate join, T_{edge} may be stored as two copies. one is sorted by tag_1 and id_1 and the other is sorted by tag_2 and id_2. Such that T_{edge} is avoided to be sorted twice in each iteration.

The second phase is to determine which pattern is frequent and which subgraph to contract. To obtain accurate solution, full space search should be applied.

Heuristic Algorithm. When graph is large, brute-force algorithm is not practical because its complexity. In this subsection, a heuristic algorithm is presented with brute-force algorithm as the base.

At first, we give a sketch of heuristic rules. Then we will discuss them in detail. Heuristic rules includes three aspects, including limiting the smallest size of subgraph and the frequency of pattern, eliminating useless patterns during processing and the combination of these two strategies.

Size and Frequency Limitation. The intuition of this rule is to filter the infrequent patterns and too small pattern.

The value of minimal numbers of nodes and edges of FPs are set to be n_{min} and e_{min}, respectively.

The smallest frequency of FP is set to be k_{min}. Based on kmin the values of the numbers of nodes and edges of FPs belonging to one FP have upper bound, $n_{max} = \dfrac{|V_G|}{k_{min}}$ and $e_{max} = \dfrac{|E_G|}{k_{min}}$. With nmin and emin, the upper bound of frequency is $k_{max} = \min imal \left\{ \dfrac{|V_G|}{n_{min}}, \dfrac{|E_G|}{e_{min}} \right\}$.

The intuition of this rule is that infrequent and too small patterns will not make the value of Formula 2 large. The filtering is performing in the end of each iteration.

Useless Pattern Filter. During the heuristic algrithm, a function f is used to judge the useful property of a pattern. f is defined as following.

$$f(p) = \max(|V_p| \cdot \alpha + |E_p| \cdot \beta) \cdot |M_p'| - \alpha \cdot |M_p'| - \sum_{g \in M_p} \log(|V_p|) \cdot (in_g + out_g)$$

$$\text{subject to} : \forall g \in M_p , g \text{ has pattern } p$$

$$M_p' \subseteq M_p$$

$$\forall g_1, g_2 \in M_p' , g_1 \cap g_2 = \phi$$

f represents the maximum possible contribution of a graph set M_p to the whole object of Problem1. This heuristic rule can be described as following:

If *patterns A and B with A $\subseteq$ B and f(A) < f(B)* is met. All subgraphs matching A can be ignored. It can be proved that given the set M_p, computing f(Mp) is an NP-Hard problem.

Theorem 3. Given a graph set S, computing f(S) is an NP-Hard problem.

Since during processing, we should computer function f frequently. For this problem, we can use greedy algorithm to estimate the value of f. The pseudo algorithm is shown in Algorithm 3, where H is a hash set to storing counted edges.

Algorithm 3. Computing f(M_p)

sort graphs in M_p with values of $(|V_p| \cdot \alpha + |E_p| \cdot \beta) - 1 - \log(|V_p|) \cdot (in_g + out_g)$

Sorted graphs are in Sequence T

$H = \emptyset$

$Result = 0$

$M'_p = NULL$

for $i = 1$ to T **do**

    if any edge in T[i] is not in H **then**

        $Result=Result+ (|V_p| \cdot \alpha + |E_p| \cdot \beta) - 1 - \log(|V_p|) \cdot (in_g + out_g)$

        insert all edges of $T[i]$ to H

        add $T[i]$ to M'p

Return $Result$

The Combination of Two Rules. The combination of two rules is that a parameter $value_{accept}$ is given. If $f(p)$ value of a pattern p is larger than $value_{accept}$. It shows that the contribution of pattern p is large enough to be considered as frequentpattern. When a pattern p is considered as frequent pattern, all the subgraphs matching it are selected to be contracted. And the values of n_{max} and e_{max} are reset as

$$n_{max} = \frac{|V_G| - \sum_{foreachg \in M_p} |V_g|}{k} \text{ and } e_{max} = \frac{|E_G| - \sum_{foreachg \in M_p} (|E_g| + in_g + out_g)}{k}$$

where M'_p is the set of subgraphs selected with Algorithm 3. The modification of n_{max} and e_{max} shows that when some subgraphs have been chosen to be contracted, these subgraphs and their incoming/coming paths will not be considered.

Determine of Subgraphs to Be Contracted. When the subgraphs are generated, the selection of subgraph to contracted is performed with greedy algorithm on the $f()$ value of each patterns. When a pattern is selected, all its subgraphs are selected to be contracted and other subgraphs overlapping with selected subgraphs are eliminated.

3   Experimental Results

We tested the Heuristic Algorithm (with the threshold = 1) on different datasets to compare their performance. "P2p-Gnutella08" is Gnutella peer to peer network from August 8 2002. "cA-GrQc" is a collaboration network of Arxiv General Relativity. "wiki-Vote" is a Wikipedia adminship vote network till January 2008. "cit-HepTh" is the paper citation network of Arxiv High Energy Physics Theory category. "email-Enron" is email communication network from Enron. "email-EuAll" is email network of a large European Research Institution. "soc-Epinion1" is a who-trust-whom online social network of a general consumer review site Epinions.com. "Amazon0302" is Amazon product co-purchaisng network from March 2 2003. "roadNet-PA" is road network of Pennsylvania. "roadNet-TX" is road network of Texas. All the details are shown in the table.

Table 1. Experiments on varity datasets

Dataset	Nodes/Edges	Compression ratio (after compression/before compression)(10^{-3})	Time(s)
p2p-Gnutella08	6301/20777	23.264	0.091
cA-GrQc	5242/28980	15.281	0.099
wiki-Vote	7115/103689	4.947	0.128
cit-HepTh	27770/352807	1.146	0.202
email-Enron	36692/367662	1.363	0.208
email-EuAll	265214/420045	1.142	0.211
soc-Epinion1	75879/508837	0.979	0.246
Amazon0302	262111/1234877	0.378	0.639
roadNet-PA	1088092/3083796	0.145	1.923
roadNet-TX	1379917/3843320	0.114	2.526

As we can see in the table above, when the size of data gets larger, the time is increasing basically. And the results of compression are distinguished. Although there must be some deviations in the test. On the whole, the algorithm has shown great potential in compression of large graph-structured data.

4   Related Work

When it comes to storage oriented graph-structured compression techniques, what we should do is to reduce the space of data storaging and bandwidth of data exchanging. Because Graph-structured that has self-describing characteristic is similar with XML data. And some research about XML have been done.

In literature[1], XML data has been put into distinct containers in which data is compressed. It is based on the dictionary compression: at first, scan the XML data to establish a semi-dynamic dictionary for frequent symbol string, and the second time of scanning is to replace symbol string. But data and digit is stay in binary-coded. In literature[2], a idea that save memory overhead of structure compression is put forward. It says that the idea can save about 18% memory at the same compression ratio. In literature[3], lempel-ziv—replace redundant that emergent succeed with XML data that emergent at the first time—is used to compress structure redundancy. In literature[4], based on probability model of XML structure, the method is to do some serialization syntax encoding with XML document With a Prototype system it named Exalt, it just need to scan the document once. In literature[5], a method is raised to cut off the memory space of loading XML document. It says to delete duplicated subtree when it is loading XML document, which proved to be an approximation algorithm of linear time.

In literature [6], after a variety of information redundancy analyzed, a method of dictionary compression is achieved with the structure named DDOM. In literature [7], it has given a evaluate method considering compression ratio and compression speed. And it contrast the efficiency of XMill, XMLPPM, XML_zip, XGrind and so on. In literature [8-10], several compression questions about XML index structure are discussed.

But all of these methods are just keen on compress tree structure effectively, not on graph structure, and that is what we should try to do.

5   Conclusion

Many challenges arise in graph structure compression. This paper studies storage oriented graph-structured compression techniques. For the graph given, many subgraph will be generated. A based on graph traversal, the frequently patterns (fp) can be found. With join the fp patterns,a new fp pattern is produced. Followed by this loop until the threshold, the it result in compression results. Analysis and experiments show that the algorithms have high performance. Future work contain how to design effective structure for quering and more efficient global algorithm.

References

[1] Skibinski, P., Grabowski, S., Swacha, J.: Effective asymmetric XML compression (J/OL). Software- Practice & Experience,
     http://www.interscience.wiley.com/journal/98517568/issue
[2] Kinno, A., Yukitomo, H., Nakayama, T., Takeshita, A.: Recursive Application of Structural Templates to Efficiently Compress Parsed XML. In: Lowe, D.G., Gaedke, M. (eds.) ICWE 2005. LNCS, vol. 3579, pp. 296–301. Springer, Heidelberg (2005)
[3] Adiego, J., Navarro, G., de la Fuente, P.: Lempel- Ziv compression of structured text. In: Proceedings of the 2004 IEEE Data Compression Conference, Snow - bird, UT, USA, pp. 112–121. IEEE Computer Society, New York (2004)
[4] Toman, V.: Compression of XML data. Charles University, Prague (2003)
[5] Busatto, G., Lohrey, M., Maneth, S.: Efficient Memory Representation of XML Documents. In: Bierman, G., Koch, C. (eds.) DBPL 2005. LNCS, vol. 3774, pp. 199–216. Springer, Heidelberg (2005)
[6] Neumüller, M.: Compression of XML data. University of Strathclyde, Glasgow (2001)
[7] Augeri, C.J., Mullins, B.E., Baird, L.C., et al.: An analysis of XML compression efficiency. In: Proceedings of the 2007 Workshop on Experimental Computer Science, San Diego, California, USA, pp. 73–84. ACM Press, New York (2007)
[8] Bao, X.Y., Tang, S.W., Wu, L., et al.: ArithRegion—an index structure on compressed XML data. Acta Scientiarum Naturalium Universitatis Pekinensis 42(1), 103–109 (2006)
[9] Jin, Y.Z., Bao, X.Y., Song, Z.S.: ArithBi~+—an XML index structure on reverse arithmetic compressed XML data. Computer Science 32(11), 119–123 (2005)
[10] Bao, X.Y., Tang, S.W., Yang, D.Q.: Interval~+—an index structure on compressed XML data based on interval tree. Journal of Computer Research and Development 43(7), 1285–1290 (2006)

Privacy Preserving Reverse Nearest-Neighbor Queries Processing on Road Network

Xin Lin , Lingchen Zhou, Peng Chen, and Junzhong Gu

Computer Science and Technology Department, East China Normal University,
Shanghai, China
xlin@cs.ecnu.edu.cn

Abstract. In recent years, with the popularity of Location Based Service (LBS) and recommendation system, spatial data query has become a hot study area. Reverse nearest neighbor (RNN) query is one of the most important queries in spatial database. It plays an important role in decision-making system, recommended system and frameworks like so on. In many cases, users do not want to disclose the specific location information to the system. It requires a certain extent anonymous of user information. Compared to the common Euclidean space, Road Network is more practical. However, in previous studies, there is no RNN queries base on road network taking into account the protection of user privacy. In this paper, we propose a novel algorithm- RN-BRNN(Road Network - Bichromatic Reverse Nearest Neighbor) query algorithm, which considering both the road network bichromatic RNN query and user location privacy protection. RN-BRNN algorithm establishes a special Voronoi Cell based on the road network, queried points anonymity, and probability calculus of obtained RNN. Extensive experimental results show that the algorithm maintenances the same time-complexity with the Euclidean space, and improved precision greatly.

Keywords: Road Network, Reverse Nearest Neighbor Query, Privacy Protection.

1 Introduction

With the popularity of Location Based Service and corresponding recommendation service, spatial data searching becomes a hot area in lots of studies. Nearest neighbor query(NN query) is one basic search method in spatial database. It returns the nearest point of the query point. The reverse nearest neighbor query (RNN query) returns the points which have query points as their NN[1] as an extension of NN query. In Fig.1, p_1 is the NN of q, but due to the NN(p_1) is p_2, so p_1 is not RNN(q); although p_3 is not NN(q), its NN is q, so p_3 is a result of RNN(q). Reverse nearest neighbor query has been widely applied in recommended system and decision-making system. For example, to build a new gas station, it needs to be the nearest neighbor of more users than the competition gas stations. In another way, if it has more RNN users as its potential customers, it will win more profit. There are two types of RNN queries, the

Z. Bao et al. (Eds.): WAIM 2012 Workshops, LNCS 7419, pp. 19–28, 2012.

monochromatic queries and the bichromatic queries [1]. In the monochromatic RNN, all points are of the same type. Thus, a data object o is considered as a reverse nearest neighbor to a query object q if there does not exist another data object o' where $dist(o, o') < dist(o, q)$. In the bichromatic RNN, there are two distinct object types $\mathcal{A}$ and $\mathcal{B}$. Thus, a data object of type $\mathcal{B}$, o_B, is considered as a reverse nearest neighbor of a query object of type $\mathcal{A}$, q_A, iff there does not exist another object of type $\mathcal{A}$, o'_A, where $dist(o_B, o'_A) < dist(o_B, q_A)$.

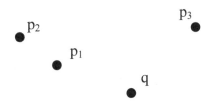

Fig. 1. Nearest Neighbor Query and Reverse Nearest Neighbor Query

LBS provides useful information according to its users' locations, this may lead to the leak of users' location privacy. Hence, in the relevant query, the user's location needs to be anonymous [2], to prevent attackers from getting the user's exact location.

Most existing studies focus on Euclidean space cases, while road network is a more accurate approximation of the reality than Euclidean space. In real life, people should reach a place along the existing paths. Two close points in Euclidean space may be far away from each other since there are some obstacles (e.g. rivers) lying between them. Thus search results in Euclidean space are often not very satisfactory. To be more specifically, there are three points: q, p_1, p_2 in Fig 2. In Euclidean RNN case, p_1 is RNN(q), but in road network, $dist(q, p_1)$ is 11, $dist(q, p_2)$ is 5, so p_1 is not RNN(q), it is RNN(p_2).

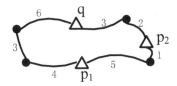

Fig. 2. Euclidean Space and Road Network

In this paper, we address the challenge of users' location privacy protection on road network based RNN query. The proposed algorithm adopts the special Voronoi in road network to define related region. Query points will be anonymity in a certain degree to protect the user's location privacy. Finally, RN-BRNN algorithm will use probability and statistics, derived reverse nearest neighbor query.

The outline of this paper is as follows: Section 2 discusses preliminaries. The problem is formally defined in Section 3. Section 4 describes the privacy protected Road Network RNN processing. Extensive experimental evaluations are reported in Section 5. Finally, concluded with a summary of contributions and directions of future work in Section 6.

2   Preliminaries

In the spatio-temporal database, the common efficient index approach for voronoi diagram is VoR-Tree[3] which based on R-Tree. For the nearest neighbor case, VoR-Tree is more suitable because it is focus on Voronoi Diagram[4][5]. Voronoi Diagram is a kind of very important geometric structure in RNN query. [4] has pointed out that dividing the searching space evenly into k sections can help us to definite corresponding Voronoi Cell of a query point q.

Due to the purpose to protect user's privacy, we use uncertain database to simulate anonymous queried objects. Cheng proposed the UV-Diagram[6] which aims to build a Voronoi Diagram for uncertain data. BRNN-UD algorithm[7] provided a strategy to solve the RNN query on uncertain dataset. To compute search results in uncertain dataset, probability computing is very important. T. Bernecker mentioned an approach to speed up similarity queries on uncertain data[8].

Road network is a new scenario on spatial database, which is more practical than traditional Euclidean space. [9] resolved nearest neighbor query and range query in road network, and [10] purposed the RNN query processing in spatial network.

3   Formal Definition

R-Tree is the basic index structure in spatio-temporal database. It derived from the B-Tree, with higher degree of balance. R-Tree is more suitable for MBR(Minimum Bounding Rectangle) indexing. MBR is the smallest rectangle completely enclosing a set of points. However, in road network, the access relationship and path weight are necessary. Therefore, the spatial R-Tree structure is more suitable[9].

Distance in Euclidean space and road network has different definition. In Euclidean space, it is defined as $dist(q, p) = ((x_q - x_p)^2 + (y_q - y_p)^2)^{1/2}$, in which x_q, y_q are horizontal and vertical coordinates of q, and x_p, y_p are horizontal and vertical coordinates of p. In road network, the distance of two nodes will be the shortest path weight between them.

Definition 1, RND (Road Network Distance), *is the weight of shortest path between two nodes in road network.*

Nearest Neighbor(NN) query and Reverse Nearest Neighbor(RNN) query are the basic searches in spatio-temporal database. The following are their definitions:

Suppose a point set S and a query point q are given, the definition of NN(q) is

$$NN(q) = \{r \in S | \; \forall p \in S : dist(q, r) <= dist(q, p)\} \tag{1}$$

For the RNN queries, two kinds of conditions are as follow separately:

Monochromatic case:

$$RNN(q) = \{r \in S | \; \forall \; p \in S : dist(r, q) <= dist(r, p)\} \tag{2}$$

Bichromatic case:

$$RNN(q) = \{r{\in}S| \; \forall \, p{\in}A{:}dist(r, q){<}{=}dist(r, p)\} \tag{3}$$

In several cases, bichromatic case has been called red-blue point problem. It means query point q is blue and it wants to find the red points which are nearer to it than to any other blue points.

Voronoi cell is the most common concept of RNN query in Euclidean space. Given a query point q and points set $\mathcal{A}$, there are n points in $\mathcal{A}$. For any point p' in VC(q), we have $dist(q, p')<dist(p_i, p')$, where pi $\in\mathcal{A}$. For two points p and q we define the bisector of p and q as the perpendicular bisector of the line segment pq. This bisector splits the plane into two half-planes. The half-plane that contains p was denoted by $h(p,q)$ and the half-plane that contains q by $h(q,p)$. Because $r{\in}h(p,q)$ iff $dist(r,p)<dist(r,q)$, the formal definition of VC(q) is:

$$VC(q) = \cap_{pi \, \in A}h(q, p_i), \; 1 \leqslant i \leqslant n \tag{4}$$

For simplicity, we assume, road network is an undirected weighted graph. It is a simplification of the actual road network. The weight of the edge can be it length or the time required for passing through.

Definition2. NVC(Network-based Voronoi Cell), *Let S_n denotes the nodes set and S_e denotes the edge set in search plane. The format of NVC is as follows:*

$$\{n{\in}S_n, \, e{\in}S_e|\vee p{\in}S_n{:}dist(n, q)<dist(n, p) \; \& \; dist(e, q)<dist(e, p)\} \tag{5}$$

In definition 2, the point q is one of all the query points. Its nickname is the generator of NVC. So, every node or edge in NVC will near to the generator points in road network than to any other destination.

In Euclidean space, to preserve users' privacy, the client will use a continuous closed region as the users' location instead of exact point. In road network distance scenario, such cloaking method would be different. If picking edges randomly near queried point, it costs much in processing time; if choose edges are around queried point closely, it cannot protect privacy effectively. We adopt the X-Star[2] way to build anonymous area to win the balance between efficiency and effectively. The X-Star anonymous object is the small subset in network. Its shape likes a snow with the inner network and outer terminal vertex.

4   RN-BRNN Algorithm

This section presents how to implement privacy protection bichromatic RNN query in road network. Firstly, we construct a related region to filter nodes which are far away from query point. After narrowing the search scope, we then find the possible results and indentify their probability to be the result. There is an important concept in this section, d_{Emax}[9]. It means the distance which in road network between two nodes transformed to Euclidean distance($d_{Emax} = d_N(q, p)$ from q.

4.1   Construction of Related Region

To win efficiency, we need to format a construct similarity to but simple than NVC-the Ring. It would used to filter queried objects. The method to construct the Ring is as follows:

Firstly, algorithm will determine the related region. Objects closer to query point than 1NN in the network, should be within Euclidean distance $d_{Emax} = d_N (q, 1NN)$ from q[9]. Due to this theory, we know that the points in $C_1(circle\ 1NN)$ are of course in result set; the points out of $C_k(circle\ kNN)$ are not in result set. So it needs to define the related region, which is the ring between two circles. There are several steps to finding the Ring.

1)   *Searching R-Tree to find the kNN[9] of query point q in road network, named separately as $p_1, p_2, ..., p_k$.*
2)   *Compute each network-based shortest path of each NN, named as $l_1, l_2, ..., l_k$.*
3)   *Compute each mid-point on spatial path of $l_i(0<i<=k)$ in turn, named as $mid\text{-}p_1, mid\text{-}p_2, ..., mid\text{-}p_k$.*
4)   *Sort these $mid\text{-}p_i$ in ascending order of their Euclidean distance to q, renamed as $m_1, m_2, ..., m_k$ with $m_1 = d_{Emax}(q, min(dist(q, mid\text{-}p_i)))$, etg.*
5)   *The ring between m_1 and m_k will be the related region we are looking for. In Fig3, the gray ring will be the Ring.*

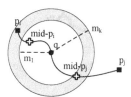

Fig. 3. The related region- the Ring

4.2   Filter Anonymous Objects

To pruning queried objects quickly, draw the MBR of Ring in A. Then the algorithm searches all the anonymous objects which intersect with MBR(Ring) to be candidates for filter in C.

4.3   Probability Computing

In this section, we rank the probability of the anonymous objects which intersect with the Ring. Assuming the intersection part of an anonymous object and the Ring is denoted by S, if the ratio of S to anonymous area is higher than threshold ρ, the anonymous area will be included in results set; Otherwise, it will be given up. Take Fig4 for example.

1)   *Object p_1 is totally in the small circle. It is undoubtedly in RNN(q);*
2)   *Objects p_2, p_3, p_4 intersects with the Ring and big circle partly. System can set*

ρ *as threshold. The algorithm will compute the ratio (denoted by t)between* *intersection and* p_2, p_3, p_4. *In Fig4, assume* $\rho = 0.6$, $t_{p2}(0.8) > \rho$, *then* p_2 *is in* *the result set;* $t_{p4}(0.3) < \rho$, p_4 *will be filtered;*

3) *Object* p_5 *is absolutely out of the Ring, so it is pruned directly.*

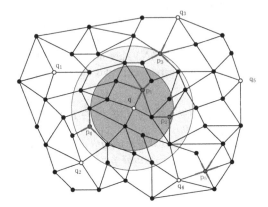

Fig. 4. The Ring

Fig. 5. Anonymous Object

About the possibility computing, we propose an algorithm. Fig 5 is the p_4 object. Point p_{41}, p_{42}, p_{43} are the three end points of p_4. Firstly, compute the ratio of number of end points who make q as its NN and all end points, supposed to be φ. Then, we will compute the average value of the three distances between end points and their separately nearest q_i, supposed to be δ. In this case, compare ρ and $\varphi*\delta$, if $\rho>\varphi*\delta$, pi will be filtered; else if $\rho<\varphi*\delta$, pi will be one of the results.

In summary, the above is the RN-BRNN query algorithm.

5   Experimental Evaluation

In this section, we evaluate the efficiency and effectiveness of privacy protected road network RNN query through experiments. The basic R-Tree structure is from libspatialindex[11]. In order to simulate the real-life conditions, we choose the real datasets of Oldenburg (OL) city and California city road networks[12]. OL dataset contains 6,104 nodes and 7,034 edges. California dataset contains 21,047 nodes and 21,692 edges. In each data set, we choose query points(q) and queried objects(p) randomly and evenly. On average, every 100 neighboring nodes will have 1 query point and 10 queried objects in them. All the experiments were conducted on Intel

Core 2.26GHz dual CPU with 2GBytes memory running on windows 7 operating system. Its program is written in Java(JDK 1.6).

To comprised, in Take-All solution(the benchmark algorithm), the area to filter anonymous objects is the MBR of kNN(q), not the Ring in NVC algorithm.

Fig6 shows the effect of increasing dataset on processing time. The processing includes both CPU time and I/O time. In OL city, the processing time difference is about 2ms. In California city, the difference is bigger to about 5ms. The reason for the differences can be announced in Fig 7.

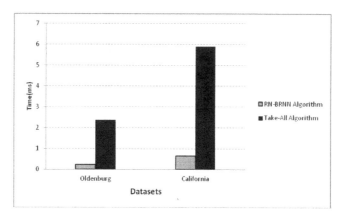

Fig. 6. OL & California Datasets

Fig7 shows the time taken by each step in NVC algorithm on OL dataset. It is clear that the main cost is spent in exactly results computing. Searching kNN is the second cost item after it. In contrast, probability computing costs least time. So, in Fig4, the MBR in Take-All solution is bigger than extension circle in NVC, so the candidates needs to be computed are more than NVC, then the processing time can differ greatly.

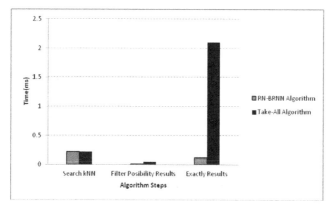

Fig. 7. Cost of Each Step

In Fig 8, Take-All solution cost increasing sharply as the parameter k grows. While, in NVC, the cost time keeps stable. The most time-consuming step is to get the exactly results. The bigger k means more candidate results, which leads to more time consumed in exact result computation. As k grows, the size of MBR will grow faster than extension circle. So the processing time will increase significantly

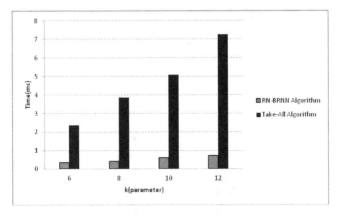

Fig. 8. Time cost by k

Compared to Fig8, Fig 9 shows the precision rate changed with parameter k. RN-BRNN define the Precision Rate to be results/candidates, which used to measuring accuracy of algorithm. Along with the increasing of the parameter k, precision rate in NVC algorithm is higher than Take-All solution in general. That's because the value of k can make a less influence on the ratio of extension circle than that on the size of MBR. So, the precision rate of RN-BRNN will higher than Take-All solution.

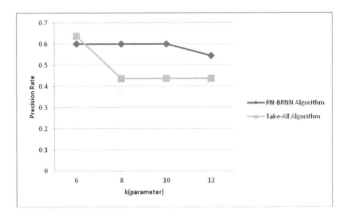

Fig. 9. Precision Rate by k

Fig10 shows the precise rate changing with probability threshold rising from 0.1 to 0.9. Precise rate is the percentage of the result anonymous objects in all candidate objects. It is clear that with the threshold goes up, the precise rate drops down. As the applications in this paper are recommended system, threshold between 0.3 and 0.5 is better.

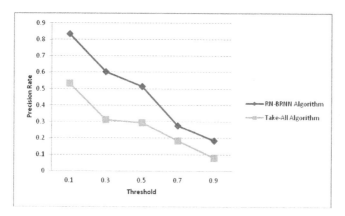

Fig. 10. Threshold and Precision Rate

6 Conclusion

In this paper we proposed the RN-BRNN query algorithm considering both privacy protection and road network distance scenario. The most innovation is the construction of the related Ring, which improved the efficiency. And then, RN-BRNN has computed the probability of the anonymous object to be the RNN result of query point to decide which points will be included in final results set. The Ring improves the precision greatly and guaranteed that the increased complexity in the tolerable range.

RN-BRNN query algorithm is very inspiring for future work. Firstly, this paper only focus undirected road network, while the algorithm based on directed road network will be more complex. Secondly, in big city, the weight of edges in road network will be dynamic in distinct time in a day, e.g., the time cost for passing though a road will be influenced by the traffic status of that time. It's an interesting extension by considering dynamic weight scenarios.

Acknowledgement. This work is supported by NSFC Grant 60903169, the Hong Kong Scholars Program and the Fundamental Research Funds for the Central Universities.

References

[1] Ioana, S., Divyakant, A., Amr, E.A.: Reverse Nearest Neighbors Queries for Dynamic Datasets. In: SIGMOD Workshop DMKD, pp. 44–53 (2000)
[2] Ting, W., Ling, L.: Privacy-Aware Mobile Services over Road Networks. In: VLDB, Lyon, France, pp. 1042–1051 (2009)

[3]   Mehdi, S., Cyrus, S.: VoR-Tree: R-trees with Voronoi Diagrams for Effecient Processing of Spatial Nearest Neighbor Queries. In: VLDB, vol. 3(1), pp. 1231–1242, Singapore (2010)

[4]   Mehdi, S., Cyrus, S.: Approximate Voronoi Cell Computation on Spatial Data Streams. In: VLDB, vol. 18, pp. 57–79 (2009)

[5]   Kefeng, X., Geng, Z., David, T., Wenny, R., Maytham, S., Bala, S.: Voronoi-based Range and Continuous Range Query Processing in Mobile Databases. Journal of Computer and System Sciences, 637–651 (2011)

[6]   Cheng, R., Xie, X., Yiu, M.L., Chen, J., Sun, L.: UV-Diagram: A Voronoi Diagram for Uncertain Data. In: ICDE, pp. 796–807 (2010)

[7]   Lingchen, Z., Xin, L., Junzhong, G., Peng, C.: Bichromatic Reverse Nearest-Neighbors Queries based on Uncertain Dataset. In: International Conference on Future Information Technology, vol. 1, pp. 294–298 (2010)

[8]   Thomas, B., Tobias, E., Hans-Peter, K., Nikos, M., Matthias, R., Andreas, Z.: A Novel Probabilistic Pruning Approach to Speed up Similarity Queries in Uncertain Databases. In: ICDE, pp. 339–350 (2011)

[9]   Dimitris, P., Jun, Z., Nikos, M., Yufei, T.: Query Processing in Spatial Network Databases. In: VLDB, vol. 29, pp. 802–813 (2003)

[10]   David, T., Maytham, S., Quoc, T.T., Wenny, R., Jong, H.P.: Spatial Network RNN Queries in GIS. The Computer Journal, 617–627 (2011)

[11]   R-Tree Index Structure, http://libspatialindex.github.com/

[12]   Dataset, http://www.cs.fsu.edu/~lifeifei/SpatialDataset.htm

Wikipedia Revision Graph Extraction Based on N-Gram Cover

Jianmin Wu[*] and Mizuho Iwaihara

Graduate School of Information, Production and Systems, Waseda University,
Fukuoka 808-0135, Japan
jianmin.wu@moegi.waseda.jp, iwaihara@waseda.jp

Abstract. During the past decade, mass collaboration systems have emerged and thrived on the World-Wide Web, with numerous user contents generated. As one of such systems, Wikipedia allows users to add and edit articles in this encyclopedic knowledge base and piles of revisions have been contributed. Wikipedia maintains a linear record of edit history with timestamp for each article, which includes precious information on how each article has evolved. However, meaningful revision evolution features like branching and revert are implicit and needed to be reconstructed. Also, existence of merges from multiple ancestors indicates that the edit history shall be modeled as a directed acyclic graph. To address these issues, we propose a revision graph extraction method based on n-gram cover that effectively find branching and revert. We evaluate the accuracy of our method by comparing with manually constructed revision graphs.

Keywords: Wikipedia revision graph, Mass collaboration.

1 Introduction

During the past decade, online mass collaboration systems have emerged and thrived on the World-Wide Web. This form of collective action involves large numbers of contributors and numerous user contents are generated. Contributors are often organized under certain projects but work independently. Typical examples include Wikipedia, Linux, Yahoo! Answers, and Mechanical Turk-based systems [3].

Wikipedia, one of the most successful systems, is an open, multilingual Internet encyclopedia written collaboratively by volunteers around the world [6][16]. Users can edit almost every article based on their knowledge, which makes Wikipedia continually changing and evolving. Since the editing process is simplified and requires little skill, articles usually possess numbers of versions that varies from tens to thousands. For each article, Wikipedia keeps all the versions and provide this edit history to the public. Other useful information including timestamps, contributors, and edit comments is also recorded.

---

[*] Corresponding author.

Z. Bao et al. (Eds.): WAIM 2012 Workshops, LNCS 7419, pp. 29–38, 2012.

Unlike what is very common in software development, Wikipedia does not maintain an explicit revision control system that manages the detailed change through revisions. The chronologically-organized edit history fails to reveal the meaningful scenarios in the actual evolution process of Wiki articles, including reverts, merges, vandalism and edit wars, which have been a target of research in [1][11]. A revision graph is a directed acyclic graph where nodes are revisions and directed edges represent parent-child relationship between revisions. Accurate reconstruction of the revision graph from an edit history is fundamental for wiki-style mass-collaborative media, having a wide range of applications from quality evaluation[12][13], trust[14], reputation[1], plagiarism detection[5], and cooperation and conflict between editors[11].

Also, the nature of the evolution process of wiki articles, which slight changes accumulate in long development, suggests that our method can be applied to the retweeting behavior research in social media mining.

In this paper, we propose a method to extract revision graphs from linear edit histories based on n-gram cover. First, we use n-gram distribution to denote revisions of the given articles with timestamps and find how a revision's n-gram distribution can be formed by previous revisions'. Secondly, we utilize smallest n-gram diff score to decrease the complexity in the n-gram cover computation.

This paper is organized as follows. In Section 2 we introduce the background of our research. Section 3 describes basic definitions regarding our problem, and explains algorithms used to extract a revision graph. In Section 4 we show experimental evaluation of our method and the results. Finally, concluding remarks and future work are discussed in Section 5.

2   Background

2.1   Mass Collaboration System and Wikipedia

A mass collaboration system enlists a crowd of users to collaborate to build a long-lasting artifact that is beneficial to the whole community[3]. Different kind of mass collaboration systems diverse in many ways such as the way they recruit and retain users, the way to combine user contributions to solve the target problem. But they are all lack of a framework to trace and evaluate contributors' work systematically.

Wikipedia, launched in January 2001, poses more than 20 million articles in 2012 [16]. Wikipedia employs an open, collaborative editing model where contributors can edit articles using wiki markup to format the text and add other elements like images and tables, and Wikipedia would keep articles with their modifications as edit history with additional information stored, such as timestamps, edit comments and contributors' information [17]. Articles can be improved immediately.

The success of Wikipedia also leads to the popularity of wiki technologies. Many websites provide a wiki engine for users to contribute their content via a web browser to form a Wikipedia-like knowledge base, which suggests that our research can be applied in a wide variety of cases.

2.2   Related Work

Detecting highly similar documents in a large collection is a common topic in computer-assisted plagiarism detection [5]. Many documents that are published on the Internet could be copies or plagiarisms of other documents. Since a plagiarism may not be identical to the original document, using conventional search techniques is difficult to distinguish plagiarized documents from those that are simply on the same topic. Existing techniques that can be used to address these problems include fingerprinting, a technique developed specially for detecting duplicates[4]. But all these contributions are based on a large collection of dissimilar documents, unlike Wikipedia's edit history.

Mikalai Sabel [8] proposed to represent revision history as a tree of versions in Wiki. In this work, revisions are arranged to a directed tree according to sentence-level edit distance, which is too raw to trace slight changes. Secondly, since [8] lacks quantitative evaluation of the proposed method, its accuracy is not warranted. Since version merges are far less frequent than branches from our observation, like [8] we focus on reconstruction of tree histories in this paper. Saccol et al.[9] proposes a machine learning-based XML version detection method, but their algorithm is limited to the problem on determining whether a given pair of documents are in a version relationship or not. Our problem is more complex because we need to determine a parent out of a set of revisions.

Cao et al. [2] proposed a version tree reconstruction method for Wikipedia articles based on keyword clustering. This method uses tf-idf (term frequency and inverted document frequency) score to cluster similar revisions and then largest common subsequences are used for more precise comparison, which is closer to string matching problem.

3   Revision Graph Extraction Method

3.1   Revision Graph Extraction

In most cases, contributors edit a Wikipedia article based on the current revision. If everyone follows this practice, the edit history should appear as a linear sequence of revisions. However, branches occur when contributors have different opinions on editing. For example, a contributor might restore the article to the previous revision if he thinks the content added in the current revision is inappropriate. Considering the nature of editing behavior, we define a *derivation relationship* as a pair of revisions that the latter revision, the child revision, is edited based on the former one, the parent revision. After all the derivation relationships are found, a revision graph that represents the revision evolution process can be constructed.

Revision Graph. A *revision graph* is a directed acyclic graph that shows the derivation relationships among revisions. We denote all the revisions with nodes, which are indexed with revision ids. An edge from revision R_i to revision R_j is constructed when R_i is one of parents of R_j. Here, Parents must be direct, in the sense that there is no intermediate revision R_j' which was created from R_i and R_j was created from R_j'.

While considering the revision graph extraction in Wikipedia, these assumptions about edition behaviors should be held:

1.   A contributor edits an article based on the existing revisions.
2.   A revision can be totally identical with one of its previous revisions.

The main task of our method for revision graph extraction is to identify the derivations of each revision. An intuitive solution is to check each previous revision for verbatim text overlaps and evaluate their edit distance in order to find the most similar one, if we treat the past of the current revision as a derivation collection. This approach can utilize many approximate string matching methods. However, this approach suffers from expensive computational complexity of $O(mn)$, where m and n are   the total numbers of the characters in the comparing revisions, not to mention that we have to perform this pairwise comparison $N(N-1)/2$ times for N revisions. Another shortcoming has been found in a series of cases that it could lead to a wrong result when identifying relationships among a small set of revisions by edit distance.

In our method, we adopt a hybrid model of bag-of-word analysis and n-gram model for revision comparison. Ukkonen[10] proposed using n-gram for text filtering, which is the problem of discarding areas of the text that cannot contain a given pattern. Also, [7] surveys other applications of n-gram to approximate string matching. Revisions can be treated as a collection of words with different occurrence so that we can compare revisions by their word frequency distribution with low computational complexity.

Notice that traditional bag-of-word analysis loses all the sequence order of words, which does harm to reflect the syntax difference between revisions like rephrasing. Here we introduce n-gram model in order to keep the string order partially. Instead of using words directly, we construct the frequency distribution for all the n-grams in a revision, where n is fixed during the entire comparison process. The larger n we use, the more precise comparison can be achieved by longer word sequences.

We compare revisions by their n-gram distributions. Given the fact that the revision documents in the collection of a single article's edit history are highly similar, it is more necessary to distinguish slight differences among them than to evaluate their similarities. We develop n-gram diff to represent the difference between revisions. It is an n-gram distribution of all the different part, while differentiating the contribution of edit behaviors like add, modify and remove. We also develop a measure to quantify the difference. If the difference between two revisions is small enough, we can find their n-gram distributions overlapped, which is like covering an n-gram distribution by the other.

A revision can be constructed by merging portions of multiple ancestors. To find a merge we need to detect a set of revisions whose n-gram distribution can cover the n-gram distribution of the current revision. However this problem involves the set cover problem, which is proved to be NP-hard. However, as we exclude merges in this paper, we just need to find a revision whose n-gram distribution indicates that the revision is the parent of the concerned revision.

3.2   Definitions

Here are the formal definitions that our method is based on.

N-Gram. A (word-level) *n-gram* is a contiguous sequence of n words from the revision text. The text of a revision is also treated as a single word sequence, where Wikipedia Markup Language symbols is removed. In our method, the size "n" starts from 2.

N-Gram Distribution. For a revision R, the *n-gram distribution* of R is the frequency distribution of all n-grams appearing in R. Each entry $f(t, R)$ in the distribution contains an n-gram t and its frequency in R.

N-Gram Diff Score. Let R_1 and R_2 be revisions. The n-gram diff score of R_1 and R_2, denoted by $DS_n(R_1, R_2)$, is

$$DS_n(R_1, R_2) = \sum_{t \in G} |f(t, R_1) - f(t, R_2)|,$$

where $G = \text{ngram}(n, R1) \cup \text{ngram}(n, R2)$, $\text{ngram}(n, R)$ is the set of n-gram of revision R, and $f(t, R)$ is the frequency of n-gram t in R.

We use this measure to evaluate the degree how two n-gram distributions are different. N-grams from base distribution that first appear in the collection are excluded.

Unique N-Grams. The *unique n-grams* UQ of a revision R is the set of all the n-grams in R that never appear in revisions earlier than R. The unique n-grams cannot be covered by any previous revision.

N-Gram Cover. Given a collection D of revisions not later than revision R, an *n-gram cover* of R is the set $S \subseteq D$ of revisions such that:

$$\text{ngram}(n, R) \subseteq UQ \cup \bigcup_{R_i \in S} \text{ngram}(n, R_i)$$

where UQ is the unique n-grams of R.

3.3   Algorithm

Basically, our algorithm performs in two stages, (1) base source detection by minimum n-gram diff score and (2) n-gram cover computing. The size parameter n is fixed during the whole process, so all the n-gram distribution can be compared by the same scale. Results that are generated with different values of n will be treated separately.

1. From the second oldest revision to the latest revision, compare every revision with all of its previous revision and calculate their n-gram distribution diff.
2. For all the n-gram diffs of a revision, choose the lowest k n-gram diff scores as candidates.
3. From the corresponding revisions of candidates, select the latest revision in time order.

4   Experimental Evaluation

4.1   Data Set

We collect articles from Wikipedia randomly to construct a test data set. Articles are filtered by the following criteria:

- English language
- Having text volume more than one page
- Having 200+ revisions

Table 1. Ground Truth Statistics

Article Title	Total # of revisions	# of Branches	Avg. run-length
Racism	10,896	23	4.26
2006 Israel–Gaza conflict	2,456	12	8.00
PhpBB	1,312	37	2.67
Edith Wharton	1,114	16	6.06
Federal republic	717	33	2.99
Sarkar Raj	592	15	6.45
Grade inflation	456	24	4.08
Natal chart	346	11	8.70
Muhammad Naguib	283	8	11.76
Clarinet Concerto	256	12	8.00

In order to keep our selection non-biased, we use the function of "Random Article" provided by Wikipedia and 10 articles are collected. We retrieve the first 200 revisions of each article from the official export page [18] and save all these 2000 revisions as XML files.

For the purpose of the building ground truth, we manually review the derivation relationship underlying these revisions based on human experience. After reviewing the collected revisions, 10 revision graphs are extracted for corresponding articles.

Table 1 shows the total number of branches manually discovered, and average running length of linear paths (i.e. avg. run-length) in the graphs. It can be surmised that controversial topics can cause more branches, which leads to shorter average run-length.

4.2   Data Processing

4.2.1   Content Extraction

First we use the open source XML parser Dom4j [15] to extract the required revisions from the original Wikipedia edit history file. Both revision text contents and edition information are exported, including timestamps, contributors and comments.

The exported revisions are indexed by time order, starting from 0. We also remove punctuations and wiki markup language operators.

4.2.2   N-Gram Distribution Generation

For each cleaned revision, we split the cleaned text of the revision into words. Then we concatenate these unigrams into n-gram as their original order in revision text, while the "n" is on demand.

We count the occurrence for the entire n-grams for each revision R and then construct a hash map and each entry maps an n-gram to its occurrence in R.

4.2.3   Inverted Index Construction

We construct another hash map for all the n-grams appear in the edit history. Each n-gram is mapped to a list of revision ids, which indicate that it has appeared in revisions with these IDs. We scan each n-gram distribution by time order so that the list of n-gram can have an ascending order.

4.2.4   First Appearance Index Construction

N-grams that first appear in a new revision cannot be covered by any previous revision. While computing a cover for a revision, these n-grams should be excluded. We construct an index for them with inverted index simultaneously.

Table 2. Example of Part of First appearance index of Edith Wharton

Revision ID	first appeared 2-grams
23	go to, promotion of, that use
24	For every, denotes the, follows from
25	number of, if and

4.3   Result Analysis

We compare the result revision graphs that generated by our method with the manually constructed graphs and evaluate the degree of how they match. Also we introduce the result of keyword clustering method (Cao et al. [2]) as a reference.

We first conduct a direct comparison by checking each revision's parent from the results and evaluate how much percent of revisions have the right parent in each graph. The higher percentage the result is, the better accuracy can be achieved.

As shown in Figure 1, both methods have accuracy around 90%, where the n-gram cover method performs slightly better in most of the case.

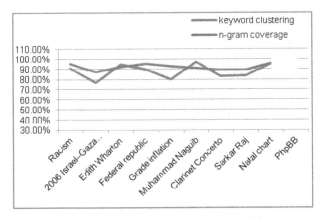

Fig. 1. Direct comparison of derivation revision

However, the direct comparison fails to evaluate the errors that happen in the branching nodes. We extend to a reachability comparison considering the errors' influence on the descendants. Given two revision graphs G_1, G_2 on the same revision set D, the accuracy of G_2 on G_1 is defined as follows:

$$C(G_1, G_2) = \frac{2|G_1^+ \cap G_2^+|}{|D|^2}$$

where G_1^+, G_2^+ are the transitive closures of G_1, G_2, $|D|^2/2$ is half the number of all the node pairs, since all revision graphs are DAGs and half of the node pairs are not reachable. We compute the accuracy of generated revision graphs on ground truth graphs and evaluate how much percent of right reachability to their descendants that all the revisions have. Branching errors that happen in an early stage, which correspond the different reachability in the higher level of the graph has a larger change in the accuracy result than those in lower levels, which reflects the purpose that those errors have greater influence.

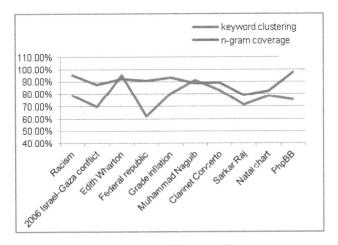

Fig. 2. Reachability comparison

As shown in Figure 2, when setting the n-gram size n to 2, our method can achieve a significant improvement than the keyword clustering method in general, especially in the articles about controversial topics. In our experiments, the adoption of larger n, i.e. longer subsequence, does not improve accuracy.

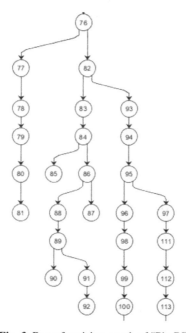

Fig. 3. Part of revision graph of "PhpBB"

Given the fact that most of the parents' n-gram diff score rank within 5 in the candidates, the parameter k in the algorithm is set to 10 in our experiment.

However, our method falls behind in the articles that have fewer branches ("Edith Wharton", "Muhammad Naguib"). It is clear that more investigations will be required before a complete understanding of how this phenomenon occurs.

5   Conclusion and Future Work

In this paper we proposed a revision graph extraction method based on n-gram cover that effectively finds key edit evolution in Wikipedia edit history. We use n-gram distribution to denote revisions of the given articles with timestamps and find how a revision's n-gram distribution can be formed by specific previous revisions'. We utilize smallest n-gram diff score to decrease the complexity in the n-gram coverer computation.

For our future work, we would focus on the problem of detecting merges by revising our n-gram distribution diff and n-gram cover algorithm.

References

1. Adler, T.B., de Alfaro, L.: A Content-driven reputation system for the Wikipedia. In: WWW (2007)
2. Cao, Z., Iwaihara, M.: Wikipedia version tree reconstruction by clustering revisions through keywords. IEICE Technical Report DE2011-32 (2011)
3. Doan, R.R., Halevy, A.Y.: Crowdsourcing systems on the World-Wide Web. Commun. ACM 54(4), 86–96 (2011)
4. Heintze, N.: Scalable document fingerprinting (extended abstract). In: Proc. USENIX Workshop on Electronic Commerce (1996)
5. Hoad, T., Zobel, J.: Methods for Identifying Versioned and Plagiarised Documents. Journal of the American Society for Information Science and Technology 54 (2003)
6. Lih, A.: Wikipedia as participatory journalism: Reliablesources? Metrics for evaluating collaborative media as a news resource. In: Proc. Int. Symp. Online Journalism (2004)
7. Navallo, G.: A Guided Tour to Approximate String Matching. ACM Computing Surveys 33(1) (2001)
8. Sabel, M.: Structuring wiki revision history. In: WikiSym, pp. 125–130 (2007)
9. de Brum Saccol, D., Edelweiss, N., de Matos Galante, R., Zaniolo, C.: XML version detection. In: Proc. ACM DocEng 2007, pp. 79–88 (2007)
10. Ukkonen, E.: Approximate String Matching with q-grams and maximal matches. Theor. Comput. Sci. 1, 191–211 (1992)
11. Viégas, F.B., Wattenberg, M., Dave, K.: Studying cooperation and conflict between authors with history flow visualizations. In: Proc. ACM CHI 2004, pp. 575–582 (2004)
12. Wang, S., Iwaihara, M.: Quality Evaluation of Wikipedia Articles through Edit History and Editor Groups. In: Du, X., Fan, W., Wang, J., Peng, Z., Sharaf, M.A. (eds.) APWeb 2011. LNCS, vol. 6612, pp. 188–199. Springer, Heidelberg (2011)
13. Wöhner, T., Peters, R.: Assessing the quality of Wikipedia articles with lifecycle based metrics. In: Proc. 5th Int. Symp. Wikis and Open Collaboration (2009)
14. Zeng, H., Alhossaini, M., Ding, L., Fikes, R., McGuinness, D.L., Computing Trust from Revision History. In: Proc. Int. Conf. Privacy, Security and Trust (2006)
15. Dom4j, http://dom4j.sourceforge.net/
16. Wikipedia, http://en.wikipedia.org/wiki/Wikipedia
17. Wikipedia Editing,
    http://en.wikipedia.org/wiki/Wikipedia:How_to_edit_a_page
18. Wikipedia edit history export pages,
    http://en.wikipedia.org/w/index.php?title=
    Special:Export&action=submit

Wireless Nerve: Invisible Anti-theft System in Wireless Sensor Network*

Meirui Ren[1,2,**], Jinsheng Duan[1], Hao Qu[1], Xinjing Wang[1], and Lei Du[1,2]

[1] School of Computer Science and Technology, Heilongjiang University,
Harbin, China 150080
[2] Key Laboratory of Database & Parallel Computing, Heilongjiang Province,
Harbin, China 150001
{renmeirui1972,nicole883776}@sina.com,
{microsko,dulei_hd}@163.com, 377740292@qq.com

Abstract. There are limitations in traditional anti-theft technologies. All the technologies have a common ground, that is: the devices are exposed to the space, also easily found and destroyed by adversaries. This paper demonstrates a Hidden Anti-theft System (HAS) based on wireless sensor network. The sensors can be hidden in walls and furniture. The intruder influences the transmitting link, and RSSI of the link changes. According to the changes, the system can detect the movement in monitoring area, and keep invisible. HAS is applied in indoor environment. Experiment results show that HAS achieves very low false positive and no false negative.

Keywords: Sensor networks, Anti-theft technology, Shadowing effect.

1 Introduction

A wireless sensor network (WSN) consists of several sensor nodes with limited storage, computation and communication. Motion detection is one of the popular applications. WSN for anti-theft significantly improves the effectiveness of traditional anti-theft ways (such as using cameras, sensor door locks and infrared devices) which have very limited detecting capability. WSN can detect

---

* This work is supported by Program for New Century Excellent Talents in University under grant No.NCET-11-0955, Programs Foundation of Heilongjiang Educational Committee for New Century Excellent Talents in University under grant No.1252-NCET-011, Program for Group of Science and Technology Innovation of Heilongjiang Educational Committee under grant No.2011PYTD002, the Science and Technology Research of Heilongjiang Educational Committee under grant No.12511395, the Science and Technology Innovation Research Project of Harbin for Young Scholar under grant No.2008RFQXG107 and No.2011RFXXG014, the National Natural Science Foundation of China under grant No.61070193, 61100032, 60803015, Heilongjiang Province Founds for Distinguished Young Scientists under Grant No.JC201104, Heilongjiang Province Science and Technique Foundation under Grant No.GC09A109, Basal Research Fund of Xiamen University under Grant No.2010121072.
** Corresponding author.

Z. Bao et al. (Eds.): WAIM 2012 Workshops, LNCS 7419, pp. 39–44, 2012.
© Springer-Verlag Berlin Heidelberg 2012

possible motions by measuring distance changes [1], sensing light [2] or vibra-
tion. However, these traditional approaches have inevitable limitations. Taking
infrared or ultrasonic for example, monitoring system fails when there are obstacles between the transmitter and receiver. Moreover, devices in these system
require exposure to the space, i.e., keep visible. More devices will be deployed
when there are obstacles in monitoring area. Furthermore, it is easy for theft to
find and destroy the devices.

We introduce a Hidden Anti-theft System (HAS) based on wireless sensor
network to solve these deficiencies. The electromagnetic wave emitted by sensor
node can go through obstacles with intact packet, only Receive Signal Strength
Index (RSSI) attenuates [3][4][5]. The attenuation will not influence communication unless the obstacle is extremely large (e.g. large and thick concrete
wall/building). According to the character of electromagnetic wave, we deploy
the device into wall or furniture. The system works and keeps invisible for any-
one.

This demo shows that HAS works in a indoor monitoring area, and detects
motion by RSSI attenuation. We call the system Wireless Nerve due to its sen-
sitivity. Experiment achieves no false negative and very low false positive.

2   Architecture and Implementation

In system framework, agent and base station are two separate parts. Agent
is a sensor node which works in system and monitors several links. In agent
part: synchronization is the communication mechanism; testing packet can be
collected and analyzed; stable state can be formed; abnormal data sample will
be detected. Base station part consists of settings and control, GUI program and
security. User operates system via GUI program.

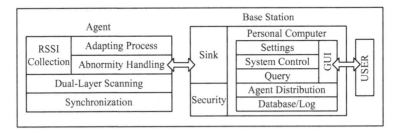

Fig. 1. Architecture of HAS

Synchronization. As we know, two sensors will likely collide when they send
packets in the same time slot. Time scheduling is required, and different sensors
broadcast testing packets in different time slot. There are n sensors in system,
the network scanning cycle $T_{cycle} = n \cdot t_{slot}$. One sensor contains one Timer to
remind itself to broadcast. The interval of the Timer is T_{cycle}. One sensor sends
packet only once in a scanning cycle.

Adapting Process. When the system has been launched, adapting process starts. In this stage, motion in monitoring area is prohibited. When adapting process ends, stable states (ranges) of every link are formed. We define: normal sample which is in the stable range; abnormal sample which is beyond the range.

Links and Agents. Sensors are deployed in the monitoring area, and user appoints several wireless links to be monitored. One agent can monitor several links. A particular link must be monitored by one and only one agent. Sensors in system are all agents. Agent analyzes the RSSI when the packet is from the monitoring links. Agent alarms when it receives an abnormal sample.

GUI. User operates the system via Java GUI program installed on PC. The GUI is shown in Figure 2. A sink node is connected to PC. After deploying sensors in monitoring area, user can draw the layout on screen according to the deployment. Parameters and settings are input by user, and sent to agents via sink by pressing SETINIT. Base station sends n packets and 1 parameter packet to sensors when there are n sensors in network. The sensors blink green LED for $n + 1$ times. Then user presses LAUNCH button to start adapting process. Sensors blink green LED continuously. In this stage, people mustn't enter into monitoring area. Sensors blink blue LED after adapting process ends. The system starts scanning stage. Sensor sends alarm packets if human motions occur in monitoring area. Sink receives the alarm packet, and forwards to GUI program. The corresponding link are colored red on screen. Alarm sound rings to notify user that there may exist intrusion in monitoring area.

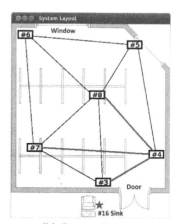

(a) System Control   (b) System Layout

Fig. 2. Graphic User Interface

There are other functions in GUI program. The system stops scanning when user presses RESET button. Pressing pause stops receiving alarm packets, though agents are sending alarm packets. User can modify the layout by adding or removing nodes or edges on system layout window.

Security. We design a super node which user holds. After pressing user-button, a particular packet with a secret key is sent to system. System can recognize identification of user. User can enter into monitoring area and the system will not alarm. Any counterfeit packets from other node (not super node) will not work, because the secret key is held by only the super node and system. When user exits the area, he presses the user-button again, the system resumes. System alarms as soon as there is motion in area. This is the locking mechanism in HAS framework.

Implementation. The system project consists of 4 parts: sink (base station), sensor node (agent), super node and GUI on PC. All programs on sensor node are implemented in Ubuntu10.10 + TinyOS2.x [6]. GUI program is developed in JDK1.6. There are 1 sink, n sensors, 1 super node and 1 PC in this demo. Sensors are set node ID and deployed in the area.

3 Demo Scenario

Demo Setup. We conducted experiments on a real testbed, shown in Figure 3 (the room area is $7.25 \times 7.45m^2$). Our deployment uses TelosB [7] motes of Crossbow company. The sensors transmit in 2.4GHz with IEEE 802.15.4 protocol (sensors are shown in red circle). Each of the sensors is bound on a fixed stick with height of $0.9m$. The monitoring area requires: indoor room or hall, people is far away from the area (distance is about $5m$), the stick with sensor is absolutely stable. Layouts are shown in Figure 2(b). Assume that the monitoring area can cover the whole room . The edges in layout are selected by users.

Fig. 3. Testbed: Key Laboratory of Database and Parallel Computing, College of Heilongjiang Province, Room 329, in Heilongjiang University

Parameter Settings. In Figure 2(a), some parameters need to be setup. *Sleep-Time* means time solt length t_{slot}: sensors broadcast in turn, the broadcasting interval is t_{slot}. *FitNumber M*: every sensor ends adapting process when it collects M packets from a particular link. *SendPower*: the transmitting power of wireless communication. *SendPower* can be dragged horizontally to set the parameter. It is used for further research.

Experiment Result. After deploying the sensors and connecting devices, user presses SETINIT and LAUNCH. The adapting process starts. In this stage, any one mustn't enter or approach the monitoring area. System scans monitoring area when blue LED of sensors blink. Experiment shows two metrics to evaluate the system. Response time: people enter in area at t_1 (time), system alarms at t_2, response time is t_2-t_1. False positive means the rate that system alarms when no body is in monitoring area.

Layout1 (6 nodes) and layout2 (4 nodes) are in the same room. Figure 4 shows the response time with human speed at $1m/s$ and $2m/s$ in layout1 and layout2. For a fixed layout, response time decreases as speed increases. For a fixed layout and speed, response time is proportional to *SleepTime*. In figure 5, k is an inner parameter. False positive is measured with *SleepTime*=$100ms$ and $M = 30$. The result shows that false positive rate is 0.0083%. False positive declines extremely when M increase in 5~15.

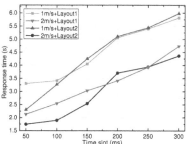

Fig. 4. Response Time   **Fig. 5.** False Positive

4   Conclusion

Traditional anti-theft approaches have an inevitable shortcoming: the exposure of devices. HAS can keep invisible and communicate through obstacles. This demo detects human motion effectively by analyzing the abnormal RSSI. User enters into monitoring area with a super node and the system will not alarm. HAS achieves low false positive and no false negative, which indicates that it can be applied in monitoring environment with high secure demand such as museums or open indoor areas.

References

1. Song, H., Zhu, S., Cao, G.: SVATS: A Sensor-Network-Based Vehicle Anti-Theft System. In: IEEE INFOCOM, pp. 2128–2136 (2008)
2. Mao, X., Tang, S., Xu, X., et al.: iLight: Indoor device-free passive tracking using wireless sensor networks. In: IEEE INFOCOM, pp. 281–285 (2011)

3. Woyach, K., Puccinelli, D., Haenggi, M.: Sensorless Sensing in Wireless Networks: Implementation and Measurements. In: 4th International Symposium on IEEE Modeling and Optimization in Mobile, Ad Hoc and Wireless Network (2006)
4. Goldsmith, A.: Wireless Communications. Cambridge University Press (2005)
5. Weisman, C.J.: The Essential Guide to RF and Wireless, 2nd edn. Prentice Hall PTR Press (2002)
6. Hill, J., Szewczyk, R., Woo, A., Levis, P., et al.: TinyOS: An open operating system for wireless sensor networks. In: Proc. of the 7th Int'l Conf. on Mobile Data Management (MDM 2006). IEEE Computer Society, Nara (2006)
7. TelosB Mote, http://www.xbow.com/pdf/Telos_PR.pdf

A Spatial-temporal Model for the Malware Propagation in MWSNs Based on the Reaction-Diffusion Equations

Zaobo He and Xiaoming Wang[*]

School of Computer Science, Shaanxi Normal University, Xi'an 710062, China
hezaobo@126.com, wangxm@snnu.edu.cn

Abstract. Mobile wireless sensor networks (MWSNs) have important applications in many fields. However, MWSNs are becoming attacked targets due to its large-scale applications. The focus of this work is to develop a modeling framework and mathematical model for characterizing the process of malware propagation in MWSNs from two aspects of time and space. It is important to understand malware's potential damages, and to develop counter-measures. Firstly, we develop a formal model for describing the process of malware propagation in MWSNs based on the reaction-diffusion equations. Then, we derive the threshold for predicting whether a malware propagates or not in MWSNs as time passes. Finally, we propose the spatial pattern based method to analyze and predict the spatial distribution of nodes infected by malwares as time passes. Both theoretical analysis and extensive simulation results show that the movement speed of nodes and the communication radius of nodes have a significant effect on the malware propagation in MWSNs.

Keywords: MWSNs, malware propagation, reaction-diffusion equation, equilibrium point, stability, spatial pattern.

1 Introduction

1.1 Background

Recently, mobile wireless sensor networks (MWSNs) have got a lot of attention, due to the fact that they have important applications in many fields, such as intelligent transportation, community health monitoring, and animal activity monitoring [1]. However, MWSNs are becoming attacked targets owing to the characteristics of its large-scale applications. Among them, injecting malwares into some nodes, especially mobile nodes, has become a serious security threat for MWSNs. Wireless malwares are malicious software or data packets, which are designed for damaging nodes, or blocking regular communications among nodes, even damaging the integrity of regular data packets in MWSNs. When malwares are injected into one or more nodes,

---

[*] Support by National Natural Science Foundation of China (NSFC) under Grant No.60970054, and 61173094, the Scientific Research Foundation for the Returned Overseas Chinese Scholars of Ministry of Education of China.

Z. Bao et al. (Eds.): WAIM 2012 Workshops, LNCS 7419, pp. 45–56, 2012.

malwares will propagate through wireless communication protocols among nodes in MWSNs. Different from the malware propagation on the Internet or WSNs [2, 3], the mobile speed, the communication range, and the initial space distribution of nodes significantly impact the process of malware propagation in MWSNs. Meanwhile, due to the mobility of nodes in MWSNs, characterizing the interacting relationship of the factors influencing the process of malware propagation becomes more difficult. Therefore, investigating the spatial-temporal dynamics of malware propagation can achieve the efficient analysis and prediction of the spatial-temporal behavior characteristics of malware propagation in MWSNs. Moreover, based on the dynamic models of malware propagation, we can effectively evaluate different prevention or control policies for malware propagation.

1.2   Related Work

For a long time, based on the differential equation, the difference equation, the Markov Chain and the cellular automaton, many formal models have been proposed to analyze the propagation of diseases or information [4]. However, these models assume all the individuals are immobile, and they just describe the temporal dynamics of the diseases or malwares diffusion, rather than represent the spatial distribution of diseases or malwares in different states with time. In particular, no related work is conducted for modeling the process of the malware propagation in MWSNs so far. Different from the existing models on the malware propagation on the Internet or wireless networks, our model focuses on the spatial-temporal dynamics of the malware propagation in MWSNs to reveal the spatial-temporal distribution laws of nodes in different states (susceptible, infected) based on the reaction-diffusion equations.

1.3   Contribution

The goal of this work is to develop a modeling framework and mathematical model that can characterize the process of malware propagation in MWSNs from the two aspects of time and space. Our main contributions are presented as follows.

1. By analyzing the mechanism of malware propagation in MWSNs, we quantify the process of malware propagation based on the Susceptible-Infected model (SI model) in epidemic theory. To the best of our knowledge, this is the first time to study the process of malware propagation based on the reaction-diffusion equations.
2. Based on the analysis of the spatially homogeneous system, we validate the threshold of the radius which determine the malware propagates continuously or not as time passes. Thereby, we can set up reasonable measures to control malware propagation in MWSNs.
3. We discuss the spatial distribution of infected nodes as time passes. We find out different parameters lead to different patterns, especially, under some parameters which can guarantee a target wave emerge, then the breakup of target waves lead to chaos patterns.

2   Spatial-temporal Model for the Malware Propagation in MWSNs

2.1   Malware Propagation Model

MWSNs are in a large 2-D region with area Φ, and W nodes are deployed in the region in a random mode at the initial time. Each node roams randomly with an average velocity $\overline{V}$ in the region. An infected node propagates multi-copies of the malware to its susceptible neighbors while transmitting data or control messages to them through a wireless broadcast communication protocol.

Let $S(x,y,t)$, $I(x,y,t)$ and $D(x,y,t)$ be the number of susceptible nodes, infected nodes and dead nodes on the location (x, y) at instant t respectively, and U is the area of the small local region that the location (x, y) represents. Therefore, when an infected node sends a malware packet through a wireless broadcast protocol, the number of susceptible nodes which is infected is λ, and we have $\lambda=\pi r^2 S(t)/U$, and r denotes the communication radius of nodes.

We use A to represent the ratio of adding regular nodes (susceptible nodes) into MWSNs in order to maintain the number of the live nodes stable. η_1 represents the ratio at which susceptible nodes become dead nodes per unit time, and η_2 represents the ratio at which infected nodes become dead nodes per unit time. Owing to infected nodes need to propagate multi-copies of the malware to its susceptible neighbors continuously, its life cycle is shorter than susceptible nodes, so $\eta_2>\eta_1$. As a result, at any instant t the change ratio of the node number may be modeled by a directed graph.

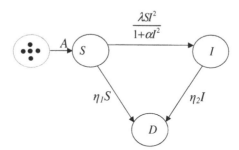

Fig. 1. Node state transition relationship

2.2   Spatial-temporal Model

In this section, we propose the formal model describing the process of malware propagation in MWSNs according to the node state transition relationship in Fig. 1. We have the following model of malware propagation in MWSNs, and it is the dynamic system described by a group of partial differential equations based on the Susceptible-Infected model [5],

$$\begin{cases} \dfrac{\partial S}{\partial t} = A - \dfrac{\lambda SI^2}{1+\alpha I^2} - \eta_1 S + \mu \nabla^2 S \\ \dfrac{\partial I}{\partial t} = \dfrac{\lambda SI^2}{1+\alpha I^2} - \eta_2 I + \mu \nabla^2 I \end{cases} \tag{1}$$

The term $\nabla^2 = \partial^2/\partial x^2 + \partial^2/\partial y^2$ is the Laplace operator which is used to describe the change ratio of the number of nodes owing to the mobile behavior on the location (x, y), and μ is diffusion coefficient, other parameters with the above. $\lambda SI^2/(1+\alpha I^2)$ is the nonlinear incidence rate, where αI^2 measures the infection force of the malware and $1/(1+\alpha I^2)$ measures the inhibition effect from the crowding effect of the infected node. The initial conditions of model (1) are $S_0 > 0$, $I_0 > 0$ and its boundary conditions are

$$\dfrac{\partial I}{\partial n}\bigg|_{(x,y)} = 0, \quad \dfrac{\partial S}{\partial n}\bigg|_{(x,y)} = 0$$

Among them, $(x, y) \in \partial\Omega$ and Ω is the whole space region and n is the boundary normal vector.

2.3   Model Analysis

2.3.1   Dimensionless Transformation
For the simplicity of analysis, we transform system (1) into a dimensionless system, with the scaling

$$S = \sqrt{\dfrac{\eta_2}{\lambda}}X, I = \sqrt{\dfrac{\eta_2}{\lambda}}Y \text{ and } t = \dfrac{1}{\eta_2}\tau . \tag{2}$$

Substituting the equations (2) into the system (1), and still using the variables S, I, t instead of X, Y, τ, we have

$$\begin{cases} \dfrac{\partial S}{\partial t} = B - \dfrac{SI^2}{1+mI^2} - kS + d\nabla^2 S \\ \dfrac{\partial I}{\partial t} = \dfrac{SI^2}{1+mI^2} - I + d\nabla^2 I \end{cases} , \tag{3}$$

where

$$B = \dfrac{A}{\eta_2}\sqrt{\dfrac{\lambda}{\eta_2}}, \quad m = \alpha\dfrac{\eta_2}{\lambda}, \quad k = \dfrac{\eta_1}{\eta_2}, \quad d = \dfrac{\mu}{\eta_2}\sqrt{\dfrac{\eta_2}{\lambda}} .$$

2.3.2   Existence of Equilibrium Point
The equilibrium points are very important to predict whether the malware will continue to propagate or die out as time passes. Due to spatial diffusion system has the same constant equilibrium state solutions with nondiffusible system. We first consider a spatially homogeneous system. We have

$$\frac{dS}{dt} = B - \frac{SI^2}{1+mI^2} - kS \quad,$$ (4a)

$$\frac{dI}{dt} = \frac{SI^2}{1+mI^2} - I \quad.$$ (4b)

Theorem 1.

1. There always exists a malware-free equilibrium $E_0(S_0,I_0)$, and $S_0=B/k$, $I_0=0$;
2. There exists no positive equilibrium if $B^2-4k(1+km)<0$;
3. There exists one positive equilibrium if $B^2-4k(1+km)=0$, denoted by $E^*(S^*,I^*)$;
4. There exists two positive equilibrium if $B^2-4k(1+km)>0$, denoted by $E_1(S_1,I_1)$, $E_2(S_2,I_2)$, where

$$S_2 = \frac{1}{I_2} + mI_2, \quad I_2 = \frac{B+\sqrt{B^2-4k(1+km)}}{2(1+km)}.$$

Proof.
To find the equilibrium points, from system (4), let $dS/dt=0$ and $dI/dt=0$. From (4b), we have

$$I = 0 \quad,$$ (5)

or

$$S = \frac{1+mI^2}{I} \quad.$$ (6)

Substituting the equation (6) into the equation (4a), we have

$$f(I) = B - k\frac{1+mI^2}{I} - I = 0 \quad.$$ (7)

From the equation (5), we obtain $S_0=B/k$, *i.e.* the zero equilibrium point which is denoted by $E_0(B/k, 0)$. Analyzing the existence of the solutions of the equation (7), we can acquire the above conclusions easily.

2.3.3   Stability of Equilibrium Point

By analyzing the stability of the equilibrium points, we can predict that the malware propagation in MWSNs will continue or die out as time passes. If the positive equilibrium points are stable, different types of nodes coexist at this equilibrium point as time passes. On the other hand, if malware-free equilibrium E_0 is stable, the type of infected nodes may finally die out as time passes. We have the following conclusions about the stability of equilibrium point.

Theorem 2.

Case 1: If system (4) has no positive equilibrium point, E_0 has global asymptotic stability;

Case 2: If system (4) has the only equilibrium point E^*, E^* is a non-hyperbolic singular point;

Case 3: If system (4) has two equilibrium points E_1 and E_2, E_1 is a saddle point which is instable. The stability of E_2 is determined by different conditions.

Proof.

According to the stability theory of ordinary differential equation, we need to get the Jacobi matrix of system (4) which is presented as follows

$$J = \begin{bmatrix} J_{11} & J_{12} \\ J_{21} & J_{22} \end{bmatrix}, \tag{8}$$

where

$$J_{11} = -\frac{I^2}{1+mI^2} - k, \ J_{12} = -\frac{2SI}{(1+mI^2)^2}, \ J_{21} = \frac{I^2}{1+mI^2} \ \text{and} \ J_{22} = \frac{2SI}{(1+mI^2)^2} - 1.$$

Case 1: Since the Jacobi matrix of system (4) at its malware-free equilibrium $E_0(S_0,I_0)$ is

$$J(E_0) = \begin{bmatrix} -k & 0 \\ 0 & -1 \end{bmatrix}.$$

Owing to the eigenvalue of Jacobi matrix at E_0 is less than zero, the malware-free equilibrium point E_0 has local asymptotic stability according to the Routh-Hurwitz rule [6]. Meanwhile, to prove $E_0(B/k, 0)$ has global asymptotic stability, we need to prove there exists no periodic orbit or limit cycle. According to the Bendixson-Dulac criterion [6], system (4) has no periodic orbit or limit cycle when there exists no positive equilibrium point, *i.e.*, on the condition of $B^2-4k(1+km)<0$. It means when system (4) reaches its equilibrium point $E_0(B/k, 0)$, the numbers of susceptible nodes and infected nodes will respectively converge to B/k and 0 as time passes. This means the malware propagation will be extinct in the MWSNs no matter how many infected nodes at the initial instant t_0.

Case 2: It is easy to prove E^* is a non-hyperbolic singular point, but the discussion of its stability is complex. Due to the limitation of space, we just consider the case of the condition of $B^2-4k(1+km)>0$.

Case 3: From the equation (7), we have

$$\frac{df}{dI} = \frac{k-(1+km)I^2}{I^2}. \tag{9}$$

From the equation (9), we know that the value of $f(I)$ gets maximum when $I = \sqrt{k/1+km}$. Thus, we obtain

$$I_1 < \sqrt{\frac{k}{1+km}} < I_2. \tag{10}$$

From the equation (6) and (8), we can obtain the Jacobi matrix of system (4) at its positive equilibrium point. In order to study the stability of system (4) at its positive equilibrium point, according to the stability theory of ordinary differential equation, we need to discuss the trace of the Jacobi matrix which is denoted by tr and the value of the determinant of the Jacobi matrix which is denoted by det at E_1 and E_2, as follows

$$tr = J_{11} + J_{22}$$
$$= -\frac{1+m+km}{1+mI^2}(I + \sqrt{\frac{1-k}{1+m+km}})(I - \sqrt{\frac{1-k}{1+m+km}}) \quad , \quad (11)$$

and

$$det = J_{11}J_{22} - J_{12}J_{21}$$
$$= \frac{1+km}{1+mI^2}(I + \sqrt{\frac{k}{1+km}})(I - \sqrt{\frac{k}{1+km}}). \quad (12)$$

From the equation (10) and (12), we have $det(E_1)<0$, $det(E_2)>0$. According to Routh-Hurwitz rule, E_1 is a saddle point which is instable. Therefore, we just consider the situations of $tr(E_2)$. According to the Routh-Hurwitz rule, E_2 has local asymptotic stability when $tr(E_2)<0$ otherwise E_2 is instable.

2.3.4   Spatial Perturbation Analysis

Based on the above stability analysis for the ordinary differential system (4), we obtain the equilibrium points and its stability. However, the problem we need to discuss is a diffusion system which is a partial differential system. Therefore, it is necessary to execute the spatial perturbation analysis to system (3) based on the conclusion of the spatially homogeneous system (4). By the spatial perturbation analysis to system (3), we can observe the impact of the mobile behavior of nodes on the spatial distribution of nodes in different states in MWSNs. We only focus on the positive equilibrium points E_2 of system (3), and propose a small spatial-temporal perturbation to E_2. Let

$$S = S_2 + u(\bar{\upsilon},t), \quad I = I_2 + v(\bar{\upsilon},t) \quad , \quad (13)$$

where $\bar{\upsilon}$ is a two-dimension spatial vector, $u(\bar{\upsilon},t)$ and $v(\bar{\upsilon},t)$ are two spatial-temporal perturbations on the variables S_2 and I_2. Substituting the equations (13) into system (3), and unfolding by the Tailor series formula, ignoring the high order terms, we obtain the following linear perturbation equations.

$$\begin{cases} \dfrac{\partial u}{\partial t} = J_{11}u + J_{12}v + d\nabla^2 u \\ \dfrac{\partial v}{\partial t} = J_{21}u + J_{22}v + d\nabla^2 v \end{cases} \quad , \quad (14)$$

where J_{11}, J_{12}, J_{21} and J_{22} is the values of the equation (8) at E_2.

We expand the perturbation variables on the Fourier space as follows,

$$u(\bar{v},t) = u_0 e^{\omega t} e^{i\kappa\bar{v}} , \quad v(\bar{v},t) = v_0 e^{\omega t} e^{i\kappa\bar{v}} . \tag{15}$$

Substituting the equations (15) into the equation (14), we have the following characteristics polynomial.

$$\begin{vmatrix} \omega - J_{11} + \kappa^2 d & -J_{12} \\ -J_{21} & \omega - J_{22} + \kappa^2 d \end{vmatrix} = 0 . \tag{16}$$

The equation (16) can be rewritten as follows,

$$\omega^2 - tr(\kappa)\omega + \det(\kappa) = 0 , \tag{17}$$

where

$$tr(\kappa) = J_{11} + J_{22} - 2d\kappa^2 , \tag{18}$$

$$\det(\kappa) = d^2\kappa^4 - \kappa^2 d(J_{11} + J_{22}) + J_{11}J_{22} - J_{12}J_{21} . \tag{19}$$

Thus, the solutions of the equation (17) are as follows.

$$\omega(\kappa) = \frac{tr(\kappa) \pm \sqrt{tr^2(\kappa) - 4\det(\kappa)}}{2} . \tag{20}$$

From the equations (15) and (20), we notice the facts: (a) when the real part of $\omega(\kappa)$ is less than zero, the values of $u(\bar{v},t)$ and $v(\bar{v},t)$ become zero as time passes, and this means system (3) finally reaches its stable state; (b) when one of the $\omega(\kappa)$ is greater than zero, the values of $u(\bar{v},t)$ and $v(\bar{v},t)$ become infinite as time passes, and this means system (3) will lose its stability, and finally go into a heterogeneous state. Assume ω_κ is the function of the control parameter θ, i.e. $\omega_\kappa = \omega_\kappa(\theta)$. At the critical point, if $\omega_\kappa(\theta_c) = 0$ and $\partial\omega_\kappa/\partial\theta \neq 0$, so $\theta = \theta_c$ is one of dynamics bifurcation point. In the bifurcation point, system exchange their stability and equilibrium states of the model intersect. To the Hopf bifurcation, it is space-independent and breaks the temporal symmetry of a system, which gives rise to oscillations that are uniform in space and periodic in time. Also the Turing bifurcation breaks spatial symmetry, leading to the formation of patterns that are stationary in time and oscillatory in space. However, in our model, the diffusion coefficient is equal, so the Turing pattern does not exist [7]. Simultaneously, the essential condition of Hopf bifurcation occurs when

$$tr(E_2) = J_{11} + J_{22} = -\frac{-1 + k + I_2^2(1 + m + km)}{1 + mI_2^2} = 0, \tag{21}$$

From equation (21), combined with the process of dimensionless transformation, we have the following function,

$$G(A, \alpha, r, \eta_1, \eta_2) = 0 . \tag{22}$$

Thus, the equation (22) is the essential condition that the Hopf bifurcation occurs concerning the main network parameters of system (1). Importantly, through the equation (22), we can predict and control the spatial-temporal distribution of nodes in different states to make better decisions for preventing or controlling the malware propagation in MWSNs.

3   Simulation Results

To solve differential equations by computer, we transform it from a continuous problem to a discrete problem. The spacing between the lattice points is defined by the lattice constant Δh. In the discrete system, the Laplacian operator describing diffusion is calculated using finite differences. The time evolution is also discrete, i.e. the time goes in steps of Δt. The time evolution can be solved by using the $rk45$ method. In this paper, we set $\Delta h=1$, $\Delta t=0.01$ and $M = 200$. All our numerical simulations employ the Neumann (zero-flux) boundary conditions.

Based on the previous analysis, from Theorem 1, we have the critical value that determine whether the system (1) exists the positive equilibrium point or not, and we denoted the critical value by r_+. From the equation (22), we have the critical values of the essential condition of Hopf bifurcation, denoted by r_c. For more clearly, we state the following conclusions.

Case 1: If $r<r_+$, the positive equilibrium does not exist, i.e., there only exists malware-free equilibrium E_0. It means the malware propagation will be extinct in MWSNs no matter how many infected nodes at the initial instant t_0. Therefore, in this case, the dynamical behavior is uniform in different space region;

Case 2: If $r_c<r$, the positive equilibrium E_2 exists and it is local stable under spatial perturbation. It means the malware propagation will converge to the equilibrium point E_2 in MWSNs. Therefore; in this case, the dynamical behavior is also uniform in different space region;

Case 3: If $r_+<r<r_c$, the positive equilibrium E_2 exist but it is instable under spatial perturbation. Therefore, the spatial perturbation executed to E_2 will lead to inhomogeneous distribution of infected nodes or susceptible nodes, i.e., the dynamical behavior is not uniform in different space region.

3.1   Uniform Dynamical Behaviors in Different Space Region

In this section, we will study the impact of different diffusion coefficient and different communication radius at the above case 1 and case 2. Due to the dynamical behaviors in different space region is uniform in this condition, we just consider the dynamical behaviors on the location (100,100) which can reflect the global dynamical behaviors. We let $A=0.2$, $\alpha=3$, $\eta_1=0.0298$ and $\eta_2=0.3298$. From Theorem 1, we have $r_+=12.2669$. From the equation (22), we have $r_c=14.9806$. Firstly, we let $r=8(r<r_+)$, $16(r_c<r)$ and give different diffusion coefficient μ which correspond to the diffusion coefficient d of dimensionless system (3). Secondly, we let the diffusion coefficient $d=1$ and give different communication radius. The simulation results of the system (3) on the location (100,100) are shown in Fig. 2.

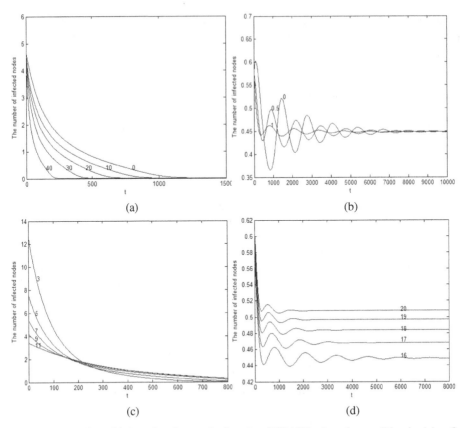

Fig. 2. The number of infected nodes on the location (100,100) when the condition is: (a) $r=8$, $d=0,10,20,30,40$ respectively (b) $r=16$, $d=0,0.5,1$ respectively (c) $d=1,r=3,5,7,9,11$ respectively (d) $d=1,r=16,17,18,19,20$ respectively

From figure (a) and (b), we have the conclusion that the diffusion behaviors of nodes are in favor of converging to the equilibrium point for the system (1). It means that the diffusion behaviors of nodes make the distribution of malware is easy to reach a stable state no matter the stable state is malware-free equilibrium state or positive equilibrium state. From figure (c), the convergence speed to malware-free equilibrium increases with the increase of the communication radius r at first, then after a period of time, the larger of r, the slower convergence speed to malware-free equilibrium. From (d), with the increase of r, the convergence speed to positive equilibrium is increasing.

3.2 Inhomogeneous Space Distribution

In this section, we will study the impact of different communication radius at the above case 3. We still let $A=0.2$, $\alpha=3$, $\eta_1=0.0298$ and $\eta_2=0.3298$. Due to the limitation of space, in the following we just simulate system (3) with the different values of r to observe the spatial distribution of infected nodes as time passes. The initial conditions is a random perturbation at E_2 in all region and let $r=14.3(r_+<r<r_c)$. The various sets of simulation time are t=0, 20000, 40000, 60000, 100000 and 500000 respectively. The simulation results are shown in Fig. 3.

Fig. 3. The number of infected nodes in different apace region. Different colors represents different distribution values, where $r=14.3$, and the number of iterations is: (a) $t=0$, (b) $t=20000$, (c) $t=40000$, (d) t=60000, (e) t=100000, (f) t=500000

From Fig. 3, we observe the target wave when $r=14.3$. Then the target wave breaks towards the core from far field. The evolution of system lead to chaos patterns when the iteration time is large enough. Therefore, we can know the distribution of infected nodes from two aspects of time and space, which is important to develop corresponding counter-measures. Based on the distribution of malware in different space region, we can take protection action in the area of severe infection to prevent the further propagation of malware.

4   Conclusion

In this paper, we have proposed a spatial-temporal model for investigating the spatial-temporal dynamics of malware propagation in MWSNs based on the reaction-diffusion equations. Making use of the proposed model, we have investigated the equilibrium points of malware propagation in MWSNs. By analyzing the stability of the equilibrium points, we derived the threshold of whether a malware will continue to propagate or not in MWSNs. In particular, we have investigated the spatial distribution characteristics of infected nodes as time passes through the spatial-temporal perturbation analysis on the variables of the proposed model. Both the theoretical analysis and the numerical results reveal that the spatial-temporal dynamic characteristics of malware propagation in the MWSNs are closely related to the network parameters. In this paper, we mostly observe the impact of different communication range of nodes and different diffusion coefficient for the malware propagation. Moreover, the influence factors of the malware prorogation are more than that, such as the mobile speed, the link reliability between one-hop nodes and the sending rate of packets, which need further discussion and investigation.

References

1. Rezazadeh, J., Moradi, M., Ismail, A.S.: Mobile Wireless Sensor Networks Overview. International Journal of Computer Communications and Networks 2(1), 17–18 (2012)
2. Wang, X., Li, Q., Li, Y.: EiSIRS: a formal model to analyze the dynamics of worm propagation in wireless sensor networks. Journal of Combinatorial Optimization 20(1), 47–62 (2010)
3. Wang, X., Li, Y.: An improved SIR model for analysing the dynamics of worm propagation in wireless sensor networks. Chinese Journal of Electronics 18(1), 8–12 (2009)
4. Anagnostopoulos, C., Hadjiefthymiades, S., Zervas, E.: An Analytical Model for Multi-epidemic Information Dissemination. Journal of Parallel and Distributed Computing 7(1), 87–104 (2011)
5. Das, P., Mukherjee, D., Hsieh, Y.: An S-I Epidemic Model with Saturation Incidence: Discrete and Stochastic Version. Int. J. Nonlinear Anal. Appl., 1–9 (2011), http://www.ijnaa.com
6. Sontag, E.D.: Mathematical Control Theory: Deterministic Finite Dimensional Systems. Springer, New York (1998)
7. Andresen, P., Bache, M., Mosekilde, E.: Stationary Space-periodic Structures with Equal Diffusion Coefficients. Physical Review E 60(1), 297–301 (1999)

The Application of a Node-Localization Algorithm of Wireless Sensor Network in Intelligent Transportation System[*]

Long Tan and Liyan Chen

School of Computer Science and Technology, Heilongjiang University, 150080 Harbin, China
Tanlong01@163.com

Abstract. The emergence of Intelligent Transportation System (ITS) has brought a revolution to the vehicle transportation which makes human, vehicles, roads united and harmonic and establishes a wider range, fully efficient, real time and accurate information manage system. In this paper, intelligent transportation technology of node-localization in wireless sensor network (WSN) has been proposed and a framework of ITS is designed and realized which provides a system model for the application of ITS.

Keywords: Intelligent transportation system, Node-localization, Localization algorithm, wireless sensor network.

1 Introduction

With rapid advances in computing and communication electronics, an Intelligent Transportation System(ITS) is feasible in the 21st century. ITS has become a center of transportation control and increase safety, security and efficiency of transport systems and decreases environment impacts. Wireless Sensor Network of ITS is based on is frequently used for testing, sensing, collecting and processing information of monitored objects and transferring the processed information to users. Thus, ITS has become an emerging global phenomenon benefiting both public and private sectors. For example, the wireless sensors and transceivers onboard individual cars can communicate with other cars, or with the road. The information collected by individual vehicle, as well as those distributed by a central transportation control communication system can help to reduce accidents and time wasted in traffic congestion.

In ITS, the actual positions of sensors are required for many applications. Most localization algorithms available in the literature are anchor based or they estimate the position of the sensors at the cost of losing accuracy. In this paper we propose a node localization technique based on radiation-frequency and ultrasonic signal which is proved to be more efficient and reliable.

In the paper, we focus on the study of the node-positioning technique of WSN in ITS which provides the technical guarantee for the positioning of transportation

---

[*] Supported by the Science Technology Research Project of Heilongjiang educational office under Grant No.11551349.

vehicles (2) the study of the prototype of ITS based on WSN, which provides a system model for the application of ITS.

2   Related Work

2.1   Intelligent Transportation System

Intelligent Transportation Systems (ITS) is a broad range of diverse technologies applied to transportation to make systems safer, more efficient, more reliable and more environmentally friendly, without necessarily having to physically alter existing infrastructure [1]. In recent years, the study of ITS has gained more and more attention. In China, after the Conference of ITS in Central Europe in 1997, ITS has been decided as an important part of China's Technological Development Strategies.

A guiding committee of ITS is founded in 2000 to conduct China's ITS developmental strategies. In 2002, China's Ministry of Science and Technology decided it as the key project of "The tenth five-year project of science and technology" which will come to its maturing stage in 2020 and become a harmonious intelligent integration of human, vehicle and road.

In the meantime, many developed countries have begun to develop the intelligent transportation systems. For instance, the United States of America implemented the plan of intelligent vehicle initiative[2]; Europe proposed the road safety action program (RSAP)[3], and Japan suggested a super intelligent vehicle system, etc.

In ITS, Vehicles can be equipped with a multitude of sensors to monitor its environment, and the collected data can be used to send alerts to the driver or serve as input to the onboard electronic system. Sensors installed on the roads can detect traffic flow, congestion, among others, and use that information to control the traffic lights, and also send the data to a control room. When the vehicle system and the roadway system are combined, then several other applications will become possible. Communication is vital to the ITS, which allows the roads and the vehicles to share information. When the road detects a slow traffic flow, it can send that information to other vehicles to take another faster route. With the vehicle and the road sensors working together, existing roads can accommodate higher traffic volume and at the same time lower congestion.

In ITS, Intelligent vehicles and roads can function on their own, or they can be connected to each other to harness the power of an interconnected system. There are several application types[4], such as Vehicle to Vehicle (V2V), Vehicle to Infrastructure (V2I) and vise versa, and infrastructure to infrastructure (I2I).

In V2V, When vehicles are equipped with the localization device, the exact position of the vehicle can be detected, and that positional data can be shared with other vehicles in the proximity. When the driver attempts to change lanes, the vehicle's central processor could check the positions of adjacent vehicles and provide a warning if a blind spot accident looks imminent. The warning could be audible or, with a more complex system, the vehicle could command the brakes or steering wheel[5].

In V2I, it requires the roads to be equipped with sensors and transceivers. As it is an ambitious project to implement such intelligent system on all roads, it makes sense to start with the major roads in major cities. If the vehicle can talk with the road and the vehicle's location is got, the vehicle's route to the destination site can be transmitted to all the upcoming intersection points, so the traffic lights can all be changed just before the emergency vehicle approaches. This will minimize the congestion and wait time for the emergency vehicle, and at the same time reducing traffic disruptions.

With I2I, the entire transportation communications network will be complete. Any car can talk to any other car on any road, any car can talk to any road at any point, and any road can talk to any other road.

In short, the localization information plays an important role in the above ITS applications. In the paper, a localization algorithm is designed and realized based on the wireless sensor network.

2.2   Localization Algorithms

The node localization refers that each node in the sensor network localizes its position in a space coordinate system. Localization may be classified broadly in two categories: range-based and range free, according to whether they depend on the accurate sensor distance or not.

1. Range-Based Localization

Range based techniques depend on absolute distances among sensors. The distances may be estimated by received signal strength or by time-off-flight of communication signal between two sensors. The range- based techniques include RSSI, TOA and TDOA, etc.

RSSI (Received Signal Strength Indicator)[6]: is an indication of utilizing the signal strengths to determine the locations of the unknown nodes. In RSSI, received signal power level assessment is a necessary step in establishing a link for communication between wireless nodes.

TOA(Time of Arrival)[7]: is the travel time of a signal from a single transmitter to a remote single receiver. By the relation between light speed in vacuum and the carrier frequency of a signal the time is a measure for the distance between transmitter and receiver.

TDOA(Time Difference on Arrival)[8]: measures time difference between departing from one and arriving at the other station compared to the TOA and changes it into distance between transmitter and receiver. This technique has been widely utilized in WSN localization models.

2. Range Free Localization

Range free localization techniques estimate positions of sensors from proximity information available from different nodes rather than providing actual locations. Now solutions of range-free localization are being pursued as a cost-effective alternative to more expensive range-based approaches.

Although many range-based localization techniques have been developed, some can't be applied because of the high expense and the limitations of hardware[9]. So researchers are engaging in developing the range-free localization techniques.[10]

The range-free localization techniques include DV-Hop, Smorphous, MDS-MAP and APIT. These network models are all composed of reference nodes and unknown nodes, which do not need any infrastructure [11].

In this paper, an efficient localization algorithm is designed which can help localize transportation vehicles in ITS.

3   The Node-Localization Algorithm in WSN Based on RF and Ultrasonic Signals

3.1   ITS Based on WSN

The ITS in WSN includes the road reference sensor, the vehicle sensor and the intelligent monitoring system. The road reference broadcasts its position by different communication radius and the vehicle sensor having received the information senses its position, the lane and direction automatically and sends the information to the intelligent monitoring system which dynamically assigns the traffic signals, no-go or no-stop signals to avoid traffic congestion.

Figure 1 is the intelligent crossroad traffic made up 4 bidirectional roads. In the figure, Both vehicle 1 and vehicle 2 can estimate their locating lanes and their speed according to the localization information and communicating radius received from the road sensor and send the information to the traffic monitoring system.

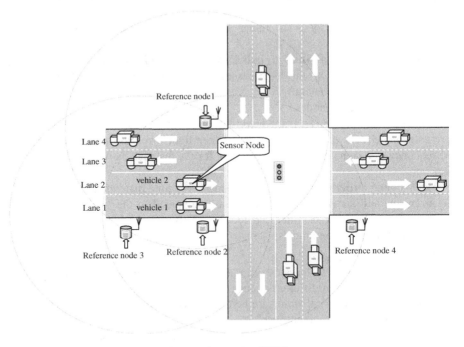

Fig. 1. ITS based on WSN

In figure 2, the localization of the vehicle sensor is based on the wireless signals and the ultrasonic distance information. In the figure, dotted wire indicates the road reference sensor, and solid wire indicates the distance from the road reference sensor

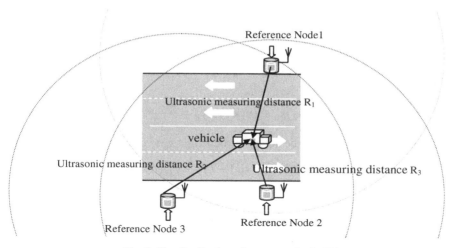

Fig. 2. The distribution of sensor nodes in ITS

measured by the ultrasonic wave, by which the precise location of the vehicle can be figured out.

3.2 Algorithm of Car-Carried Sensor

This algorithm is based on ultrasonic distance measurement and triangle locating. As shown in figure 3, If the coordinate of node M is (xu, yu), and three coordinates of road reference nodes A,B, C are (x1, y1), (x2, y2), (x3, y3), then the distance to the three nodes are d1、d2、d3.

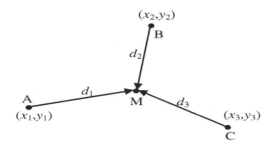

Fig. 3. The triangle locating

Then we have

$$\begin{cases} d_1 = \sqrt{(x_1 - x_u)^2 + (y_1 - y_u)^2} \\ d_2 = \sqrt{(x_2 - x_u)^2 + (y_2 - y_u)^2} \\ d_3 = \sqrt{(x_3 - x_u)^2 + (y_3 - y_u)^2} \end{cases} \quad (1)$$

After squaring, we can get (2)

$$d_1 = \sqrt{(x_1 - x_u)^2 + (y_1 - y_u)^2}$$

$$d_2 = \sqrt{(x_2 - x_u)^2 + (y_2 - y_u)^2} \qquad (2)$$

$\Rightarrow$

$$d_1^2 = x_1^2 - 2x_1 x_u + x_u^2 + y_1^2 - 2y_1 y_u + y_u^2$$
$$d_2^2 = x_2^2 - 2x_2 x_u + x_u^2 + y_2^2 - 2y_2 y_u + y_u^2$$

$\Rightarrow$

$$(x_1 - x_2)x_u + (y_1 - y_2)y_u = \frac{d_1^2 - d_2^2}{2}$$

The coordinates of node M can be got by introducing x_u to equation (1)

$$x_u = \frac{(y_1 - y_2)(d_3^2 - d_1^2) - (y_1 - y_3)(d_2^2 - d_1^2)}{(x_1 - x_3)(y_1 - y_2) - (y_1 - y_3)(x_1 - x_2)}$$

$$(3)$$

$$y_u = \frac{(x_1 - x_2)(d_3^2 - d_1^2) - (x_1 - x_3)(d_2^2 - d_1^2)}{(y_1 - y_3)(x_1 - x_2) - (x_1 - x_3)(y_1 - y_2)}$$

The distance information of three nodes d_1 、 d_2 and d_3 have to be known in order to get the coordinates of M. in this paper, the distance between sensors is computed by the time difference between wireless signal and the ultrasonic signal. As we know, the ultrasonic distance measurement is measured by the speed of ultrasonic wave transmitted in the air and get the actual distance by computing the time difference between the starting and arriving time. The formula is

$$L = C \times T \qquad (4)$$

L is the distance, C is the speed of ultrasonic wave and T is the time difference.

The speed of ultrasonic is affected by the air density which is in turn affected by the air temperature. The relationship between speed of ultrasonic and the air temperature is as

$$C = C_0 + 0.607 \times T \qquad (5)$$

In the equation, C0 indicates that the ultrasonic speed is 332m/s when the temperature is zero, and 0.607 is the proximity factor and T represents the actual temperature[12]. The temperature has to put into consideration when ultrasonic distance reaches 1mm. For example, the ultrasonic speed is 332m/s when the temperature is zero, and it is 350m/s when the temperature is 30. The speed difference of 18m/s is caused by the temperature change.

From the above analysis, we can get that the distance between two objects can be formulated by the temperature and the sending and receiving time of ultrasonic. The speed of the wireless signal is 299792458m/s, and so sending the ultrasonic signal after Δt time of wireless signal, the receiving end can computing the distance according to the time difference.

Suppose the sending time of wireless signal is t_{s0}, the sending time of ultrasonic signal is $t_{s0} + \Delta t_s$, the receiving time of wireless signal is t_{r0}, and the receiving time of ultrasonic signal is $t_{r0} + \Delta t_r$, then the distance between the sending and the receiving is

$$d = C \times [(t_{r0} + \Delta t_r) - (t_{s0} + \Delta t_s)] \qquad (6)$$

Then we can get

$$d = C \times [(t_{r0} + \Delta t_r) - (t_{s0} + \Delta t_s)]$$
$$= C \times (t_{r0} - t_{s0} + \Delta t_r - \Delta t_s)$$
$$= C \times \frac{d}{3 \times 10^8} + C \times (\Delta t_r - \Delta t_s)$$
$$= \frac{3 \times 10^8 \times C}{3 \times 10^8 - C} \times (\Delta t_r - \Delta t_s)$$
$$\approx C \times (\Delta t_r - \Delta t_s)$$

In the equation, C is the speed of sound and the distance is computed by the sending and the receiving time of ultrasonic waves.

3.3   The Model of Vehicle Sensor Localization

There are two kinds of sensors in the localization process: the reference sensor and the localizing sensor. The former one can communicate wirelessly and sending ultrasonic waves and the latter can communicate wirelessly and receiving ultrasonic waves and sensing the temperature.

3.3.1   The Algorithm of Reference Sensor

This section is decribing the localization algorithm of reference nodes. The mechanism of CSMA/CA of protocol of IEEE802.11 is adopted in order to avoid the conflicts between reference sensors. It can broadcast its location information in wireless network if a busy channel has occurred. If the broadcasting is successful, the reference node will be in the waiting state for sending ultrasonic waves at the prescribed time intervals. Because the wireless sensor is working in an event-driven way, the data-collection and topology operation can be conducted at the same time.

The carry-out of the algorithm of reference sensor is described as follows:

The carry-out of the algorithm of reference sensor

---

1. Node Initiating:

   Initiating the wireless modular, ultrasonic modular, sensor modular and the variables. Get the location in the node distribution area by GPS.

2. Program Initiating:

   The wireless modular, the sensor modular and the other modulars come into the working state.

3. CSMA/CA channel monitoring and conflict detecting

4. IF   *channel is busy or conflicts*   THEN

5.   Goto 3

6. ELSE

7. Delaying Δt_s time to ensure other sensor nodes to send the ultrasonic signals .

---

8.   Broadcasting the location message :   broadcasting the location message   of itself (coordinates X, Y) including the radius, node id, the
node type and the information type.

9.   Sending ultrasonic signals after time Δt_s .

10.   Waiting and collecting the localization result sent by the localizing
sensor and transmit it to the server .

11. ENDIF

3.3.2   The Algorithm of Localizing Sensor Node

The localizing sensor keeps in the state of waiting for the wireless signals at the beginning stage and does not conduct the localization algorithm until the wireless receiving event is triggered. The sensor is in the state of receiving and responds at any
time to the localization request of the sensor.

The carry-out process of the algorithm is described as follow:

The algorithm of localizing sensor node

1. Node Initiating:

Initiating the wireless modular, ultrasonic modular, sensor modular and
the variables.

2. Waiting state :

The sensor comes into the, and waits for the localization information received
by the wireless equipment, and other tasks can be conducted during the time.

3. Information filtering :

The localizing sensor files   information   by *MsgFilter* function.

4. IF *receiving the message package*   THEN

/* A reference node will send the ultrasonic signal.*/

5.   The sensor node collects the ultrasonic signals and the temperature data.

6.   The localization computing will be conducted after three reference nodes
have received the information

7. ENDIF

8. GOTO Step 2.

/*Regardless of the result, the nodes will come into the wireless waiting state,
and wait for more localization messaged sent by either reference sensors or
the localized sensors.*/

The algorithm of the localizing sensor is realized in the flow chart of Fig.3. The process will come back after the localization of the coordinates for more broadcasting message. The second time of carrying out the process after receiving new localizing message is the correction of the location coordinate estimated by the first time.

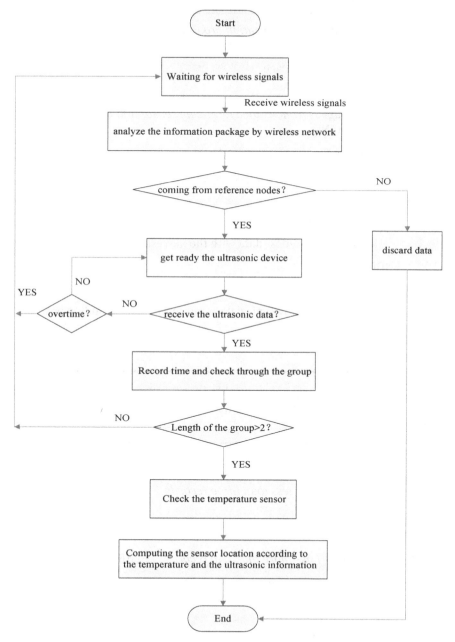

Fig. 3. The flow chart of localizing sensors

The valid message package is saved in ***msgTable*** after the invalid message is discarded and analyzing the ***msgTable*** by ***PositEsti*** function. The MsgFilter Function is describe as follows:

Function: MsgFilter

---

1. IF *the message is the localization broadcasting information and   ultrasonic data is not received within the given time*   THEN

2.   Discarding the message

3. ENDIF

4. IF *the message is from its own localizing sensor and ultrasonic data has existed*   THEN

5.   Discarding the information package

6. ENDIF

7. FOR *i* from *0* to *4*

8.   IF *msgTable[] is full*   THEN

9.   Replacing the oldest data with the newest one

10.   ENDIF

11.   IF *msgTable[] is not full*   THEN

12.   Wireless and ultrasonic data is put in the corresponding position

13.   ENDIF

14. ENDFOR

The processing of Analyzing msgTable [] and getting the current coordinate of the sensor is done by ***PositEsti*** function. The ***PositEsti*** Function is describe as follows:

Function: PositEsti()

---

1. IF   *msgTable[] has less than three records*   THEN

2.   No action.

3. ENDIF

4. IF   *msgTable[] has more than two records*   THEN

5.   collect the temperature T and compute the sound transmission speed C_0.

6. $C = C_0 + 0.607 \times T$;

---

7. $d \approx C \times (\Delta t_r - \Delta t_s)$;

8. ENDIF

4   The Analysis of the Performance of the Vehicle Sensor Algorithm

The TinyOS platform designed by UC Berkeley is used as the test platform in the paper. All the data is got by TOSSIM of TinyOS 1.1.7. TOSSIM is a stimulator of a discrete event in the wireless sensor network of TinyOS. A certain number of sensor nodes are produced at random in the range of 50m×50m in TOSSIM. The coordinates (x,y) of every sensor is produced at random and the biggest is communication radius is 120m. The reference nodes number accounts for 60% of the total number. We have a comparison test with the dv-hop algorithm, including the amount of the sensor information and the average localization error of sensor nodes. The total amount of the information packages is defined as the total amount of times of sending and receiving information packages by all the sensors. The average error refers to the distance between the real coordinates and the estimated ones, which can be got by distance dividing sensor numbers.

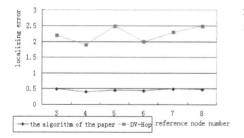 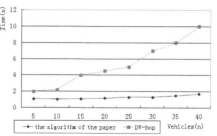

Fig. 4. Comparison of localization accuracy   **Fig. 5.** Comparison of localization time

From Figure 4, we can see that the localization accuracy varies within 0.5 meters, which is much smaller than dv-hop algorithm. The theoretical accuracy of the ultrasonic measurement algorithm can be pinned down within centimeters, but because of the block of the vehicle, the accuracy falls in 0.5 meter, which can meet the demand of the intelligent transportation system.

The localization time is the major index of measuring the time quality of the algorithm. From Figure 5, we can see the average localization time varies within 1-2 seconds, while the localization time of DV-hop algorithm getting longer with the increase of the vehicles, which is caused by the more hops.

5   Conclusion and Future Work

The main ideas of the paper is the localization techniques of the wireless sensor nodes in ITS, and a localization technique of the wireless sensor nodes based on ultrasonic distance measurement is proposed which supports the technique for localizing the transportation vehicles. In our future work, more work will be done in the systematic structure of ITS by combining V2V, V2I and I2I, etc.

References

1. Li, L., Liu, Y.-A., Tang, B.-H.: SNMS: an intelligent transportation system network architecture based on WSN and P2P network. The Journal of China Universities of Posts and Telecommunications 14(1), 65–70 (2007)
2. Wang, S.: Cooperative driving based on inter-ve-hicle communications experimental platform and algorithm. In: IEEE Conference Oil IlLS, Beijing, China, pp. 5073–5078 (2006)
3. European Commission. European transport policy for 2010: Time to decide, EB/OL (2011), http://ec.europa.eu/transport/rsap_en.html
4. Hung, C., Yarali, A.: Wireless Services and Intelligent Vehicle Transportation Systems. In: CCECE 2011, pp. 63–68 (2011)
5. Murray, C.: The Biggest Thing in Safety (cover story). Design News 63(8), 58–60 (2008)
6. Bahl, P., Padmanabhan, V.N.: RADAR: an In-building RF-based user location and tracking system. In: Proceedings of the Nineteenth Annual Joint Conference of the IEEE Computer and Communications Societies, INFOCOM 2000, vol. 2, pp. 775–784. IEEE (2000)
7. Girod, L., Estrin, D.: Robust Range Estimation Using Acoustic and Multimodal Sensing. In: Proceedings of the IEEE/RSJ International Conference on Intelligent Robots and Systems, vol. 3, pp. 1312–1320 (2001)
8. Savvides, A., Han, C.-C., Strivastava, M.B.: Dynamic Fine-Grained Localization in Ad-Hoc Networks of Sensors. In: Proceedings of the 7th Annual International Conference on Mobile Computing and Networking, Pages, pp. 166–179 (2001)
9. Bahl, P., Padmanabhan, V.N.: RADAR: An In-Building RF-Based User Location and Tracking System. In: IEEE InfoCom 2000, vol. 2, pp. 775–784 (March 2000)
10. Doucet, A., Godsill, S., Andrieu, C.: On Sequential Monte Carlo Sampling Methods for Bayesian Filtering. Statistics and Computing 10(3), 197–208 (2000)
11. Nagpal, R., Shrobe, H.E., Bachrach, J.: Organizing a Global Coordinate System from Local Information on an Ad Hoc Sensor Network. In: Zhao, F., Guibas, L.J. (eds.) IPSN 2003. LNCS, vol. 2634, pp. 333–348. Springer, Heidelberg (2003)
12. Caballero, F., Merino, L., Gil, P., Maza, I., Ollero, A.: A probabilistic framework for entire WSN localization using a mobile robot. Robotics and Autonomous Systems 56(10), 798–806 (2008)

Parallel Network Virtualization Resource Management System[*]

Juan Luo[**], Lei Chen, Shan Fu, and Renfa Li

School of Information Science and Engineering, Hunan University
Changsha, Hunan, China
juanluo@sina.com, chenleixyz123@126.com, 490717971@qq.com,
lirenfa@vip.sina.com

Abstract. Network virtualization is an important solution for overcoming the internet ossification. Designing excellent and general resource management system is one of challenges. A kind of parallel distribution management system is proposed in the paper, which can well regulate resource management and allocation. By using resources distributors to allocate resources for virtual network and providing resources reservation and locking mechanism, it greatly improves the virtual network distribution rate and the performance of the whole system under the premise of guaranteeing the distribution independence of each virtual network. Simulation results show that, without influencing the performance of various virtual network distribution algorithms, the parallel distribution mechanism can greatly improve the distribution rate of virtual networks and the utilization rate of physical nodes and links.

Keywords: Network Virtualization, Resource Management, Parallel Assignment, Resource Mapping.

1   Introduction

Network virtualization is the key technology for solving Internet ossification [1]. It is considered to be crucial component of the next generation of Internet system structure[2]. With the development of network technology, new technology and new service such as distinguish service, IP multicast, routing safety and so on, emerge endlessly and develop rapidly [3]. Thus, the current single Internet architecture can't satisfy the demands of new technology deployment [4-5]. While the old system structure can not meet new application requirements, the new architecture is difficult to be evaluated and deployed. Internet ossification emerges accordingly.

For conquering Internet ossification, network virtualization technology does not use new system structure to replace the existing one. Instead, it constructs a platform

---

[*] This work is partially supported by Hunan Provincial Natural Science Foundation of China (11JJ5039); The Project-Sponsored by SRF for ROCS, SEM; Hunan young core teacher project; Young teacher project from Hunan University.
[**] Corresponding author.

Z. Bao et al. (Eds.): WAIM 2012 Workshops, LNCS 7419, pp. 69–77, 2012.

based on the existing system structure to meet the demands of new network system structure deployment. This platform can simultaneously deploy several virtual networks (VN) with parallel operation, and each virtual network can use different system structure.

The most important thing of parallel operation of multiple virtual networks is reasonable sharing physical network resources and effective distribution and management of resources on demands. To achieve the effective distribution of resources on demands, firstly it needs advanced resource allocation algorithm, then it requires highly efficient system to manage resource. In the past, people focused on the research of advanced algorithm of resources allocation [6-7], resource management framework rarely got people's attention. As one of the key factors of achieving effective resource allocation on demands, the resources distribution management system is also very important. It is not only the bridge of coordination between terminal users and infrastructure suppliers, but also it guarantees smoothly running of resource allocation algorithm.

2   Related Work

Zhang and other researchers[8] put forward QoS oriented network management framework which was based on network virtualization. The management structure put emphasis on virtual routing, using multi goods flow theory and stratification mechanism to improve efficiency of the management structure. The article used multi goods flow theory to balance the load of whole physical network, and it adopted stratification mechanism to increase the throughput and efficiency of network. However, this paper did not consider resources management. It limited the resources management in the routing level without considering the aspects such as resource discovery, collection and so on. In addition, the management structure in the article was only suitable for the Qos-guaranteed network environment with poor generality. Marquezan and some researchers[9] studied resource management of distributed network virtualization and put forward a distributed self-organizing model to allocate and manage the resources. Through the distributed self-organizing model, it greatly improved the efficiency of resource management and distribution. Because of the complexity of heterogeneous virtual network and the resource allocation, the traffic between the physics nodes in the distributed model was extremely huge and the stability of the model was poor.

Barachi[10] and some other researchers put forward a service oriented framework of resource management which was virtualized in network. The structure presented a service model to classify the virtual network service according to the priority level. Meanwhile, it distributed and managed resource on the basis of its rank. As a result, it improved the efficiency of resources management and allocation. The structure used priority theory to redefine the management and the distribution order of virtual network, but it didn't express resource allocation efficiency very well. For service provider of virtual network, Che [11] and other investigators put forward a network virtualization access framework. The framework also used hierarchy to divide priority level according to the customer's demands and properties of network. Aimed at priority, it managed and distributed the resources in virtual network. Similar to Barachi [10] and others, the access mechanism in the paper made use of the priority thought, but it hadn't considerably improved the efficiency of resources management and distribution. According to the next generation network, Mobile network, Hoffmann [12] came up with a resource management structure. It applied to any application in Mobile network, but it was only for Mobile network and did not study resource management and distribution efficiency.

In literature [13], based on multi region and level in virtual network management structure, a virtual network management framework was proposed. The framework divided the physical area into multiple independent areas by making use of regional division theories, then it allocated an independent management unit for each region, which increased the efficiency of management framework. The framework made greatly improvement in efficiency and generality of resources allocation. Nevertheless, although the framework made division in the physical network, it didn't give division standard. Since physical area was as a whole and relationship between the nodes was close, making division in area was very difficult.

In this paper we propose a concurrent network virtualization and resource management system. The system adopts the concurrent processing mechanism with concurrent distribution and resource management of multiple virtual networks. Furthermore, through using division theories, the system divides the complex virtual network into sub division, forming a small star network topology. By using concurrent mechanism, it further improves the network resources mapping and the efficiency of distribution.

3   Concurrent Resource Management Architecture Design

Concurrent resource management system is composed of four modules: strategy scheduling module, resources mapping module, state database and local database, as shown in Fig.1. Strategy scheduling module monitors virtual network request in real time and divides the virtual network into multiple star sub-network. Then the module sends the divided network set to resources mapping module and assigns the corresponding mapping algorithm (network set including the sub-network and coordination network). According to the number of virtual network, strategy scheduling module can dynamically add and reduce the number of resources mappers in resource mapping module, so as to dynamically adjust resource concurrent allocation and management rate. Without destroying the relationship among nodes in physical network, improve the efficiency of resource allocation and mapping. According to virtual sub-topology allocated by the strategy management module, resources mapping module concurrently distributes several virtual sub-topology to resources mapping device. Meanwhile, it designates resources mapping algorithm correspondingly. On the basis of specific mapping algorithm, each resource mapper carries through the resources mapping, which refers the current state of network that is in state database. Local database module stores all the resources allocation mapping situation of virtual network in order to manage and schedule strategy manager conveniently.

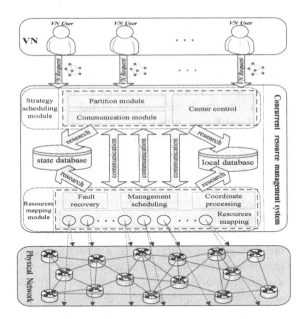

Fig. 1. Resource management system structure

Strategy scheduling is the core processor module of the whole system. Central control module is the coordinator between network partition and communication module, monitoring the whole resource allocation and management situation in real time. First of all, the central control module monitors the arrival of the virtual network. When the virtual network arrives, the central controller accepts and stores it in the waiting sequence of network partition modules, waiting the management of network partition module. Another task of central control module is regularly monitoring the waiting sequence length of network partition module. Network segmentation module is responsible for topology division of waiting sequence in the virtual network. Resources mapping module is responsible for the physical resource allocation of arrived virtual topology. It is consisted of management scheduling, coordinated processing, fault recovery and resources mapping.

Management scheduling is commander and coordinator of the whole module, saving all of the relevant information of resources mapping including abundant resources mappers and detailed information of mappers (quantity of sub topology waiting for mapping, mapping algorithm, the finish situation of virtual network, etc.). Management scheduling is also in charge of collaboration of resources mapping, coordinate processing and fault recovery.

State database stores state information of current physical network, including total quantity of resources on nodes and link, amount of used and remaining resources, locked state, etc. According to the current state of network, resources mapper and coordination processor carry through resources mapping of virtual network by using concurrent and mechanism resources mapping algorithm.

Local database has stored all information of the resource allocation, management and the communication between modules since the network initializes. Through the local database, concurrent system can monitor network operation, the physical resource allocation, network load, network throughput rate and utilization efficiency of node and link, etc. In addition, system can make analysis and statistics of various data so as to greatly reduce the difficulty of resource management, improve the management efficiency and promote the performance of concurrent system.

4   Concurrent Distribution Mechanism Design

Resources mapping module contains several resources mapping device which can allocate resources concurrently. Strategy scheduling module divides virtual topology, then resources mapping module assigns sub topology and mapping algorithm for several resources mapper and conducts resources mapping concurrently.

If concurrent mechanism is adopted, there will necessarily cause chaos of mapping order of virtual network and destroy the topology of the original relationship between nodes and link. Therefore, this article applies reservation mechanism to reserve physical resources in virtual network in order to ensure the original relationship between nodes and link in the virtual network topology. In the physical network, the nodes have a locked mark which means the node is occupied or not. When resources mapper occupies physical node, it will lock nodes and link which is directly connected with the physical node. Thus, it can avoid the occupation of other resources mapper and reserve resources for virtual topology before mapping completes. Virtual network forms multiple star topologies after strategy scheduling division. Since the links between the topological nodes are close, the traffic is great and the nodes need to map to physical concentrated area, ensure the integrity of topological relationships, improve resources mapping rate and enhance performance of the resource allocation.

Concurrent distribution mechanism mainly divided into two parts: sub topology allocation and coordinated topology allocation.

Virtual network is divided into sub topology by topology division. Several resource mappers manage resource mapping of sub topology concurrently. Mapping processes are as follows:

(1)Select position of first physical node. According to the specified resource mapping algorithm, physical node is selected and judged whether locked or not. If locked, the next node will be visited (if all nodes are locked, wait for a short time to restart mapping); Otherwise, it takes up the physical node, and reserves resources according to the second step.

(2)On basis of resources reserve mechanism, lock the nodes and links which directly connect physical node, then sign network number correspondingly. Only the resources mapper which is distributed network number can conduct the locked physical resources mapping. When Lock finishes, then turn into the third step.

(3)Judge whether locked resources meet the demands of sub topology or not. If not, around the locked nodes look for the unlocked to lock them, expand the scope of locking until the locked resource meet the demands in network, then turn into the forth step.

(4)Do the mapping according to the resources algorithm, then turn into the fifth step.

(5)Relieve all locked resources and store mapping information to the management scheduling module.

After all the sub topology mapping in the same virtual network is completed, it needs to manage the resources mapping of coordination topology. Once management scheduling module monitors the finish of all sub topology mapping in virtual network, it will notice the coordination mapping module to coordinate the topology mapping. The system has more than one processor to conduct multiple coordination topology mapping. Coordination processes are as follows: according to the information of sub topological mapping stored in management scheduling module, look for unlocked link resources in physical network for resources mapping on the basis of the resources mapping algorithm. If the physical network can not meet demands of topology coordination, coordination topology will be moved to the end of waiting sequence to wait for the next distribution, until mapping is completed. The process of concurrent distribution mechanism is shown in Fig.2.

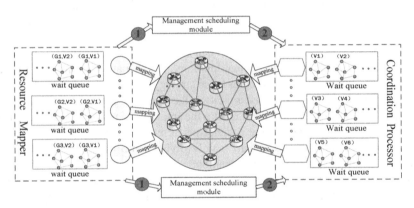

Fig. 2. Concurrent distribution mechanism

5   Simulation Results

This paper uses the cloud computing simulation platform-CloudSim2.0[14] as tools to simulate the algorithm. Several aspects of a variety of algorithm are simulated, such as virtual network adoption rate, average node utilization rate, average link utilization ratio, the biggest node load, the largest link load and so on. The simulation compares performance difference before and after adopting the concurrent distribution mechanism. We compares three virtual network resource allocation algorithms, G-SP[4], D-ViNE [15], PG-VNE [16].

Fig.3 compares average nodes utilization rate of the three kinds of algorithms before and after using concurrent distribution mechanism. We can see the concurrent distribution mechanism greatly improves the average nodes utilization rate. Through the concurrent distribution mechanism, several resources mapper simultaneously conduct virtual network resources mapping, which increase the distribution number of virtual network per minute, accelerate the resources allocation rate and improve the average utilization rate of physical network node. In addition, when the system is initializing, the average utilization rate of network nodes using concurrent distribution mechanism inferior to that of nodes not using mechanism.

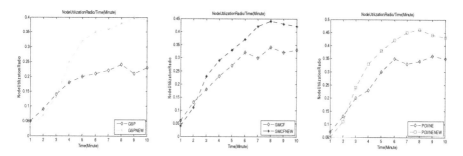

Fig. 3. Node average utilization rate

Fig.4 compares average link utilities rate of three kinds of algorithm before and after using concurrent distribution mechanism. From the picture, when the system initializes, the number of resources mapper is small and concurrent distribution granularity is poorer. It causes low resource allocation rate by adopting concurrent mechanism. Meanwhile, the system cost generated by resource reserving and locking mechanism leads that the link utilization rate is below that of common resources distribution algorithm. Nonetheless, with the rising number of arrived virtual network and resources mapper, the increasing size of concurrent granularity and continuous improvement of sources mapping and distribution rate, link average utilization rate will be significantly superior to that of ordinary algorithm. And, with the continuous optimizing of resources allocation algorithm, link average utilization rate constantly improves through applying allocation mechanism.

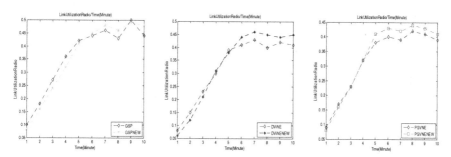

Fig. 4. Average link utilization rate

Fig.5 describes the link load in physical network at different moments. The concurrent distribution mechanism does not improve link load conditions, but lead to even more intense. Concurrent distribution uses resources reserving and locking mechanism. On one hand, it ensures the virtual network distribution rate. On the other hand , it increase the area link load, causes the link load less balanced and leads to the link utilities fluctuation even more intense.

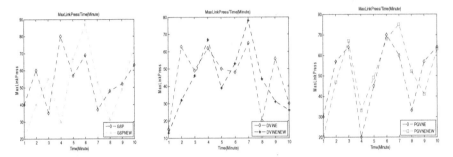

Fig. 5. Maximum link load

6   Conclusion

According to the lack of resource management system, poor universality of utilization, resource allocation efficiency and coordination between resource management and resource allocation and so on, this paper puts forward a concurrent resource allocation and management mechanism. The system contains a set of concurrent distribution mechanism, adopting several resources mappers to manage concurrent resources mapping. As a result, it enhances the resources mapping rate, improves virtual network acception rate and perfects the resource management performance. Concurrent distribution makes use of resources reserving and locking mechanism for virtual network to reserve resources, ensure the monopoly of the physical node, improve the efficiency of resource allocation and optimize the resources distribution performance. Concurrent distribution mechanism is suitable for most centralized resource allocation algorithm with great generality. In addition, resource management system has designed a set of complete communication and fault handling system. Communication mechanism safeguards coordination services among each module in the concurrent system. Failure mechanism guarantees the stable operation of virtual network to improve the robustness. However, all virtual networks are equal in the concurrent systems. It does not take into account the special needs of each network. Therefore, virtual network conducting priority division will be the next work.

References

1. Anderson, T., Peterson, L., Shenker, S., Turner, J.: Overcoming the Internet impasse through virtualization. IEEE Computer Magazine 38(4), 34–41 (2005)
2. Lv, B., Huang, T., Wang, Z.-K., Chen, J.-Y., Liu, Y.-J., Liu, J.: Adaptive scheme based on status feedback for virtual network mapping. The Journal of China Universities of Posts and Telecommunications 18(5), 63–72 (2011)

3. Armstrong, W.J., Arndt, R.L., Boutcher, D.C., Kovacs, R.G., Larson, D., Lucke, K.A., Nayar, N., Swanberg, R.C.: Advanced virtualization capabilities of Power5 systems. IBM Journal of Research and Development 49(4/5), 523–532 (2005)
4. Zhu, Y., Ammar, M.: Algorithms for assigning substrate network resources to virtual network components. In: Proceedings of IEEE INFOCOM, pp. 1–12. Institute of Electrical and Electronics Engineers Inc., Barcelona (2006)
5. Liu, J., Huang, T., Chen, J.-Y., Liu, Y.-J.: A new algorithm based on the proximity principle for the virtual network embedding problem. Journal of Zhejiang University - Science C 12(11), 910–918 (2011)
6. Yu, M., Yi, Y., Rexford, J., Chiang, M.: Rethinking virtual network embedding: Substrate support for path splitting and migration. ACM SIGCOMM Computer Communication Review 38(2), 17–29 (2008)
7. Fiedler, M.: On resource sharing and careful overbooking for network virtualization. International Journal of Communication Networks and Distributed Systems, 232–248 (2011)
8. Zhang, Y., Wang, C., Gao, Y.: A QoS-Oriented Network Architecture based on Virtualization. Education Technology and Computer Science, 959–963 (2009)
9. Clarissa, C.M., Lisandro, Z.G., Giorgio, N., Brunner, M.: Distributed Autonomic Resource Management for Network Virtualization. In: Network Operations and Management Symposium (NOMS), pp. 463–470. IEEE (2010)
10. May, E.B., Nadjia, K., Rachida, D.: Towards A Service-Oriented Network Virtualization Architecture. In: Kaleidoscope: Beyond the Internet? - Innovations for Future Networks and Services, pp. 1–7. IEEE Computer Society, Pune (2010)
11. Che, Y., Yang, Q., Wu, C., Lianhang, M.: BABAC: An access control framework for network virtualization using user behaviors and attributes. In: IEEE GreenCom, pp. 747–754. IEEE Computer Society, Hangzhou (2010)
12. Marco, H., Markus, S.: Network Virtualization for Future Mobile Networks General Architecture and Applications. In: 2011 IEEE International Conference on Communications Workshops, pp. 1–5. Institute of Electrical and Electronics Engineers Inc., Kyoto (2011)
13. Lv, B., Wang, Z., Huang, T., Chen, J., Liu, Y.: A Hierarchical Virtual Resource Managemen Architecture for Network Virtualization. In: IEEE WICOM, pp. 1–4 (2010)
14. Calheiros, R.N., Ranjan, R., Beloglazov, A., DeRose, C.A.F., Buyya, R.: CloudSim: a toolkit for modeling and simulation of cloud computing environments and evaluation of resource provisioning algorithms. Journal Article (JA), 23–50 (2011)
15. Chowdhury, N.M.M.K., Rahman, M.R., Boutaba, R.: Virtual network embedding with coordinated node and link mapping. In: The 28th Conference on Computer Communications, IEEE INFOCOM 2009, pp. 783–791. Institute of Electrical and Electronics Engineers Inc., Rio de Janeiro (2009)
16. Gao, X.J., Yu, H.F., Anand, V., Sun, G., Hao, D.: A New Algorithm with Coordinated Node and Link Mapping for Virtual Network Embedding Based on LP Relaxation. In: Procs of SPIE - The International Society for Optical Engineering, pp. 152–153. SPIE, Shanghai (2011)

The Analysis of Priority-Based Slotted CSMA/CA Algorithm in IEEE 802.15.4 Sensor Network

Yang Zao, Liu Yan, and Li Renfa

College of Information Science and Technology,
Hunan University, China
chaozao1987@163.com, liuyan@hnu.edu.cn, lirenfa@vip.sina.com

Abstract. IEEE 802.15.4 standard provides a MAC and PHY layer specification for a low rate, low power, low complexity and short range Wireless Personal Area Network (WPAN). The basic access mechanism used by the standard is the CSMA/CA. In this paper, Markov chain model is used to reflect the characteristics of the IEEE 802.15.4 medium-access control protocol, and proposed an enhanced priority-based slotted CSMA/CA algorithm. The simulation results used OMNET environment show that proposed algorithm can achieve higher channel access rate.

Keywords: Mac layer protocol, CSMA/CA, 802.15.4.

1 Introduction

With rapid improvements in wireless technologies, wireless sensor networks (WSN) are attracting growing attention form both research communities and vendors. WSNs have been widely deployed in our lives, connecting sensor nodes in a simple and efficient manner with low power and low cost has become an important issue. A WSN essentially requires that a common communication link between nodes, so that the performance of WSN is critically dependent on the performance of MAC layer protocol. The IEEE 802.15.4 standard for low-rate wireless personal area networks has been introduced to achieve above requirements [1].

IEEE 802.15.4 networks can operate in a nonbeacon-enabled mode. In this mode, nodes in the network communicate with each other according to an unslotted CSMA/CA protocol. And in the beacon-enabled mode, nodes' communication according to a slotted CSMA/CA protocol based on a super frame structure. Each super frame consists of an active period and an inactive period. The active period consists of a beacon period, a contention access period (CAP), and a contention free period (CFP). During the inactive period, the coordinator and the nodes shall not interact with each other and may enter a low-power mode [2].

In the original CSMA/CA algorithm of IEEE 802.15.4, a node starts transmitting packets only when it has enough time in the current superframe for a successful transmission. However, if the remaining time in the current superframe is not

Z. Bao et al. (Eds.): WAIM 2012 Workshops, LNCS 7419, pp. 78–86, 2012.

sufficient then the packet transmission is deferred to a new superframe. This problem was addressed in the revised version of the new standard. In this paper, we propose an enhanced Priority-based slotted CSMA/CA algorithm, and establish the Markov chain model of this improved-mechanism to observe the throughput performance of the algorithm, and the Markov chain model reflects the characteristics of the IEEE 802.15.4 MAC protocol. Through simulations, we verify that the effective of our algorithm.

The remainder of the paper is organized as follows. In Section2, we will present a brief overview for CSMA/CA algorithm. The enhanced Priority-based slotted CSMA/CA algorithm is illustrated in Section3. Section 4 presents a set of experimental results obtained by the developed algorithm. Finally, we make a summary of accomplished work and describe future work in Section 5.

2   CSMA/CA Algorithm of 802.15.4

IEEE 802.15.4 publish standard for the Physical (PHY) and MAC layer of low rate wireless sensor networks. The standard uses slotted CSMA/CA as its access mechanism. In the slotted CSMA-CA channel access mechanism, each time a node wishes to transmit a packet during the Contention Access Period (CAP). Each node maintains the value of 3 parameters: Number of Backoffs (NB), Contention Window (CW) and Backoff Exponent (BE). NB is always initialized to 0 for a new packet transmission and it denotes the maximum attempts the CSMA/CA algorithm can make, while attempting the current packet transmission. CW defines the length of the contention window, which is the number of backoff periods for which the medium should be sensed idle before attempting the packet transmission. CW is initialized and reset to 2 each time the channel is found busy. BE defines the random backoff period that a device should wait before attempting to assess the channel [3].

The CSMA/CA algorithm is divided into six steps. In step 1, the values of three parameters (NB, CW, BE) are initialized according to the IEEE 802.15.4 standard. The boundary of backoff slot is set after initialization of parameters in step 1. When algorithm goes to step 2, the delay of random backoff period is selected in the range from 0 to $2^{BE}-1$. After algorithm starts counting down the number of backoff periods, algorithm moves to step 3.

A node starts transmission of packets only when it has enough backoff periods available in the current superframe for a successful transmission.   Then the algorithm will perform two Clear Channel Assessment (CCA), transmit the packet. This decision is made in step 3. If the CAP has enough backoff periods, algorithm will proceed to step4. If the channel is assessed to be busy then it proceeds to step 5. In step 6, it will decrement the CW by 1 and will check if two CCAs are performed, if not then it moves to step 4 and on next backoff boundary channel is assessed again. More details refer to [3, 4].

3   Priority-Based CSMA/CA Algorithm

In this section, we introduce the Priority-based Slotted   Algorithm.

In some specific monitoring network, when the nodes detect an emergency event , they must send the datagram to the sink node as soon as possible. So we must distinguish the priority of all datagram need to transmission[5]. Those datagram contains emergency events should be given privileged access to transmission. And other datagram don't own emergency event should make a corresponding move to delay transmission[6].

3.1   Algorithm

Figure 1 shows the main procedure of Priority-based Slotted CSMA/CA Algorithm. .

The major difference between this advanced algorithm and the standard Slottd CSMA/CA Algorithm is step1 and step4. In step 1, we add a judgment flow to distinguish the priority of datagram. To the high priority datagram ,CW will be initiated to 1 .As for the low priority datagram, CW still remain to 2.After the CCA of steps in addition ,we add another judgment flow. Give the CW on the same settings in step1.And to the high priority datagram, we made its NB plus 1and low priority plus 2.

After this setting change, the high priority datagram have a CW smaller than the low priority datagram. It decides the high priority datagram can made its CW=0 before the low priority datagram, then it has prior authority to access the channel. This mechanism can also made the NB of low priority datagram reach its maximum limit before the high priority datagram. That means if the channel is in busy state for a long time, the MAC can get rid of the low priority datagram because its NB across the high limit. And the high priority datagram can still remain its current state to contest for the servitude of the channel.

Both improvement methods on two parameters can made the high priority datagram have much more chance to access the channel and start to transmission. The mechanism to abandon low priority datagram after a certain number of trying to access the channel and keep the high priority database to compete make channel release the pressure of node's competition. Reduce the energy consumption of the whole network on the precondition of keep the network's communication efficiency. In general, the   Priority-Based Slotted CSMA/CA algorithm just makes a bit change in initiate the two parameter and the method to dispose them after the CCA compare to the standard CSMA/CA algorithm.

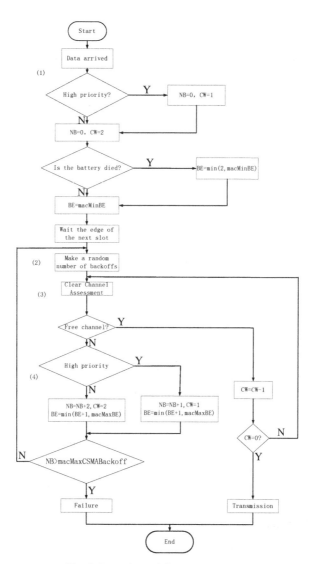

Fig. 1. Procedure of the new algorithm

3.2   Markov Model Analysis

In Figure 2, it is the Markov model of Slotted CSMA/CA Algorithm without considering node sleep mode. BO_{ik} is the backoff state of node, here $i = NB \in [0, m]$, $k \in [0, W_{i-1}]$, $W_i = W_0 2^i$. W_i means the size of the BE when BE is i. $CCA_{i,1}$ and $CCA_{i,2}$ means the results of the first and the second CCA is busy. T_l is the state that data frame is transmitting, $l \in [1, L]$, L is the slot's quantity which the node need when it transmitting data frame, and l is

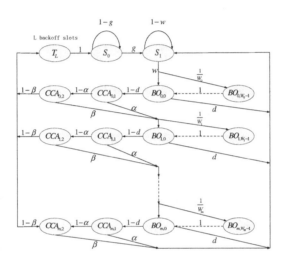

Fig. 2. Markov model of Slotted CSMA/CA Algorithm

The slot node which node current locate. S_0 is the sleep mode which has no data frame need to transmit. S_1 means there is a data frame wait in the node to transmit but the node is still in sleep mode. α and β means the probability the first and the second CCA results is busy. g is the transition probability that node transfer its state from S_0 to S_1 when there is a data frame arrived and wait to transmit. w is the probability that the node transfer its state from S_1 to a random backoff state that wait the channel become free.

On the basis of the Markov Model in Figure 2 and the principle of Markov Chain, we can infer that all of the node's state show's in the figure has a smooth probability distribution[7]. The transition probability of each state is show in the next equation system.

$$
\begin{cases}
P\{BO_{i,k} \mid BO_{i,k+1}\} = 1, i \in [0,m], k \in [0, W_i - 2] \\
P\{BO_{0,k} \mid BO_{i,0}\} = w[(1-d)(1-\alpha)(1-\beta)g \\
\quad +d]/W_0, i \in [0, m-1], k \in [0, W_0 - 1] \\
P\{BO_{i,k} \mid BO_{i-1,0}\} = (1-d)[\alpha + (1-\alpha)\beta]/W_i, \\
\quad i \in [1,m], k \in [0, W_i - 1] \\
P\{BO_{0,k} \mid BO_{m,0}\} = w[(1-d)(1-\alpha)(1-\beta)g + \\
\quad (1-d)[\alpha + (1-\alpha)\beta] + d]/W_0, k \in [0, W_i - 1] \\
P\{BO_{0,k} \mid S_1\} = w/W_0, k \in [0, W_0 - 1] \\
P\{S_1 \mid S_0\} = g \\
P\{S_0 \mid T_l\} = 1, l \in [1, L] \\
P\{T_l \mid BO_{i,0}\} = (1-d)(1-\alpha)(1-\beta), i \in [0,m], l \in [1, L] \\
P\{CCA_{i,1} \mid BO_{i,0}\} = 1 - d, i \in [0,m] \\
P\{CCA_{i,2} \mid BO_{i,0}\} = (1-d)(1-\alpha), i \in [0,m]
\end{cases}
\tag{1}
$$

Then from those transition probability of the nodes, we can infer the equation system of the model's probability of stability

$$
\begin{cases}
\pi_{BO_{0,k}} = (W_0 - k)/W_0 \left\{ \left(\sum_{i=1}^{m} \pi_{BO_{i,0}} \right) [(1-d)(1-\alpha)(1-\beta)g+d] + \pi_{BO_{m,0}}(1-d)[\alpha+(1-\alpha)\beta] \right\} \\
\pi_{BO_{i,0}} = \pi_{BO_{i-1,0}}(1-d)[\alpha+(1-\alpha)\beta] \triangleq \pi_{BO_{i-1,0}} q = ... = \pi_{BO_{0,0}} q^i, i \in [1,m] \\
\pi_{BO_{i,k}} = (W_i - k)/W_i \left\{ \pi_{BO_{i-1,0}}(1-d)[\alpha+(1-\alpha)\beta] \right\} \\
\quad = (W_i - k)\pi_{BO_{i,0}}/W_i, i \in [1,m] \\
\sum_{i=1}^{m} \pi_{BO_{i,0}} = \pi_{BO_{0,0}}(1-q^{m+1})/(1-q) \triangleq \tau \\
\sum_{i=1}^{m} \sum_{k=0}^{W_i-1} \pi_{BO_{i,k}} + \sum_{i=1}^{m} \pi_{CCA_{i,1}} + \sum_{i=1}^{m} \pi_{CCA_{i,2}} + L\pi_{T_i} + \pi_{S_0} + \pi_{S_1} = 1
\end{cases}
\tag{2}
$$

We assume the quantity of the high priority node and the low priority node is N_1 and N_2. p_0 is the probability of the node receives new data frame which is waiting to transmit ,and the probability of data frame arrives node obey Poisson distribution. We set the arrive rate of Poisson distribution is λ_q ,then $p_0 = \lambda_q / N_q$.

Here N_q is the quantity of node, q means the sort of nodes, when $q=1$, the node is high priority, and when $q=2$, the node is low priority. $q \in \{1, 2\}$

We use $\{s(t), \omega(t)\}$ to indicate the node's state in moment t. $s(t)$ indicate NB and $\omega(t)$ indicates BE. Then we made:

$$
e_{i,k} = \lim_{t \to \infty} p\{s(t) = i, \omega(t) = k\}
\tag{3}
$$

$e_{i,k}$ indicates the probability of stability when $\{s(t)=i,\omega(t)=k\}$, we can infer that:

$$e_{i,0}=\alpha e_{i-1,0} \tag{4}$$

And on the basis of the character of Markov Chain,:

$$e_{i,k}=e_{i,0}(W_i-k)/(W_i-W_{i-1}),i\in(0,m),k\in(0,W_i-1)$$

$$e_{-1,-1}=e_{0,0}(1-p_0)/p_0 \tag{5}$$

$$\sum_{k=0}^{L-1}e_{-2,k}+\sum_{i=0}^{m}\sum_{k=0}^{W_i-1}e_{i,k}=1 \tag{6}$$

On the basis of those equation, we can use iterative algorithm to derived the probability of the node finish its first random backoff; the probability of the channel's state is busy after the first and the second CCA; Te probability that the channel is busy after both two CCAs.

4   Experimental Results

We use the professional network simulation tool OMNeT++ (Objective Modular Network TestBed in C++) to test and verify the theoretical derivation.

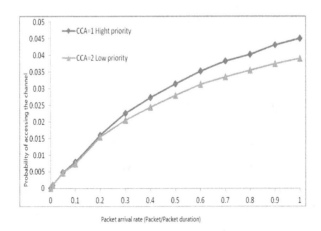

Fig. 3. The Accessing rate of two kinds of priorities datagram

In Figure 3, e can see the when the packet arrival rate (λ) is low, the difference between the node which finish 1 and 2 CCAs doesn't have obvious effect to the probability of accessing the channel. That's because most nodes is in the sleep mode when λ is low. Then along with the λ growing up, high priority datagram shows an obvious higher access rate than low priority datagram. That's because the high priority node just need 1 CCA to access the channel when the channel's state is free, but the low priority need finish 2 CCAs. That makes the throughput of high priority datagram increase obviously. But when λ growing up further, the channel is approaching it's saturation state. That makes the collide rate of data frame increase rapidly.

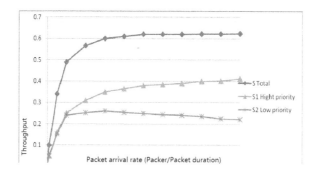

Fig. 4. The Accessing rate of two kind of priorities datagram

5   Conclusions

The paper presented a enhanced priority-based slotted CSMA/CA algorithm for IEEE 802.15.4 MAC layer for emergency response applications with low-power requirement used Markov chain model. The analysis focuses on the quality of service and priority schedule strategy. Comparison to simulation results validates our analysis.

Acknowledgements. This paper is supported in part by the National Nature Science Foundation of China under Grant 60673061.

References

1. Wireless Medium Access Control (MAC) and Physical Layer (PHY) Specifications for Low Rate Wireless Personal Area Networks (LR-WPANs), IEEE Std. 802.15.4 Specification (October 1, 2003)
2. Jung, C.Y., Hwang, H.Y., Sung, D.K., Hwang, G.U.: Enhanced Markov Chain Model and Throughput Analysis of the Slotted CSMA/CA for IEEE 802.15.4 Under Unsaturated Traffic Conditions. IEEE Transactions on Vehicular Technology 58(1), 473–478 (2009)
3. Rehamn, S.U., Berber, S., Swaim, A., Holmes, W.: Modeling the impact of deferred transmission in CSMA/CA algorithm of IEEE 802.15.4 for acknowledged and unacknowledged traffic. In: Proceedings of the 8th ACM Symposium on Performance Evaluation of Wireless Ad Hoc, Sensor, and Ubiquitous Networks, New York, USA, pp. 25–32 (2011)
4. Banerjee, S., Bozorgzaden, E., Dutt, N.: Performance analysis of beacon-enabled IEEE 802.15.4 MAC for emergency response applications. In: Proceedings of the 3rd International Symposium on Advanced Networks and Telecommunication Systems, New Delhi, India, pp. 1–3 (2009)
5. Bianchi, G.: Performance analysis of the IEEE 802.11 distributed coordination function. IEEE Communicaton Magazine 41(2), 116–125 (2003)

6. Ramachandran, I., Das, A.K., Roy, S.: Analysis of the contention access period of IEEE 802.15.4. MAC. ACM Trans. Sensor Netw. 3(1), 1–29 (2007)
7. Dam, T.V., Langendoen, K.: An Adaptive Energy-Efficient MAC Protocol for Wireless Sensor Networks. In: Proceedings of the 1st International Conference on Embedded Networked Sensor Systems (SenSys 2003), pp. 171–180 (November 2003)
8. Demirkol, I., Ersoy, C., Alagoz, F.: MAC Protocols for Wireless Sensor Networks: A Survey. IEEE Communications Magazine 44(4), 115–121 (2006)

QoI-Based Data Gathering and Routing Guidance in VANETs

Cheng Feng, Rui Zhang, Shouxu Jiang, and Zhijun Li

Harbin Institute of Technology, School of Computer Science and Technology,
92 West Dazhi Street, Nan Gang District
Harbin, China
{fengcheng,raychang,jsx,lizhijun_os}@hit.edu.cn

Abstract. The optimization of data gathering in VANETs is to save bandwidth consumption as much as possible. However, the QoI of data collected may not satisfy the requirement of routing guidance demanded by the user if it merely focuses on saving bandwidth consumption, resulting in the inaccuracy of routing guidance for the user. Therefore, we propose a framework of QoI-based data gathering and routing guidance, which can take the requirement of QoI into consideration in the process of data gathering. Firstly, the requirement of QoI is extracted from the QoS requirement of the user. Then QUERY with QoI constraints is distributed in the interested area. Finally, data of which QoI can satisfy the requirement is gathered through data aggregation by QoI-DG protocol used for routing guidance. Simulations show that our proposed solution achieves effective and efficient data gathering in VANETs.

Keywords: data gathering, data aggregation, QoI, routing guidance, VANETs.

1   Introduction

Over the last decades, the technology of Vehicular Ad-hoc NETworks (VANETs) [1] has developed rapidly. VANETs have a broad application prospect in Intelligent Transportation Systems (ITS) [2]. There are two kinds of communication equipments in VANET, vehicle nodes and Access Points (APs) by the roadside. In VANETs one vehicle node communicates with the other one through V2V [3] communication mode, while the vehicle node uses V2I [4] communication mode to communicate with APs. Vehicle nodes in VANETs use kinds of sensing devices to get information about the drivers' environment, such as speed, GPS, acceleration, available parking space and so on. Then the information is collected to APs through data aggregation and data routing. APs use such information to provide several applications (real-time navigation, safe warning of intersections, discovery of free parking places, etc.) [5] in order to make driving safer, more efficient and comfortable. The quality of information gathered through VANETs has direct effect on the performance of routing guidance application. Data collection with high QoI (Quality of Information) can make travel time estimated more accurate, and the result provided by dynamic routing guidance application is better. At the same time, the higher the QoI of data

Z. Bao et al. (Eds.): WAIM 2012 Workshops, LNCS 7419, pp. 87–98, 2012.
© Springer-Verlag Berlin Heidelberg 2012

collected can reach, the more wireless communication bandwidth consumption needs. Susceptible situations in VANETs (communication disruption, the mobility of vehicles, etc.) make the available bandwidth limited. Therefore, data gathering protocol in VANETs should cost wireless communication bandwidth as little as possible under the premise of satisfying the accuracy requirement of routing guidance application proposed by the user.

The contribution of this paper is the proposal of QoI-based data gathering in VANETs (QoI- DG) protocol. The main idea in our approach is that the process of data gathering in VANETs considers the requirement of QoI the routing guidance application needs while it decreases communication bandwidth consumption as much as possible. First, we get QoI constraints of data collected from the requirement of routing guidance raised by the user. Then we create the query of data gathering based on QoI and diffuse the query. At last we use QoI-DG protocol to collect data in the interest area to requesting AP.

The rest of the paper is organized as follows: Related works are reviewed and analyzed in Section 2. Section 3 describes the overview of QoI-based data gathering and routing guidance in VANETs. Extracting QoI constraints of data gathering from the requirement of routing guidance application is presented in Section 4. QoI-based data gathering query propagation is stated in Section 5. In Section 6, data gathering with QoI constraints, QoI-DG protocol is introduced in details. Section 7 contains the results and the analysis of an experimental study on QoI-based data gathering and routing guidance in VANETs. Finally, in Section 8, conclusions are given and some future research directions are outlined.

2   Related Works

Data gathering is one of the most important issues in VANETs which brings real-time data to applications. Many research and implementation efforts [6-16] have been involved in the data gathering. In VANETs data gathering is studied in two main aspects: routing-related and aggregation-related aspects. The routing-related research focuses on routing protocols including when and where which vehicle nodes broadcast. The other focuses on data compression and aggregation techniques.

Great efforts have been made in data routing protocol in the process of data gathering in VANETs. According to the communication mode, data routing in the process of data gathering can be classified in three sorts, which are routing only through V2I communication, both through V2I and V2V communication, and through cellular base station. Data collection in [8] only uses the communication of vehicle and roadside unit, so it does not effectively use the communication between vehicles. The approaches given by [7] and [12] are based on the idea of clustering. The nodes in the cluster send data to the cluster head, and then the cluster head uses V2I communication to forward all data to APs. [6] uses cellular network to improve the accuracy of collected data in the event detection application, but the service provided by telecommunication company costs a lot of money.

Data aggregation in the process of data gathering can decrease bandwidth consumption. At the same time, it will make data accuracy decrease and delay longer. In the aggregation-related aspect, the research is focused on data accuracy and delay. [18] and [19] choose to do data compression that is no less accuracy aggregation. [15, 16, 25] consider data accuracy in the process of data aggregation. The longer data is retained in the node, the more opportunity more data can be aggregated. But it will increase delay. [20, 21, 24] consider the tradeoff between delay and decrease of amount of data.

In summary, existing researches about data gathering did not proceed from the requirement of applications, so they did not do the optimization of communication bandwidth consumption under the premise of satisfying the demand of QoI of data collected. This makes either that QoI of data collected is higher than the requirement needed, which means waste communication bandwidth or that QoI of data collected is lower than the requirement of shortest time routing guidance algorithm, which means QoS of the application is greatly affected.

3   Overview of QoI-Based Data Gathering and Routing Guidance

While QoI constrains are used to guide the process of data gathering, extracting QoI constraints from the requirement of application is the basic step. The next step is to create QoI-based data gathering query and distribute it in the interested area. Then data is collected by QoI-DG protocol to requesting AP.

Shortest time routing guidance algorithm uses travel time that cars pass by each road to find the shortest travel time path and calculate travel time of this path. At first, we introduce the definition of QoS of routing guidance, which describes the degree of accuracy of travel time provided by shortest time routing guidance algorithm.

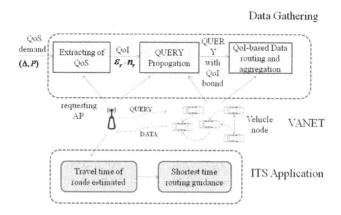

Fig. 1. The framework of QoI-based data gathering and routing guidance

Definition 1. QoS of routing guidance (δ, p): Assume the result provided by routing guidance is $^\wedge t_r$, the real travel time is t_r if $^\wedge t_r$ satisfies (1), QoS of routing guidance is (δ, p).

$$p\left(\left|\hat{t}_r - t_r\right| < \delta\right) > p \tag{1}$$

Assume QoS requirement of routing guidance given by the user is (Δ, P), first we calculate the bound of QoI of travel time estimated in the road (Δ_r, P_r) from it. The next step is to extract the amount n_r and the error bound ε_r of data that needs be collected in each road. According to QoI bound of data collected, we create the QUERY message and distribute it. Then data is gathered by QoI-DG protocol to requesting AP. Fig. 1 presents the overview of collecting the travel time information and routing guidance application in VANET environment.

4   Extracting QoI Constraints of Data Gathering

From data sensed by the sensors in the vehicle to the result calculated by the application, data is processed by three stages. The first step is data gathering, in which data is routed and aggregated to AP. The next step is road travel time estimating, which its input is inaccuracy data from vehicle nodes in each road so that travel time estimated has deviation. At last, routing guidance algorithm uses road travel time estimated to get possible travel time from departure to destination. In the whole process, QoI needs be considered, so we should get QoI requirement of road travel time estimated from the requirement of users in order to extract QoI constraints of data gathering.

Graph $G_r = (V, E)$ is described as road network, of which the node set is intersection set, the edge set as road set. Travel time passed by road r is t_r, and the length of road r is l_r. The user inputs the departure, the destination and the requirement of QoS of routing guidance (Δ, P). Shortest path algorithm can find out the shortest length path from departure to destination and its length l_{sd}^{min}, then we use [27] to calculate all paths of which the length is smaller than $(1+\tau) l_{sd}^{min}$. Assume the number of roads of which the path has the most roads in those paths is n_{sd}^{max}. (Δ, P) is divided to QoI bound of road travel time estimated (Δ_r, P_r) through (2) and (3).

$$\Delta_r = \frac{\Delta}{n_{s \to d}^{max}} \tag{2}$$

$$P_r = \sqrt[n_{s \to d}^{max}]{P} \tag{3}$$

Then we use (Δ_r, P_r) to extract error bound of data aggregation ε_r. Assume the data sensed by vehicles in the road r is $t_1, t_2, \ldots, t_k$; the data AP received is $\tilde{t}_1, \tilde{t}_2, \ldots, \tilde{t}_k$; travel time of road t is normally distributed. Assume travel time estimated of the road is $^\wedge t_r$.

$$\left|\tilde{t}_i - t_i\right| < \varepsilon_r \tag{4}$$

$$\hat{t} = \frac{\sum_{i=1}^{k} \tilde{t}_i}{k} \tag{5}$$

According to definition 1, we can get (6), and use (4) to deduce (7). Then we can get (8), and is independent and identically distributed, so we can calculate error bound ε_r.

$$p(|\hat{t}-t|<\Delta_r)=p(-\Delta_r<\hat{t}-t<\Delta_r)=p(-\Delta_r<\frac{\sum\limits_{i=1}^{k}\tilde{t}_i}{k}-t<\Delta_r)=p(-\Delta_r<\frac{\sum\limits_{i=1}^{k}\tilde{t}_i}{k}-t)p(\frac{\sum\limits_{i=1}^{k}\tilde{t}_i}{k}-t<\Delta_r) \qquad (6)$$

$$p(-\Delta_r<\frac{\sum\limits_{i=1}^{k}\tilde{t}_i}{k}-t)<p(-\Delta_r<\frac{\sum\limits_{i=1}^{k}t_i}{k}-t+\varepsilon_r) \quad p(\frac{\sum\limits_{i=1}^{k}\tilde{t}_i}{k}-t<\Delta_r)<p(\frac{\sum\limits_{i=1}^{k}t_i}{k}-t-\varepsilon_r<\Delta_r) \qquad (7)$$

$$p(-\Delta_r<\frac{\sum\limits_{i=1}^{k}t_i}{k}-t+\varepsilon_r)p(\frac{\sum\limits_{i=1}^{k}t_i}{k}-t-\varepsilon_r<\Delta_r)>P_r \qquad (8)$$

5   Distributing the QoI-Based Data Gathering Queries

In this section, we create QUERY message with QoI constraints of data gathering, and distribute it in the interested area. Firstly, requesting AP create QoI-based data gathering query. Table1 describes the structure of QoI-based data gathering QUERY. The key items in QUERY message are interested area, amount of data collected and error bound of aggregated data.

Table 1. The structure of QUERY message

Name	Meaning
AP	the ID of AP that create this query
CreateTime	The time of creating this query
lifetime	The remaining time of Query
Interested Area	The area from which data is collected
n_r	The number of data that need de collected
ε_r	Error bound that need be satisfied in data aggregation
(e_r, X_r)	The value used to set respond nodes

After creating QUERY message, requesting AP distributes this QUERY message to the vehicles in the interested area. Query Propagation deals with the diffusion of the query message from the requesting AP within the region of interest. Until now there are many researches about broadcasting protocol in VANETs. But they are different from our work, in which response nodes are chosen during query propagation in our work.

AP broadcasts a QoI-based data gathering QUERY message. When a vehicle node receives a QUERY message, it stores the query only if it is in the interested area. Then it prepares to schedule the dispatches of QUERY message. It checks the neighbor list, the information of which is got by beacon messages in cycle times. A random number between 0 and 1 is generated for each node in its neighbor list, if this number is above and beyond the value of P that is static parameter, put this node ID into set S, and decrease X_r. Then it continues diffusing the query and Set S to its neighbors. When the node receives QUERY, it checks whether its ID is in Set S. if it is, set itself as response node of QUREY. At the same time, the node checks whether QUERY is expired through lifetime of QUERY, if it is, abort this query propagation.

6   Gathering the Data with QoI Constraints

The crucial task of collecting the data and routing them toward APs is managed by our QoI-based data gathering protocol (QoI-DG). The protocol need solve three questions as follows:

- Which part of data can be aggregated? When and how data aggregation is done?
- When does node forward data, retain data and discard data?
- Which node should broadcast when several nodes in the conflict range want to broadcast data?

Therefore, QoI-DG protocol should include three parts: QoI-based data aggregation, strategy selection and wireless communication schedule. QoI-based data aggregation based on dynamic programming calculates which data can be aggregated in order to decrease the amount of data as much as possible. Strategy selection points out that which strategy node should adopt according to the current state of the node. Wireless communication schedule provides a method based on priority that deals with node communication demand conflict.

6.1   QoI-Based Data Aggregation

We define QoI-based data aggregation on node problem Π.

Problem Π: assume data set is $S=\{d_1, d_2, ..., d_n\}$, find minimal set partition $\{B_1, B_2, ..., B_m\}$ under this constraint:

$$\forall B_i, \forall d_j, \ \left|d_j' - d_j\right| \leq \varepsilon, d_j \in B_i, d_j' = \frac{\sum_{d_k \in B_i} d_k}{|B_i|} \tag{9}$$

To solve Problem Π, we propose a dynamic programming algorithm. We define $A(i, j)$ the optimal set partition number of set $\{d_i, ..., d_j\}$, $B(i, j)$ the flag that sign whether aggregation of $\{d_i, ..., d_j\}$ satisfies the constrain (if is, the value is 1; if not, the value is ∞).

We analysis QoI-based data aggregation get (9).

$$A(1,n) = \min\{(A(1,1)+A(2,n)),(A(1,2)+A(3,n)),....(A(1,n-1)+A(n,n)),B(1,n)\} \tag{10}$$

The pseudocode of this algorithm is as follows.

```
Data aggregation with error bound algorithm
Input : S={d₁,d₂, ...,dₙ}, ;
Output : {B₁,B₂, ...,Bₘ};
Algorithm:
begin
   For(i=0 to n, j=0 to n) Calculate B(i,j);
   For(i=0 to n) A(i,i)=1;
```

```
For(k=2 to n) for(i=1 to n-k+1);
   j=i+k-1;
   if B(i,j)=1 then A(i,j)=1;
   else for(y=I to j-1) q=A(i,y)+ A(y+1,j);
           if q<A(i,j) then A(i,j)=q;
end.
```

6.2   Strategy Selection

Node has three ways to deal with data, which is to forward, retain or discard. Carrying the data is convenient because it can reduce bandwidth consumption while bringing the data closer to the requesting AP. Moreover when a vehicle is carrying data, it can receive and aggregate data from other vehicles. Vehicle need to decide whether to retain the data they are carrying or to forward the data.

Whether node is fit for forwarding data depends on the moving speed of itself and its neighbors. If the moving direction of node is away from requesting AP, node must choose to forward data. If there is HELP node in its neighbors, node also should forward data. We define HELP node as follows.

Definition 2. HELP node: node j is said HELP node of node i, if and only if it satisfies the conditions as follows:

(1) The moving direction of node j is towards AP
(2) $v_j - v_i > V$, v_j is the current speed of node j, v_i is the current speed of node i, V is system parameter.

Here the information of node speed and moving direction is got by BEACON message.

It depends on available storage space whether the node discards data. When the proportion of available storage space in entire storage space is less than 20%, node aggregates the stored data. If there is no free storage space when node receives new data, data with the oldest timestamp is discarded.

6.3   Wireless Communication Schedule

When several nodes in the communication conflict field want to forward data, only one node in one time can use wireless channel. At this time which node can obtain the right to use it depends on the priority of node. At first, nodes exchange the respective priority, and node with highest priority broadcast data while other neighbor nodes are ready to receive data. The priority is determined by the distinctive feature of data set in node. We can get the bit vector that node has data through BEACON message. We define the distinctive feature of data set S in node i $DF(S)^i$.

Definition 3. Repetition frequency of data k: assume the neighbor node set of node i is N_i, the bit vector of data in node j is BV_j, BV^k_j is the value in the bit vector corresponding to data k.

$$TF^i_k = \frac{\sum_j BV^k_j}{|N_i|} \tag{11}$$

Definition 4. Distinctive feature of data set S in node i, $DF(S)^i$:

$$DF(S)^i = \frac{\sum_{k \in S} 1 - TF_k^i}{|S|} \tag{12}$$

7   Simulation Results and Analysis

In this section, we compare QoI-DG protocol and naïve approach from the aspects of effectiveness of data gathering, bandwidth consumption, and the rate of data collection. Naïve approach is in the process of data gathering without data aggregation and random data routing strategy.

7.1   Experimental Setup

We use Singapore expressway as the scene of experiment. The Singapore expressway has 11 intersections and 28 links (both directions). The structure of the expressway is shown in figure 2.

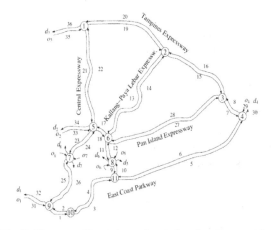

Fig. 2. Singapore Expressway (central and eastern parts)

Table 2. Simulation parameters

Parameter	Value
The number of intersections	11
The number of roads	28
The number of cars	60
Communication radius	250m
Departure	Intersection 9
destination	Intersection 2

We use traffic simulation software "Sumo" to generate the traces of vehicles. In the experiment we apply discrete time simulator. The nodes are synchronized. The communication model is disc model that when the distance between one node and the other is less than 100m, they can communicate. Table 2 lists simulation parameters.

7.2   Analysis of Results

In order to analyze the performance of QoI-DG protocol, we point out two different metrics which measure the effectiveness and efficiency of the protocol.

- Effectiveness measures the ability of QoI-DG to gather enough data under error bound in order to provide application service with enough quality.
- Efficiency is an index that measures the level of bandwidth optimization. we use the number of all the DATA message to measure it.

First, we analyze the effectiveness of QoI-DG protocol. As shown in figure 3. QoI-DG protocol guarantees that the accuracy of travel time provided by routing guidance with QoS requirement changes with the QoS requirement. Regarding the effectiveness, QoI-DG protocol can reach the requirement proposed by the user. From figure 3, we can find out the proportion of the result in the error bound is almost P. The value of P is set 90% by the user.

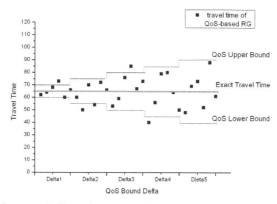

Fig. 3. Comparisons on QoS requirements and the results of QoI-based routing guidance

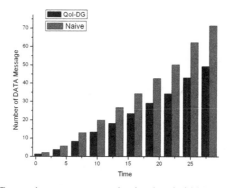

Fig. 4. Comparisons on communication bandwidth consumption

Then we compare the efficiency of the protocol with naïve approach. From figure 4, it is clear to demonstrate that QoI-DG cost the less communication bandwidth. As the time increases, the proportion of deduced bandwidth consumption is rapidly increasing.

Another aspect of the protocol assessed is the rate of data collection. Figure 5 shows the proportion of data that has been collected in the whole required data increases as the time. We can find out that the rate in the whole process of data gathering is relatively stable. Moreover, the rate is always higher that naïve approach.

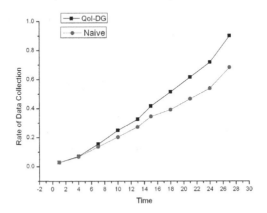

Fig. 5. Comparisons on the rate of data collection

8   Conclusion and Future Works

This paper analyses the disadvantages of current research on data gathering in VANETs, and points out that data gathering should be guided by the requirement of ITS applications. A framework for data gathering based on QoI bound was designed, in which the gathering process was divided into three steps, i.e., extracting of QoI bound, query propagation and QoI-based data routing and aggregation. The main mechanism of our solution is to use QoI bound to control the optimization of data aggregation and routing. The core work is the proposed QoI-DG protocol which can achieve least communication bandwidth consumption under the premise of satisfying the application requirement of the user. Simulations are carried out to verify the effectiveness and efficiency of QoI-DG protocol in the aspects of both satisfying the user demands and saving bandwidth consumption.

In this paper we focus on only one kind of ITS application, i.e., dynamic routing guidance. For future work, it would be interesting to add other ITS applications into it, which will mainly complicate the extraction of QoI fit for all demands. Furthermore, the delay will be another key aspect that needs to be researched.

Acknowledgement. The research is sponsored by National Natural Science Foundation of China (No. 60803148) and (No.60973124), Ph.D. Programs Foundation of Ministry of Education of China (No. 20102302110036).

References

1. Jakub, J., Yevgeni, K.: State of the art and research challenges for VANETs. In: Proceedings of the 5th Consumer Communications and Networking Conference (2009)
2. Beresford, A.R., Bacon, J.: Intelligent transportation systems. IEEE Pervasive Computing 5(4), 63–67 (2006)
3. Wang, C.X., Cheng, X.: Vehicle-to-vehicle channel modeling and measurements: recent advances and future challenges. IEEE Communications Magazine 47(11), 96–103 (2009)
4. Cho, W., Kim, S.I., Choi, H., Oh, H.S., Yong, D.: Performance evaluation of V2V/V2I communications: The effect of midamble insertion. In: Proceedings of 1st Conference on Wireless Communication, Vehicular Technology, Information Theory and Aerospace & Electronic Systems Technology, pp. 793–797 (2009)
5. Hartenstein, H., Laberteaux, K.: A tutorial survey on vehicular ad hoc networks. IEEE Communication Magazine 46(6), 164–171 (2008)
6. Lee, J., Kang, M.: Data Collection Scheme for Two-Tier Vehicular Sensor Networks. In: Pandurangan, G., Anil Kumar, V.S., Ming, G., Liu, Y., Li, Y. (eds.) WASA 2010. LNCS, vol. 6221, pp. 295–298. Springer, Heidelberg (2010)
7. Ismail, S., Mohamed, C., SidiMohammed, S.: A New Architecture for Data Collection in Vehicular Networks. In: Proceedings of IEEE International Conference on Communications, Dresden, Germany (2009)
8. Arbabi, M., Weigle, M.: Using vehicular networks to collect common traffic data. In: Proceedings of the Sixth ACM International Workshop on VehiculAr InterNETworking, Beijing, China, pp. 117–118 (2009)
9. Delot, T., Ilarri, S.: Data Gathering in Vehicular Networks: The VESPA experience. In: Proceedings of IEEE 36th Conference on Local Computer Networks, pp. 797–804 (2011)
10. Mathur, S., Kaul, S., Gruteser, M., Trappe, W.: ParkNet: A Mobile Sensor Network for Harvesting Real Time Vehicular Parking Information. In: Proceedings of the 2009 MobiHoc S3 Workshop, New York, USA, pp. 9–11 (2009)
11. Christian, L., Scheuermann, B., Christian, W.: Data aggregation and roadside unit placement for a VANET traffic information system. In: Proceedings of the Fifth ACM International Workshop on VehiculAr Inter-NETworking, San Francisco, California, USA, pp. 58–65 (2008)
12. Chang, W.R., Lin, H.T., Chen, B.X.: TrafficGather: An Efficient and Scalable Data Collection Protocol for Vehicular Ad Hoc Networks. In: Proceedings of the 5th IEEE Consumer Communications and Networking Conference, pp. 365–369 (2008)
13. Zarmehri, M.N., Aguiar, A.: Data Gathering for Sensing Applications in Vehicular Networks. In: Proceedings of 2011 IEEE Vehicular Networking Conference, pp. 222–229 (2011)
14. Fernandes, H., Boukerche, A., Pazzi, R., Samarah, S.: Efficient Data Gathering and Position Dissemination Protocols for Heterogeneous Vehicle Ad hoc and Sensor Networks. In: Proceedings of the 5th IEEE GCC Conference & Exhibition, pp. 1–4 (2009)
15. Wischhof, L., Ebner, A., Rohling, M., Halfmann, R.: SOTIS - a self-organizing traffic information system. In: Proceedings of The 57th IEEE Vehicular Technology Conference, New York, NY, USA, pp. 2442–2446 (2003)
16. Nadeem, T., Dashtinezhad, S., Liao, C., Iftode, L.: TrafficView: traffic data dissemination using car-to-car communication. ACM Mobile Computing and Communications Review (MC2R), Special Issue on Mobile Data Management, 6–19 (2004)

17. Picconi, F., Ravi, N., Gruteser, M., Iftode, L.: Probabilistic Validation of Aggregated Data in Vehicular Ad-hoc Network. In: Proceedings of the 3rd International Workshop on Vehicular Ad Hoc Networks, New York, NY, USA (2006)

18. Ibrhim, K., Weigle, M.: Accurate data aggregation for VANETs. In: Proceedings of the Fourth ACM International Workshop on Vehicular Ad Hoc Networks, Montréal, Québec, Canada, pp. 71–72 (2007)

19. Khaled, I., Michele, C.W.: CASCADE: Cluster-based Accurate Syntactic Compression of Aggregated Data in VANETs. In: Proceedings of the Global Communications Conference, New Orleans, LA, USA (2008)

20. Yu, B., Gong, J., Xu, C.: Catch-Up: a data aggregation scheme for VANETs. In: Proceedings of the Fifth ACM International Workshop on VehiculAr Inter-NETworking, San Francisco, California, USA, pp. 49–57 (2008)

21. Yu, B., Xu, C., Guo, M.: Adaptive Forwarding Delay Control for VANET Data Aggregation. IEEE Transactions on Parallel and Distributed Systems 23(1), 11–18 (2012)

22. Christian, L., Bjorn, S., Martin, M.: A probabilistic method for cooperative hierarchical aggregation of data in VANETs. Journal of Ad Hoc Networks 8(5), 518–530 (2010)

23. Han, Q., Du, S., Ren, D., Zhu, H.: SAS: A Secure Data Aggregation Scheme in Vehicular Sensing Networks. In: Proceedings of 2010 IEEE International Conference on Communications, pp. 1–5 (2010)

24. Palazzi, C.E., Pezzoni, F., Ruiz, P.M.: Delay-bounded data gathering in urban vehicular sensor networks. Journal of Wide-Scale Vehicular Sensor Networks and Mobile Sensing 8(2), 180–193 (2012)

25. Schoch, E., Dietzel, S., Bako, B.Z., Karql, F.: A Structure-free Aggregation Framework for Vehicular Ad Hoc Networks. In: Proceedings of the 6th ACM International Workshop on Vehicular Ad Hoc Networks, Beijing, China, pp. 79–88 (2009)

26. Eichler, S., Merkle, C., Strassberger, M.: Data Aggregation System for Distributing Inter-Vehicle Warning Message. In: Proceedings of 2006 31st IEEE Conference on Local Computer Networks, pp. 542–544 (2006)

27. Aneja, Y.P., Aggarwal, V., Nair, K.P.K.: Shortest chain subject to side constraints. Networks 13, 295 (1983)

PTL: Partitioned Logging for Database Storage on Flash Solid State Drives

Robin Jun Yang and Qiong Luo

Hong Kong University of Science and Technology
{yjrobin,luo}@cse.ust.hk

Abstract. We propose Partitioned Logging (PTL), a storage layout for databases on flash solid state drives. In PTL, we replace data writes with logging, and put data and logs into separate blocks. Moreover, we group data blocks into partitions so that updates on each partition are appended as log entries to one log block. This way, we can tune the partition size to balance the read and write performance based on the hardware and workload characteristics. We have implemented PTL in PostgreSQL, which involves moderate changes to the buffer manager, the storage manager, and the transaction manager. We have also developed an analytical model to determine the PTL parameter values. We have evaluated PTL using the standard TPC-C benchmark as well as homegrown workloads. The empirical results match our analytical analysis, and show a considerable improvement over both the traditional storage and a leading flash-based database storage scheme.

1 Introduction

Recent studies [1,2] have suggested that flash solid state drives be a high-performance storage device for transactional database workloads such as TPC-C [3]. However, flash memory has an inherent read-write asymmetry, which may make existing database techniques underutilize the hardware capability in fast access. In this paper, we propose a new database storage scheme called PTL (Partitioned Logging) for flash SSDs. In PTL, we take a log-structured approach [4] so that updates on data pages are replaced with writing logs sequentially. Such a logging approach generates sequential writes, which are much faster than random writes; however, it also bears two disadvantages: (1) reading a page from disk requires applying logs to the original data page to construct the up-to-date version; and (2) merging of logs into data pages on disk and discarding the obsolete logs must be performed as necessary to reclaim the space.

To address these two drawbacks, we design PTL to group data pages into partitions with each partition having a separate PTL log block. This way logging is not done sequentially for the entire database but for each partition. To improve the PTL log writing performance, we design an in-memory PTL log buffer in which each partition has exactly one buffer page to accumulate its PTL log entries. The buffer size is dynamically adjusted to the database size with the number of pages equaling the number of data partitions. When a PTL log buffer page is full, it is flushed to the corresponding PTL log block on disk. When a PTL log block is full, it is merged into the original data pages of the corresponding partition. By setting three parameters, data partition size,

Z. Bao et al. (Eds.): WAIM 2012 Workshops, LNCS 7419, pp. 99–108, 2012.

PTL log page size, and PTL log block size based on hardware and workload characteristics, we can balance the read and write performance as well as to adjust the frequency of merging and space consumption.

We have developed an analytical model to estimate the performance of PTL. We have implemented PTL in PostgreSQL and evaluated it using standard TPC-C as well as custom workloads of different read-write ratios and spatial distributions in data accesses. Our analytical model matches well with the real execution time results. It also facilitates PTL to adapt to changing workloads through checkpoint-time adjustment. Furthermore, PTL outperforms both the original PostgreSQL and IPL [5], a state-of-the-art flash-based database storage scheme on two types of flash disks as well as a hard disk. The performance improvement varies by workloads and types of storage devices. On an Intel X-25M flash SSD under TPC-C, the speedup of PTL is two times over the original PostgreSQL and five times over IPL.

2   Background and Related Work

In this section, we review related work on optimization techniques that addressed the read-write asymmetry of flash SSDs.The NAND flash memory used in most flash SSDs has a physical limitation that in-place update is not supported. As a result, manufacturers added the FTL[6] (Flash Translation Layer) inside an SSD to alleviate this problem by directing writes to clean pages; however, this redirection also causes garbage collection to run more frequently. RAM caches inside an SSD can help alleviate this problem by accumulating dirty pages, but it will affect the performance of other access patterns as well as increase the manufacturing cost. Consequently, random writes remain the worst-performing access pattern on flash SSDs. A few new file systems [7,8] have been proposed to solve the random write problem by adopting log-structured approaches [4]. However, the garbage collection problem becomes more severe in these file systems, especially on flash SSDs of a large capacity. Within a database system, most of the work focused on how to modify existing data structures and algorithms to utilize the asymmetric I/O performance of flash SSDs to achieve a better overall performance. There has been initial work on optimizing DBMS components such as query processing [9], buffer management [10], and indexing [11,12] for flash-based databases. Some researchers were interested in using flash SSDs for Key-Value Stores (KV-stores), e.g., FlashStore[13] and SkimpyStash [14]. However, none of them proposed a solution for OLTP (Online Transactional Processing) applications. Recently PDL (Page-Differential Logging)[15] has been proposed as a database-independent approach to storage on flash; however, it requires direct access to physical flash pages and is not applicable to flash SSDs today[16]. The work that is most related to ours is IPL [5]. However, writes at such a fine granularity as appending a few bytes to a flash page perform poorly on current flash SSDs due to their block-based interface. Furthermore, given the size limit of one erase block and the one-page size of the IPL log region within each erase block, merge operations occur frequently - whenever the log region is full.

Table 1. Notations

Parameter	Description
S_{pt}	**Size of a partition in bytes**
S_{lp}	**Size of a PTL-log page in bytes**
S_{lb}	**Size of a PTL-log block in bytes**
S_B	Size of an erase block of the flash SSD in bytes
E_B	Time of erasing an erase block on the flash SSD
S_D	Size of the database in bytes
S_{dp}	Size of a data page in bytes
S_{memory}	Size of the PTL-log buffer in bytes
S_{disk}	Total size of PTL-log blocks in bytes
R_{SEQ_P}	Bandwidth of sequential reads with a unit size S_P
R_{RND_P}	Bandwidth of ran. reads with a unit size S_P
W_{SEQ_P}	Bandwidth of seq. writes with a unit size S_P
W_{RND_P}	Bandwidth of ran. writes with a unit size S_P
W_{S-RND_P}	Bandwidth of semi-ran. writes with a unit size S_P in a region of the size S_{disk}
n_{merge}	Total number of merge operations
n_{lpr}	Number of PTL-log page reads
n_{lpw}	Number of PTL-log page writes
n_{dpr}	Number of data page reads
n_{dpw}	Number of data page writes

3   Design and Implementation of PTL

We present the design and PostgreSQL-based implementation of Partitioned Logging (PTL) in this section. Notations used throughout the paper are summarized in Table 1. The goal of PTL is to achieve a high overall performance through balancing read and write operations. To achieve this goal, we take a log-structured approach with tunable parameters on the granularity of the PTL logging IO as well as the size ratio of the PTL logs to the data pages.

Figure 1 illustrates the overall organization of PTL. In PTL, all data pages on disk are grouped into **partitions**. The partition size is a multiple of the erase block size of the flash SSD. As in traditional systems, data pages are brought into the main memory resident database buffer pool for reading and writing. Furthermore, writes to data pages on disk are transformed to recording **PTL-logs** in the **PTL-log buffer**. The PTL-log buffer page size is a multiple of the flash page size. Each PTL-log buffer page is responsible for one data partition, and the number of PTL-log buffer pages is equal to the number of data partitions. As a result, the PTL-log buffer size is associated with the database size in number of partitions and no page replacement occurs in the buffer. When a PTL-log buffer page becomes full, it will be flushed to append to the **PTL-log block** on disk that corresponds to the data partition. The size of a PTL-log block is also a multiple of the flash erase block size. By changing one or more of the three parameters - the partition size, the PTL-log buffer page size, and the PTL-log block size, we can adjust the disk and memory space usage of PTL.

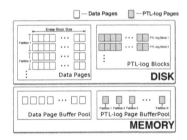

Fig. 1. Illustration of PTL Organization

Table 2. Comparison of Page Write Operations Between Original and PTL-based PostgreSQL

	Original	PTL-based
Dirty Buffer Page Replacement	Yes	No
Background Dirty Page Flush	Yes	No
PTL Log Page Flush	No	Yes
Checkpoint	Yes	Yes
Write-Ahead-Log	Yes	Yes

3.1 Database Operations and IO Operations in PTL

Because of the log-based approach used in PTL, the logic of database operations needs to be changed.

Insertion, Deletion, Update. In the database buffer pool, these operations on the data pages are done as in traditional systems so that the in-memory version is always up to date. Moreover, the changes to be applied to disk are recorded as PTL-logs in corresponding PTL-log buffer pages in the memory. If a PTL-log buffer page is full, it will be flushed to its corresponding PTL-log block on disk by the *PTL-log page write* operation. If the PTL-log block on disk is full when flushing the PTL-log buffer page, a *merge* operation will be executed to apply the logs to associated data pages.

Data Page Write. Traditionally data pages from the buffer are written to disk directly. In PTL, these small, random writes are replaced by the PTL-log page appending operations, *PTL-log page write* and the *merge* operation. As a result, all the updates on a data page are recorded in the form of PTL-logs in the corresponding PTL-log buffer page and PTL-log blocks on disk. These PTL-logs are used to construct the up-to-date data page upon a read request.

Data Page Read. When a requested data page is not in the buffer, also known as a buffer miss, the storage manager will read the page from disk to the buffer pool. Since the data page read from disk may not be the latest version, we need to apply PTL-logs associated with this data page in the memory to return the up-to-date version.

Merge. When a PTL-log block is full, a merge occurs to apply the PTL-logs in the entire block as well as those in the corresponding PTL-log buffer page to associated data pages. Then, it flushes the up-to-date data pages to disk, and vacates both the log block and the log buffer page. Since reads and writes are performed in bulk, with a suitable PTL log size ratio and IO granularity, the overall performance can be improved.

3.2 PTL Implementation

We have implemented PTL in PostgreSQL-8.3.7. Most of the modifications are made in the operations of the storage manager such as page read and page write, and the merge operation is added. The change outside the storage manager is to add the logic of transforming updates on data pages to PTL-logs and to make checkpointing apply all the logs in PTL-log buffer pages and PTL-log blocks to data pages before flushing them to the disk so that all the data on the disk will be up to date after checkpointing. The main difference between PTL-based PostgreSQL and original one is summarized in Table 2. In the design of PTL, we have no special requirement on the PTL-log record format, as long as it is sufficient for reconstructing the up-to-date version of a data page correctly. Our current implementation of PTL in PostgreSQL uses a similar format of the WAL(Write-Ahead-Logging) logs for the PTL logs. Note that the WAL transaction logs and transaction processing mechanisms are intact in the PTL-based PostgreSQL to guarantee the transaction semantics as in the original PostgreSQL.

4   Analytical Model

In this section, we develop an analytical model to estimate the performance of PTL using the IO time of each type of operations. We further model the number of each type of IO operations as a function of three parameters, (a)*the partition size*(S_{pt}), (b)*the PTL-log page size*(S_{lp}), and (c)*the PTL-log block size*(S_{lb}). We then estimate the optimal performance through these parameters. Finally, given a workload and the hardware properties, with our model, we can set suitable values for the three parameters for an optimal performance.

Table 3. IO Time of the Operations with and without PTL

Without PTL		With PTL	
Data page read (T_{dpr})	$\dfrac{S_{dp}}{R_{RND_{dp}}}$	Data page read (T_{dpr})	$\dfrac{S_{dp}}{R_{RND_{dp}}}$
		PTL-log page read (T_{lpr})	$\dfrac{S_{lp}}{R_{RND_{lp}}}$
Data page write (T_{dpw})	$\dfrac{S_{dp}}{W_{RND_{dp}}}$	PTL-log page write (T_{lpw})	$\dfrac{S_{lp}}{W_{S-RND_{lp}}}$
		Merge (T_{merge})	$\dfrac{S_{lb}}{R_{SEQ_{lp}}} + \dfrac{S_{pt}}{R_{SEQ_{dp}}} + \dfrac{S_{pt}}{W_{SEQ_{dp}}} + \dfrac{S_{lb} \cdot E_B}{S_B}$

In a conventional system, the entire data IO time (excluding transaction log IO time) consists of two components, *data page read* time and *data page write* time. In comparison, with PTL, the IO time consists of four components, *data page read*, *PTL-log page read*, *PTL-log page write* and *merge*. Detailed IO time for each type of IO operations with and without PTL is shown in Table 3. We treat PTL-log page writes as semi-random writes [17] as these pages may be written to different PTL-log blocks but within each PTL-log block, the pages are appended sequentially.

Given the IO time of each type of operations, we use Equation (1) to estimate the total IO time.

$$T_{PTL} = n_{lpr} \cdot T_{lpr} + n_{dpr} \cdot T_{dpr} + n_{merge} \cdot T_{merge} + n_{lpw} \cdot T_{lpw} \qquad (1)$$

The IO time of each type of operation can be determined by the hardware characteristics and the three PTL size and granularity parameters. In comparison, the total number of each type of operations is dependent on the workload. Fortunately, because these numbers are accumulative counts, they can also be modeled as simple functions of the three parameters despite of the workload characteristics. For ease of presentation, we use a workload with a uniform spatial access pattern on disk, or in short, a uniform workload, as an example. In summary, we model the numbers of three special types of IO operations in PTL as follows:

$$n_{lpr} = \alpha \cdot \frac{S_{lb}}{S_{pt} \cdot S_{lp}}, n_{lpw} = \beta \cdot \frac{1}{S_{lp}}, n_{merge} = \gamma \cdot \frac{1}{S_{lb}} \tag{2}$$

The constants α, β, and γ are workload-dependent. We show experimental results to validate this model in Section 5.3.

Using Equation (1), (2) and Table 3, we can derive the optimal setting of parameters shown in Equation (3) to achieve a peak performance of PTL.

$$\frac{S_{lb}}{S_{pt}} = \sqrt{\frac{\gamma \cdot R_{RND_{lp}} \cdot (W_{SEQ_{dp}} + R_{SEQ_{dp}})}{\alpha \cdot W_{SEQ_{dp}} \cdot R_{SEQ_{dp}}}} \tag{3}$$

Given that the optimal setting depends only on a ratio of the PTL-log block size and the partition size, to reduce the cost of each merge, we choose the smallest PTL-log block size, i.e., one erase block of the flash SSD in use. This choice also leads to the smallest partition size; both ensures least variation in the response time due to the occurrences of merge. In case there is a limit on the amount of disk or memory space to use, we can increase the PTL-log block size and the partition size with the ratio close to optimal.

5   Experiment Results

In this section, we experiment with our PostgreSQL-based implementation on a hard disk and two types of flash disks. We first present empirical results of running our prototype on real devices under representative workloads to validate our analytical model. We further compare the performance of PTL with the original PostgreSQL and our PostgreSQL-based IPL implementation.

5.1   Experimental Setup

All of our experiments are run on a desktop PC with a 2.5GHz AMD Phenom X4 9850 CPU and 4GB RAM. We use a 500GB 7200RPM Western Digital magnetic disk, and two flash disks, an Mtron MSD-SATA3035064 with a 64GB capacity and an Intel X25-M with a 80GB capacity as the storage devices. All three disks are connected to the PC via a SATA interface. The default OS is CentOS 5.2 Final (kernel version 2.6.18-92.el5). We use Ext3 of Linux as the default file system. We disable the file system cache to stress the disk IO performance.

For a detailed study to validate our analytical model, we use synthetic workloads with different read-write ratios and different localities of updates. To compare the overall performance, we use the OSDL Database Test Suite (DBT2)[1] to simulate workloads very similar to TPC-C. DBT2 is designed to produce a real-world OLTP workload, based on TPC-C, to stress test DBMSs on Linux without the complexity and expense of running a full TPC-C benchmark. Our PTL implementation is based on PostgreSQL 8.3.7. For IPL, we consider it a special case in PTL where the partition size is the flash erase block size and the PTL-log block size is the same as the flash page size.

5.2   Impact of Three Parameters on IO Operations

We perform code-level tracing to count the number of IO operations in Table 3 for the original PostgreSQL and the modified one with PTL under different kinds of workloads on the three disks. The overhead of such tracing can be ignored because the tracing code already exists in all IO operations and we focus on the impact of the parameters on the number of IO operations, not the time performance. To obtain an accurate count of these IO operations that are physically performed on the disk, we disable the cache of the file system for this set of experiments.

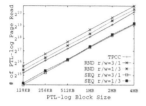
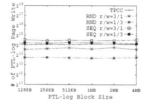
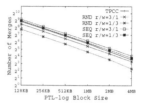

Fig. 2. PTL-log Page Reads   **Fig. 3.** PTL-log Page Writes   **Fig. 4.** PTL-log Merges

When studying the impact of one parameter, we fix the other two to their default values. The default setting of the three parameters is S_{lp} = 4KB, S_{lb} = 256KB, and S_{pt} = 512KB, obtained through our experiments. We only present the figures using the information collected on the Intel SSD for brevity. The results on the other two devices are similar. We choose five kinds of workloads with both the access pattern (random or sequential) and the read-write ratio varied. *TPCC* is generated by DBT2 with 20 warehouses, 1 connection, and 2 terminals per warehouse. The database size is 2GB and the data page size is 8KB by default. The database buffer pool is set to 32MB, a relatively small buffer size, to stress the IO performance. Each run has a warm-up period for database buffer pool.

Our results show that the numbers of IO operations as functions of the three parameters in our analytical model are well validated by the empirical results on different kinds of workloads running on three representative storage devices. For example, in Figure 2, the number of PTL-log page reads is proportional to the PTL-log block size. Figure 3 shows that the number of PTL-log page writes is not affected by the PTL-log block size. Finally, the number of merge operations is inversely proportional to the PTL-log page size, shown in Figure 4.

---

[1] http://sourceforge.net/projects/osdldbt/

5.3   Comparison between Measured and Estimated Overall Performance

With our analytical model, we can determine the optimal parameter setting. Further-more, with α, β and γ values set through calibration and the hardware-related band-width values, we can obtain the estimated overall performance through Equation (1) for various settings of the three parameters. As an example, we pick the workload *RND r/w=1/3* to derive the optimal parameter setting as well as to compare the esti-mated with the measured overall performance under various parameter settings. Figure 5 shows the comparison between analytical and experimental results on the workload. The difference between the two for a given parameter setting is 5-10%. Moreover, the peak performance happens at the point where $\frac{S_{lb}}{S_{pt}} = \frac{1}{8}$ which is close to the optimal ratio of $\frac{S_{lb}}{S_{pt}} = \frac{1}{8.69}$ given by our model.

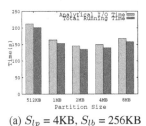 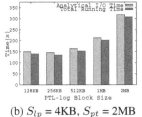 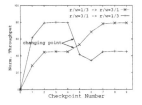

(a) S_{lp} = 4KB, S_{lb} = 256KB   (b) S_{lp} = 4KB, S_{pt} = 2MB

Fig. 5. Performance Comparison between Measured and Analytical Results

Fig. 6. Self-tuning Performance of PTL

5.4   Performance under Changing Workloads

Our implementation of PTL on PostgreSQL achieves self-tuning by re-evaluating PTL parameters at checkpoint time based on the workload information collected from the last checkpoint with little overhead (only two counters for *PTL-log page read* and *merge* are needed). Figure 6 shows a crossover point when we change the workload at a time between checkpoint 4 and 5. With the read-intensive workload changing to write-intensive, the throughput goes down because writes are more expensive on flash disks. For the same reason, the throughput increases with the workload changing in the reverse direction. As the workload information collected and parameters adjusted at subsequent checkpoints, the throughput of both workloads gradually improves.

5.5   Comparison between PTL, The Original PostgreSQL, and IPL

Finally, we compare PTL with the original PostgreSQL and IPL under different work-loads and on three types of devices. The default workload is TPC-C and the default device is Intel X-25M.

Since PTL optimizes write operations, it improves the overall performance more as the writes in the workload increase. In Figure 7a, all three have an almost identical

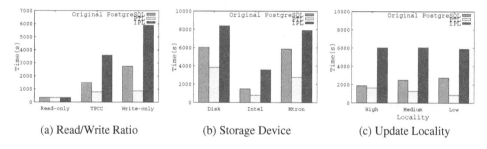

(a) Read/Write Ratio   (b) Storage Device   (c) Update Locality

Fig. 7. Comparison on Measured Overall Performance

performance for the read-only workload because the read-only workload does not generate any PTL logs, and PTL improves the PostgreSQL performance approximately 2x under TPC-C and more than 3x under the write-only workload. Due to the small log writing granularity and the fixed and tightly coupled logging structure, IPL slows down the original PostgreSQL more than two times both on TPC-C and on write-only.

Figure 7b shows that PTL can improve the performance on all the three storage devices. On the hard disk, the improvement is about 33%; on flash disks, the improvement is more, almost doubling the performance, due to the read-write asymmetry. In contrast, IPL slows down PostgreSQL considerably.

In Figure 7c, our empirical results show that the overall performance can be improved with PTL under workloads with various degrees of locality. The high-locality workload updates only 20% of the entire database, the medium updates 50%, and the low updates the entire database uniformly. The locality or skewness of the workload matters to PTL because it slightly affects the optimal parameter setting, shown in our analytical model. As a result, the performance obtained for a high-locality workload may not be optimal. In comparison, under a workload of low locality, or a workload of more uniformity in its spatial access region, PTL improves the PostgreSQL performance by about three times. In contrast, the original PostgreSQL runs slightly faster under a workload of higher locality and IPL appears insensitive to the locality.

6   Conclusion

We have presented PTL, a database storage scheme aimed for flash-based systems. PTL combines logging with partitioning, where each log block records changes about its corresponding data partition. Reads are done through reconstructing data pages by applying PTL-logs; writes are replaced with PTL-logging. We design a PTL-log buffer to accumulate PTL-log entries for each partition, and flush full PTL-log pages to their corresponding PTL-log blocks on disk. When a PTL-log block is full, it will be merged to its corresponding data partition. We have developed an analytical model to determine the size ratios of PTL-log blocks to data partitions as well as the granularity of the PTL-log page to optimize the overall performance. PTL can also adapt to dynamic workloads by re-evaluating the optimal parameters based on our analytical model at checkpoint time. We have implemented PTL as well as IPL for comparison in PostgreSQL. We have conducted experiments on a hard disk and two flash disks under a variety of workloads. The empirical results match well our analytical model. Furthermore, PTL improves the overall performance of PostgreSQL up to three times and outperforms IPL by five times.

References

1. Lee, S.-W., Moon, B., Park, C.: Advances in Flash Memory SSD Technology for Enterprise Database Applications. In: SIGMOD Conference, pp. 863–870 (2009)
2. Lee, S.-W., Moon, B., Park, C., Kim, J.-M., Kim, S.-W.: A Case for Flash Memory SSD in Enterprise Database Applications. In: SIGMOD Conference, pp. 1075–1086 (2008)
3. TPC: TPC Benchmark C Standard Specification Revision 5.9. Transaction Processing Performance Council (2007)
4. Rosenblum, M., Ousterhout, J.K.: The Design and Implementation of a Log-Structured File System. ACM Trans. Comput. Syst. 10(1), 26–52 (1992)
5. Lee, S.-W., Moon, B.: Design of Flash-based DBMS: An In-page Logging Approach. In: SIGMOD Conference, pp. 55–66 (2007)
6. Intel: Understanding the Flash Translation Layer (FTL) Specification. Intel Corporation (1998)
7. Manning, C.: YAFFS: The NAND-specific Flash File System (2002), http://www.yaffs.net
8. Woodhouse, D.: JFFS: The Journalling Flash File System. In: Ottawa Linux Symposium (2001)
9. Tsirogiannis, D., Harizopoulos, S., Shah, M.A., Wiener, J.L., Graefe, G.: Query Processing Techniques for Solid State Drives. In: SIGMOD Conference, pp. 59–72 (2009)
10. Lv, Y., Cui, B., He, B., Chen, X.: Operation-aware Buffer Management in Flash-based Systems. In: SIGMOD Conference, pp. 13–24 (2011)
11. Li, Y., He, B., Luo, Q., Yi, K.: Tree Indexing on Flash Disks. In: ICDE (2009)
12. Wu, C.-H., Chang, L.-P., Kuo, T.-W.: An Efficient B-Tree Layer for Flash-Memory Storage Systems. In: Chen, J., Hong, S. (eds.) RTCSA 2003. LNCS, vol. 2968, pp. 409–430. Springer, Heidelberg (2004)
13. Debnath, B., Sengupta, S., Li, J.: FlashStore: High Throughput Persistent Key-Value Store. PVLDB 3(2), 1414–1425 (2010)
14. Debnath, B., Sengupta, S., Li, J.: SkimpyStash: RAM Space Skimpy Key-value Store on Flash-based Storage. In: SIGMOD Conference, pp. 25–36 (2011)
15. Kim, Y.-R., Whang, K.-Y., Song, I.-Y.: Page-differential Logging: An Efficient and DBMS-independent Approach for Storing Data into Flash Memory. In: SIGMOD Conference, pp. 363–374 (2010)
16. Bonnet, P., Bouganim, L.: Flash Device Support for Database Management. In: CIDR, pp. 1–8 (2011)
17. Nath, S., Gibbons, P.: Online Maintenance of Very Large Random Samples on Flash Storage. In: VLDB (2008)

Adaptation Mechanism of iSCSI Protocol for NAS Storage Solution in Wireless Environment[*]

Shaikh Muhammad Allayear[1], Sung Soon Park[2,**], Shamim H. Ripon[1], and Gyeong Hun Kim[2]

[1] Dept. of Computer Science and Engineering, East West University, Bangladesh
[2] Anyang University, Gluesys Co, Ltd. Korea
{allayear,dshr}@ewubd.edu, sspark@anyang.ac.kr, kgh@gluesys.com

Abstract. The continued growth of both mobile appliances and wireless Internet technologies is bringing a new telecommunication revolution and it has extended the demand of various services with mobile appliances. However, in wireless environment the availability of mass storage is limited due to their limited size and weight. Although the problem can be alleviated by iSCSI (Internet Small Computer Interface) based Network Storage System, it has drawbacks in high availability and performance. To address this issue, this paper presents an architecture to adapt iSCSI protocol with traditional NAS (Network Attached Storage) cluster system with an error recovery method. To realize the access to a NAS storage system, our experiments suggest the optimal values for various parameters. The test cases show that the best values of the parameters are not always the default values specified in the iSCSI standard.

Keywords: iSCSI Protocol, Network Storage, wireless network.

1 Introduction

A NAS (Network Attached Storage) system is a specially designed device providing clients with files on a LAN. NAS has many advantages, such as sharing files in a hetero-architecture, making full use of the existing LAN architecture, easy installation, operation and management, PNP, good connection compatibility and network adaptation, low costs and so on. However, NAS supports only file I/O protocol, such as NFS and CIFS, block-level storage applications are not available on NAS. The performance of NFS and CIFS is only a fraction of the exported storage system due to their single server design, which binds one network endpoint to all files in a file system.

iSCSI is an Internet Protocol-based storage standard for linking data storage facilities, and presents storage to servers as disk targets [1] which appears to be storage attached locally to the server. iSCSI works better than NAS for most

---

[*] The research is supported by Basic Research Program through the National Research Foundation of Korea (NRF), Grants No. 2009-0075576, funded by Ministry of Education, Science and Technology, and WBS(World Best S/W) Development Project, Grants No.10040957, funded by Ministry of Knowledge Economy 2011.
[**] Corresponding author.

Z. Bao et al. (Eds.): WAIM 2012 Workshops, LNCS 7419, pp. 109–118, 2012.

applications because it provides the illusion that the networked storage being used by the server is exclusive. Whereas, CIFS or NFS-based network allows simultaneous access from multiple servers. As iSCSI presents storage space as a virtual block-level device, operating systems and applications can put their own file systems on them, which is something not possible with NAS.

In the passage of time, mobile appliances are going to be used in many areas. Due to their mobility, mobile appliances must be small in size and they use flash memory storage. However storage of multimedia data and installation of large software system is still a major challenge [2][3][4][5]. To alleviate the problems as well as to access mass storage we have developed MNAS [6], an iSCSI based NAS Cluster system, for the allocation of a mass storage space to mobile clients through network.

In this paper, we extend our earlier approach and propose the inclusion of error recovery method into the adaptation of iSCSI protocol on traditional NAS system to support block based I/O for mobile appliances. After extensive experiments we suggest optimal values for iSCSI parameters into CDMA network.

To adapt iSCSI in mobile appliances we have to address the obstacles to TCP congestion and iSCSI parameters in Wireless connections. Since the iSCSI PDU is a SCSI transport protocol over TCP/IP [1], the iSCSI Data-In PDUs are passed to the TCP layer. When the iSCSI Data-In PDU size is greater than the MSS (maximum segment size) the PDU will be further fragmented into smaller segments. Generally, the size of iSCSI Data-In PDU is larger than that of MSS. If some segments of an iSCSI Data-In PDU are lost due to high bit error rate, TCP layer would require re-transmitting the segments. The other segments of those parts must wait for being reassembled into an iSCSI Data-In PDU in TCP buffer. It decreases the performance of the system due to the reduction of the available capacity of TCP buffer.

In a wireless network with high bit error rate, more the size of iSCSI PDU is increased by the MRDSL (MaxRecvDataSegmentLength) parameter, more the segments would have to wait due to the loss of some segments of parts of an iSCSI PDU in TCP buffer. The write operation has the same results as the read operation in caused by the parameter for target. A wireless link generally becomes a bottleneck portion in an end-to-end TCP connection because of its narrow bandwidth, as compared to wired links. Thus a TCP sender's congestion controls are apt to be caused by wireless link congestions.

When congestion occurs in wireless links, there are two indications of packet loss, timeout occurring and receipt of duplicate ACKs. Though TCP can reduce the transmission amount of data segments using congestion avoidance mechanism, it still transmits data segments until detecting congestion in wireless network. If the amount of SCSI data payload, which is continuously passed to the TCP layer, is increased by increasing the value of MaxBurstLength for write operation or Number of sectors per command for read operation, and the size of an iSCSI PDU is enlarged by increasing the value of MRDSL, the sender will transmit more segments until recognizing congestion in wireless network. It causes the falling of performance of the iSCSI based mobile applications.

The rest of the paper is organized as follows. Section 2 gives an overview of iSCSI architecture. A brief description of error recovery module for iSCSI protocol is presented in Section 3. Our experimental settings and analysis of results are depicted in Section 4. The experimental analysis of sequence error and connection failure is given in Section 5. Finally, Section 6 draws the conclusion of the paper and outlines our future plan.

2   Architecture of iSCSI Module for Block I/O Service in NAS

We illustrate the software components that are of interest for data transfer through the I/O path in Fig. 1. The NAS metadata server provides the I/O interface to the user-space world. The I/O system will call in metadata server use MUVFS (or generic file system) to perform their tasks, while individual file system, such as EXT3 or NFS, plugs their code into the MUVFS handlers.

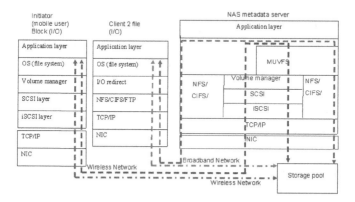

Fig. 1. Software Architecture of NAS for Block I/O and File I/O Service

When NAS offers the block I/O services, it invokes the iSCSI module, the requester is Initiator. The data read/write process is as follows: (i) The block I/O commands (SCSI commands) sent by the users in the Initiator are encapsulated into the IP data packets with the iSCSI device driver and the TCP/IP protocol stack, then transferred via the wireless network; (2) When the encapsulated packets arrive at the NAS metadata server, they are restored to the previous SCSI commands in an unpacked process, then passed to the MUVFS layer. Having been processed by the MUVFS, the previous I/O commands are packed up once again and sent to an appointed NAS via an inner network; (3) required data blocks are packed into the iSCSI protocol data units by the NAS and returned to the Initiator via a wireless Network. The way in which the mobile appliance communicates with the NAS via a high-speed IP channel is similar to that the mobile appliance communicates with the metadata server. When NAS offers the file I/O service, the data read/write process is almost the same as that in a usual NAS.

To support Unix Client, Windows Client and iSCSI mobile appliance, we developed Multi User Virtualized File System (MUVFS). MUVFS is the most important core modular in MNAS that enables the centralized management of many NAS nodes and provides a unified file system view for clients. Fig. 2 illustrates the structure of MUVFS. It consists of Unfsd, UiSCSId, Samba, the management partition, an iNAS configuration table, and a virtual partition (logic volume) map table. Unfsd is a wrapper of the global NFS daemon and provides file services for UNIX/Linux clients. Samba is used for windows clients. UiSCSId is a wrapper of the global iSCSI daemon and provides block I/O services for iSCSI clients (mobile appliances).

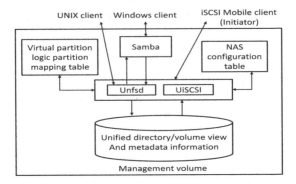

Fig. 2. Structure of Multi user Virtualization File System MUVFS

The NAS configuration table contains the system information such as hostnames, independent IP of MNAS members, export points, and so on. The virtual partition (logic volume) map table specified the data partition which stores file entities, etc.

3   Error Recovery Module in iSCSI Protocol

Earlier an error recovery module was implemented for MNAS [6]. A storage management tool was placed on top of it. The storage management tool is used to allocate server's storage to each client to pre-defined extent. Fig. 3 shows the architecture of the error recovery module. For mobile appliances, an error recovery module is very important because of high error rate in the wireless network. The following two considerations prompted the design of much of the error recovery functionality in iSCSI [7],[8]. An iSCSI PDU may fail the digest check and can be dropped, despite being received by the TCP layer. The iSCSI layer must be optionally allowed to recover such dropped PDUs.

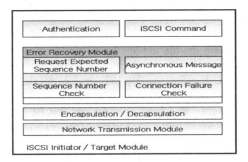

Fig. 3. Architecture of Error Recovery Module of MNAS (Mobile NAS Cluster System)

A TCP connection may fail at any time during data transfer. All the active tasks must optionally be allowed to continue on a different TCP connection within the same session.

iSCSI system is organized by iSCSI initiator and target. The initiator is used on a mobile device, PDAs in this paper. iSCSI target has one or more mass storage devices. Many kinds of errors can be occurred in wireless environment. However, iSCSI error recovery module considers the errors on iSCSI protocol layer. In this paper, the error recovery module considers following two errors.

- *Sequence Number Error*: When the iSCSI protocol transmits iSCSI Command PDU or Data PDU that has a sequence number, some PDU can be lost. So iSCSI PDU receiver (an initiator or a target) will not get the valid PDU.
- *Connection Failure*: If iSCSI target or initiator cannot communicate each other via a TCP connection, we define this situation as "connection failure".

3.1   Error Recovery Procedure

iSCSI protocol with error recovery checks the sequence number of every iSCSI PDU. If iSCSI target or initiator receives an iSCSI PDU with an out of order sequence number, then it requests an expected sequence number PDU again.

In case of connection failure, when a connection has no data communication during the engaged time, iSCSI protocol with error recovery checks all the connection status by the nop-command. We assume the multiple connections.

Sequence Number Error. When an initiator receives an iSCSI status PDU with an out of order or a SCSI response PDU with an expected data sequence number (ExpDataSN) that implies missing data PDU(s), it means that the initiator detected a header or payload digest error one or more earliest ready to transmission (R2T) PDUs or data PDUs. When a target receives a data PDU with an out of order data sequence number (DataSN), it means that the target must have hit a header or payload digest error on at least one of the earlier data PDUs. The target must discard the PDU and request retransmission with recovery R2T.

Connection Failure. At an iSCSI initiator, the following cases lend themselves to connection recovery.

- TCP connection failure: The initiator must close the connection. It then must either implicitly or explicitly logout the failed connection with the reason code "remove the connection for recovery" and reassign connection allegiance for all commands still in progress associated with the failed connection on one or more connections. For an initiator, a command is in progress as long as it has not received a response or a Data-in PDU including status.
- Receiving an Asynchronous Message that indicates one or all connections in a session has been dropped. The initiator must handle it as a TCP connection failure for the connection(s) referred to in the Message.

At an iSCSI target, the following case lend themselves to connection recovery.

TCP Connection Failure. The target must close the connection and, if more than one connection is available, the target should send an Asynchronous Message that indicates it has dropped the connection. Then, the target will wait for the initiator to continue recovery.

4   Experimental Methodology

We consider throughput as performance metrics, which is the total number of application level bytes carried over an iSCSI connection divided by the total elapsed time taken by the application, as expressed in Bytes per millisecond (B/ms). We used the system timer of PDA, which is based on WinCE in order to measure throughput. Total elapsed time was measured as the time interval from the initial time when the experiment program started generating the first byte of data to the time when the last byte of data was confirmed to have been sent (received). The elapsed time therefore includes all data transfers and all read and write commands as well as responses at all levels of the protocol stack. We perform two kinds of experiments. In the first experiment, we generate an I/O stream of 5 megabytes, and then measure throughput. Our request for a 5 megabytes I/O operation is passed through the file system and the SCSI subsystem to the iSCSI initiator as a number of SCSI commands. Then the initiator sends the commands and data to the target device. At that time, we adapt various settings of the iSCSI parameters to investigate the best parameter values for performance in wireless network. Each data point plotted in every graph of this paper was calculated as the average of 5 runs with identical parameter settings. We also perform the same experiment again with a different size of I/O stream, which is a 100 megabytes I/O stream. We perform the second experiment in order to examine the effect of the characteristics of the high bit error rate within the commercial CDMA network and the variable bandwidth of the network. Our experiment program generates read I/O bursts continuously for 15 hours, and then measures throughput every 1000 seconds. The second experiment shows that the increase of the MaxRecvDataSegmentLength parameter value cannot always bring performance improvement to iSCSI-based Mobile NAS System in unstable wireless network with high bit error rate and variable bandwidth.

4.1   Result and Analysis

Our experiment setup consists of a PDA and NAS Storage server connected on to Internet with CDMA 2000 1xEVD0 network.

The graph in Fig. 4 illustrates that the increase of the Number of sectors per command causes the performance improvement in the wireless network. It is very similar to the experiment results in a wired network. However, there are the performance falling at the Number of sectors per command value of 2048 with MRDSL of 1Kbytes and 4Kbytes. At the MRDSL values of 1Kbytes and 4Kbytes, there are 1% and 15% de-creases respectively in throughput as Number of sector per command increases from 1024 to 2048. In the wireless network, the results related to the values of MRDSL from our experiment are different from those in a stable wired network. The throughputs are better at the MRDSL values of 1Kbytes, 2Kbytes and 4Kbytes than those of 512bytes and 8Kbytes. At the Number of sectors per command value of 2048, the throughput of 2Kbytes MRDSL is increased by 48% and 52% respectively compared with those of 512bytes and 8Kbytes MRDSL. However, when MRDSL value is among 1Kbytes, 2Kbytes and 4Kbytes, it is difficult to say which one is the best value, for different MRDSL values may outperform others at different values of Number of sectors per command. For example, MRDSL of 2Kbytes can bring better throughputs when Number of sectors per command is between 256 and

512, while MRDSL of 1Kbytes is better when Number of sectors per command is around 1024. Therefore, it is not always correct in wireless network to increase the value of MRDSL to get performance improvement. Correspondingly, the default value 8Kbytes of MRDSL in standard is also not suitable for a wireless network. The reason of performance dropping at the MRDSL value of 512bytes is due to extra PDUs overhead. When MRDSL is 8Kbytes, however, the reason of performance falling is due to the increase of data segments which are transmitted continuously by TCP until detecting congestion even though congestion occurred in the wireless network with narrow bandwidth. Therefore, we suggest that you should use the parameters settings of the MRDSL of 1Kbytes, 2Kbytes and 4Kbytes with the Number of sectors per command values of 1024 or 2048 when per-forming a small file read operation in CDMA network.

In write operation, the Number of sectors per command is kept constant at 1024(512Kbytes) and the MaxBurstLength (MBL) varies from 512Bytes to 256Kbytes. We use the fixed Number of sectors per command, because the value of the parameter MaxBurstLength limits the total amount of all data segments of all PDUs which were requested by R2T. Fig. 5 shows that the throughput is increased as MBL increases from 512Bytes to 128Kbtyes in a wireless network, after that there is a slight decrease in throughput when MBL is around 256Kbytes. The same kinds of decrease in throughput happen in read operations too, which is caused by the narrow bandwidth of CDMA network. The number of iSCSI PDUs which are continuously passed to the TCP layer can be increased by increasing either the value of MaxBurstLength for write operation or Number of sectors per command for read operation, then the sender will transmit more segments until recognizing congestion in wireless network, and thus causes the performance falling of iSCSI based NAS for mobile appliances.

In a write operation, the result is more obvious because the CDMA 2000 1x EV-DO network has the narrower bandwidth for upload than that for download. Therefore it is not always true that the increase of MBL value would cause the performance improvement. The MBL value of 256Kbytes in standard is not suitable too. When the MRDSL value is 2Kbytes or 4Kbytes, the throughputs are better than that is 512bytes or 1Kbytes or 8Kbytes. At the MBL of 128Kbytes, the throughput of MRDSL of 2Kbytes is 26% and 31% increases respectively than those of MRDSL of 512bytes and 8Kbytes. However, when MRDSL is at a value of 2Kbytes or 4Kbytes, it is hard to say which one is the best value. From the figure 4 we can know that either of them may achieve better performance within a certain value range of Number of sectors per command. From these we can also see the standard value for wired network is not suitable here for a wireless network. Therefore, we suggest that you use the parameters settings of the MRDSL of 2Kbytes and 4Kbytes with the MBL of 128Kbytes when performing a small file write operation in CDMA network.

Fig. 6 shows the results from our experiment of 100 megabytes read operation. When MRDSL is at a value of 1Kbyte, the throughput is better than that at the value of 512bytes or 8Kbytes. However, when the MRDSL value is 1Kbytes, 2Kbytes and 4Kbytes, it is difficult to say which one is the best value. In the case of large size of I/O burst, such as 100 megabytes read operation, the best value for performance is also different from the standard value. When the MRDSL value is 2Kbytes, 4Kbytes and 8Kbytes, the performance is sharply dropped or increased with the increase of Number of sectors per command. At 8Kbytes of MRDSL value, there is a drastic decrease by 64% in throughput as Number of sectors per command increases from 64 to 512, after that there is an increase in throughput suddenly when the Number of sectors per command is around 1024. This experiment has different results from 5 megabytes I/O operation because the characteristics of a high bit error rate within the wireless channel, and a narrow and variable bandwidth of the wireless channels affect the large file read operation which was performed for a long time. Therefore, we perform a second experiment in order to examine the effect of the characteristics of CDMA network.. In the experiment for continuous read I/O bursts, we measure throughput every 1000 seconds in order to observe the degree of performance increasing or decreasing with respect to 1Kbytes, 2Kbytes and 4Kbytes MRDSL parameter values respectively, and then the Number of sectors per command is kept constant at 1024(512Kbytes). Among the 1Kbytes, 2Kbytes, and 4Kbytes MRDSL, there are the fewest changes in throughput of 1Kbytes of MRDSL in Fig. 7.

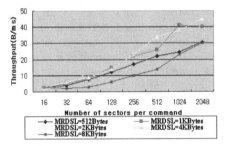

Fig. 4. Effect of Number of sectors per command on throughput for 5 Mbytes read operation in CDMA network

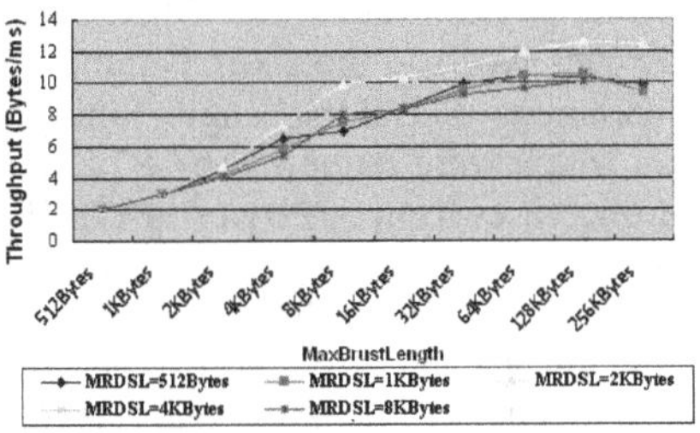

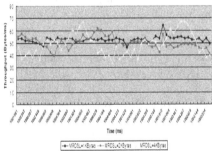

Fig. 5. Effect of MaximumBrustLength (MBL) for 5 Mbytes write operation in CDMA network

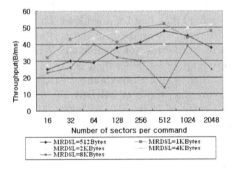

Fig. 6. Effect of Number of sectors per command on throughput for 100 Mbytes read operation in CDMA network

Fig. 7. Effect of MRDSL on throughput about 15 hours in wireless network

From the results, we found out that the smaller size of iSCSI PDU could be less affected by the characteristics of a high bit error rate within the wireless channel, and a narrow and variable bandwidth of the wireless channels. We suggest setting the MRDSL parameter value at 1Kbytes when performing a large file I/O operation while moving into CDMA network.

5   Experiments for Sequence Error and Connection Failure

PDU loss is generated factitiously for this experiment and round trip time (RTT) of PDU is about 0.4 second. We use iSCSI write Command PDUs for 1 Mbyte data in this experiment. The experiment progress is:

1. Initiator sends iSCSI PDUs to write 1Mbyte data.
2. Several PDUs would not be sent by the PDU loss rate during the transmission.
3. Target checks the PDU loss by the sequence number checking, and then it requests the PDU of the expected sequence number.

We consider an error recovery failure that target cannot receive the PDU of the expected sequence number during the time RTT × 2 + α = 1 second; where α= 0.128.

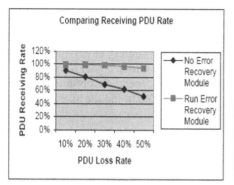
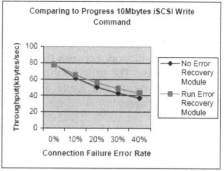

Fig. 8. Comparing the iSCSI PDU Receiving Rate. In this figure Y-axis is representing the PDU Receiving Rate and X-axis is representing the PDU Loss Rate.

Fig. 9. Comparing to Progress 10Mbytes iSCSI Write Command. In this figure Y-axis is representing the throughput and X-axis is representing the connection failure error rate.

Fig. 8 shows that PDU receiving rate is drastically decreasing without an error recovery module. However, the case of running our error recovery module just has about 8% PDU loss at 50% PDU loss rate. We assumed following factors for the experiments of connection failure:

- PDU loss is not considered.
- When iSCSI module without an error recovery has a connection failure, it reconnects after 60 seconds (Linux version 2.4.2 TCP Timeout). When iSCSI module has an error recovery, it reconnects by an error recovery module. We did not consider the time to reconnect by TCP or an error recovery module.
- Connection failure error rate is 0%, 10%, 20%, 30% and 40%. 10% Connection failure error means 1/10 connections of all connections has connection failure during the time (DefaultTime2Wait + DefaultTime2Retain = 22 seconds) [8].
- We experimented with 4 connections and connection failure can be happened concurrently.
- We experimented iSCSI write commands because iSCSI read commands have similar characters with iSCSI write commands.

Fig. 9 shows that writing progress with our error recovery module has better throughput than writing progress without an error recovery module. Increasing the connection failure error rate, the difference of throughput becomes larger. Result showed that the difference of throughput on 40% error rate was above 6 Kbytes/sec.

6   Conclusion

In this paper, we have presented a mechanism to adapt iSCSI protocol for mass storage solution in wireless environment for mobile appliances. In order to overcome the erroneous behavior in wireless environment, especially, *sequence number error* and *connection failure* we have implemented an error recovery module for iSCSI protocol based on mobile NAS cluster system. Our rigorous experiments confirmed that data can now be transferred in a stable and efficient manner than that without such error recovery module making our iSCSI appliances more reliable.

Our future research goal is to develop SMA (Session Management Agent) for adapting iSCSI protocol with massive storage system in wireless environment. The SMA system will work between iSCSI and TCP where the ultimate goal is to increase TCP throughput. The efficiency of the SMA agent will depend on optimum iSCSI PDU size. Therefore, our future works will emphasis to determine the size of iSCSI PDU for SMA agent.

References

1. SAM-3 Information Technology – SCSI Architecture Model 3, Working Draft, T10 Project 1561-D, Revision7 (2003)
2. Allayear, S.M., Park, S.S.: iSCSI Multi-connection and Error Recovery Method for Remote Storage System in Mobile Appliance. In: Gavrilova, M.L., Gervasi, O., Kumar, V., Tan, C.J.K., Taniar, D., Laganá, A., Mun, Y., Choo, H. (eds.) ICCSA 2006, Part II. LNCS, vol. 3981, pp. 641–650. Springer, Heidelberg (2006)
3. Kim, D., Ok, M., Park, M.-S.: An Intermediate Target for Quick-Relay of Remote Storage to Mobile Devices. In: Gervasi, O., Gavrilova, M.L., Kumar, V., Laganá, A., Lee, H.P., Mun, Y., Taniar, D., Tan, C.J.K. (eds.) ICCSA 2005. LNCS, vol. 3481, pp. 1035–1044. Springer, Heidelberg (2005)
4. Park, S., Moon, B.-S., Park, M.-S.: Design, Implement and Performance Analysis of the Remote Storage System in Mobile Environment. In: Proc. ICITA 2004 (2004)
5. Ok, M., Kim, D., Park, M.-S.: UbiqStor: A Remote Storage Service for Mobile Devices. In: Liew, K.-M., Shen, H., See, S., Cai, W. (eds.) PDCAT 2004. LNCS, vol. 3320, pp. 685–688. Springer, Heidelberg (2004)
6. Allayear, S.M., Park, S.S.: iSCSI Protocol Adaptation With NAS System Via Wireless Environment. International Conference On Consumer Electronics (ICCE), Las Vegus, USA (2008)
7. Allayear, S.M., Park, S.S., No, J.: iSCSI Protocol Adaptation with 2-way TCP Hand Shake Mechanism for an Embedded Multi-Agent Based Health Care Service. In: Proceedings of the 10th WSEAS International Conference on Mathematical Methods, Computational Techniques and Intelligent Systems, Corfu, Greece (2008)
8. RFC 3270, http://www.ietf.org/rfc/rfc3720.txt

Band Selection for Hyperspectral Imagery with PCA-MIG

Kitti Koonsanit[*], Chuleerat Jaruskulchai, and Apisit Eiumnoh

Department of Computer Science, Faculty of Science, Kasetsart University,
Bangkok, Thailand
sc431137@hotmail.com

Abstract. Although hyperspectral imagery provides abundant information about bands, their high dimensionality also substantially increases the computational burden. An interesting task in hyperspectral data processing is to reduce the redundancy of the spectral and spatial information without loss of any valuable details. In this paper, a band selection technique with principal components analysis, maxima-minima functional, and information gain for hyperspectral imagery such as small multi-mission satellite imagery is presented. Band selection method in the present research does not only serve as the first step of hyperspectral data processing that leads to a significant reduction of computational complexity but also an invaluable research tool to identify optimal spectral for different satellite applications. In this paper, an integrated PCA, maxima-minima functional method and information gain is proposed for hyperspectral band selection. Based on tests in SMMS hyperspectral imagery, this new method achieves good result in terms of robust clustering.

Keywords: Band Selection, Principal Components Analysis, PCA, Satellite image, Maxima-Minima Functional, Information Gain, IG.

1 Introduction

Satellite application in which such fast processing is needed is the dimension reduction of hyperspectral remote sensing data. Dimension reduction can be seen as a transformation from a high order dimension to a low order dimension. Principle Component Analysis (PCA) is perhaps the most popular dimension reduction technique for remotely sensed data [1]. The growth in data volumes due to the large increase in the spectral bands and high computational demands of PCA has prompted the need to develop a fast and efficient algorithm for PCA. In this work, the author presents on an implementation of maxima-minima functional and information gain with PCA dimension reduction of hyperspectral data. For this paper, hyperspectral data was obtained from the Small Multi-Mission Satellite (SMMS) [2] which has a ground pixel size of 100m x 100m and a spectral resolution of 115 channels, covering

---

[*] Corresponding author.

Z. Bao et al. (Eds.): WAIM 2012 Workshops, LNCS 7419, pp. 119–127, 2012.

the range from 450 nm to 950 nm. The author focuses on a collection of data taken in June 12, 2010 in the northern part of Amnat Charoen province, Thailand. The data consists of a total size of 8.86 Mbytes by 115 bands.

In this paper, the research interest is focused on comparison the effects of integrated PCA, maxima-minima functional method and information gain (PCA-MIG) of band selection on the final clustering results for hyperspectral imaging application. In the following sections, the various methods in this research were included PCA, maxima-minima functional, information gain and the proposed integrated PCA-MIG, are described, and the experimental results are presented later.

2   Methods

2.1   Dimensionality Reduction

Nowadays, hyperspectral imagery software becomes widely used. Although hyperspectral imagery provides abundant information about bands, their high dimensionality also substantially increases the computational burden. An important task in hyperspectral data processing is to reduce the redundancy of the spectral and spatial information without losing any valuable details. Therefore, these conventional methods may require a pre-processing step, namely dimension reduction. Dimension reduction can be seen as a transformation from a high order dimension to a low order which eliminates data redundancy. Principal Component Analysis (PCA) [1], [3], [9], [10] is one such data reduction technique, which is often used when analyzing remotely sensed data. The collected hyperspectral imagery data is in the form of a three dimensional image cube, with two spatial dimensions (horizontal and vertical) and one spectral dimension (from SMMS spectrum 1 to spectrum 115 in this study). In order to reduce the dimensionality and make it convenient for the subsequent processing steps, the easiest way is to reduce the dimensions by PCA.

2.2   Background on PCA

PCA is a widely used dimension reduction technique in data analysis. It is the optimal linear scheme for reducing a set of high dimensional vectors into a set of lower dimensional vectors. There are two types of methods for performing PCA, the matrix method, and the data method. In this work, the author will focus on the matrix methods. To compute PCA, the author follows the general 4 steps given below [1]:

  I. Find mean vector in x-space
  II. Assemble covariance matrix in x-space
  III. Compute eigenvalues and corresponding eigenvectors
  IV. Form the components in y-space

It has been previously shown that only the first few components are likely to contain the needed information [4]. The number of components that hold the majority of the information is called the intrinsic dimensionality, and each data image may have a different intrinsic dimensionality. PCA condenses all the information of an "N" band original data set into a smaller number than "N" of new bands (or principal components) in such a way that maximizes the covariance and reduces redundancy in order to achieve lower dimensionality as shown in the figure 1.

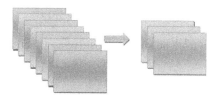

Fig. 1. PCA is a dimension reduction technique

2.3   Maxima and Minima

A further technique was integrated in this research, maxima and minima[5], known collectively as extrema, are the largest value (maximum) or smallest value (minimum), that a function takes in a point either within a given neighborhood (local extremum) or on the function domain in its entirety (global or absolute extremum). More generally, the maxima and minima of a set (as defined in set theory) are the greatest and least values in the set as shown in the figure 2. To locate extreme values is the basic objective of optimization.

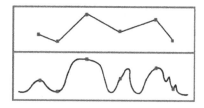

Fig. 2. Maxima and minima

Main features that classify the different objects should be extracted and preserved. The optimal bands is accordingly defined as the bands that not only maintains the major representation of the original data cube, but also maximally preserves features that separate different object classes [6]. Since the PCA method does not necessarily guarantee that the resulting transformation will preserve the classification information among different object classes, a maxima and minima method is proposed in this study to achieve a better performance to satisfy the optimal band selection criteria. The author thinks that maxima and minima value can use preserved features that separate different object classes. Maxima and minima value can calculate and explain as shown in session 2.3.1-2.3.4.

2.3.1 Theorem on First Derivative Test
Let f (x) be a real valued differentiable function. Let "a" be a point on an interval I such that f '(a) = 0.

(a) a is a local maxima of the function f (x) if

i) f '(a) = 0

ii) f '(x) changes sign from positive to negative as x increases through a.
That is, f ' (x) > 0 for x < a and f ' (x) < 0 for x > a

(b) a is a point of local minima of the function f (x) if

i) f ' (a) = 0

ii) f ' (x) changes sign from negative to positive as x increases through a.
That is, f ' (x) < 0 for x < a, f ' (x) > 0 for x > a

2.3.2 Working Rule for Finding Extremum Values Using First Derivative Test

Let f (x) be the real valued differentiable function.

Step 1: Find f '(x)

Step 2: Solve f '(x) = 0 to get the critical values for f (x). Let these values be a, b, c.
These are the points of maxima or minima.

Arrange these values in ascending order.

Step 3: Check the sign of f'(x) in the immediate neighbourhood of each critical value.

Step 4: Let take the critical value x= a. Find the sign of f '(x) for values of x slightly less than a and for values slightly greater than a.

(i) If the sign of f '(x) changes from positive to negative as x increases through a, then f (a) is a local maximum value.

(ii) If the sign of f '(x) changes from negative to positive as x increases through a, then f (a) is local minimum value.

(iii) If the sign of f (x) does not change as x increases through a, then f (a) is neither a local maximum value not a minimum value. In this case x = a is called a point of inflection.

2.3.3 Theorem on Second Derivative Test

Let f be a differentiable function on an interval I and let a is a member of I. Let f "(a) be continuous at a. Then

i) 'a' is a point of local maxima if f '(a) = 0 and f "(a) < 0

ii) 'a' is a point of local minima if f '(a) = 0 and f "(a) > 0

iii) The test fails if f '(a) = 0 and f "(a) = 0. In this case we have to go back to the first derivative test to find whether 'a' is a point of maxima, minima or a point of inflexion.

2.3.4 Working Rule to Determine the Local Extremum Using Second Derivative Test

Step 1

For a differentiable function f (x), find f '(x). Equate it to zero. Solve the equation f '(x) = 0 to get the critical values of f (x).

Step 2

For a particular critical value x = a, find f "'(a)

(i) If f "(a) < 0 then f (x) has a local maxima at x = a and f (a) is the maximum value.

(ii) If f "(a) > 0 then f (x) has a local minima at x = a and f (a) is the minimum value.

(iii) If f "(a) = 0 or infinity, the test fails and the first derivative test has to be applied to study the nature of f(a).

2.4   Information Gain

One more technique was integrated in this research; Information Gain is a measure of dependence between the feature and the class label. It is one of the most popular feature selection techniques as it is easy to compute and simple to interpret. The information gain of a given attribute X with respect to the class attribute Y is the reduction in uncertainty about the value of Y when we know the value of X, IG(X ,Y) So information gain of a feature or band X and the class labels Y is calculated as

$$IG(X,Y) = H(X) - H(X \mid Y) \tag{1}$$

Entropy (H) is a measure of the uncertainty associated with a random variable. H(X) and H(X|Y) is the entropy of band X and the entropy of band X after observing Class Y , respectively calculated as

$$H(X) = -\sum_i P(x_i) \log_2(P(x_i)) \tag{2}$$

$$H(X \mid Y) = -\sum_i P(y_i) \sum_i P(x_i \mid y_i) \log_2(P(x_i \mid y_i)) \tag{3}$$

The maximum value of information gain is 1. A feature with a high information gain is relevant. Information gain is evaluated independently for each feature and the features with the top-k values are selected as the relevant features. Information Gain does not eliminate redundant features.

2.5   PCA-MIG

At the band selection stage, several projection-based methods are studied; including principle components analysis (PCA), maxima and minima function (MM), information gain (IG) and integrated PCA-MIG methods. We integrated PCA-MIG method which is follow as

X Band Selected = PCA of Band ∩ MM of Band ∩ IG of Band   (4)

Let we take three sets PCA of band, MM of band and IG of band, (∩ = intersection) PCA of band ∩ MM of band ∩ IG of band, is the set of band selected such that the behind statement is true for the entire element x: X is a band member of band selected if one means if X is a band member of PCA and X is a band member of MM and X is a band member of IG.

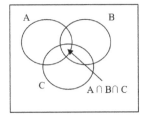

Fig. 3. The intersection set consists of elements common to three sets is a PCA-MIG

This paper presents a PCA-MIG method, which can effectively reduce the hyperspectral data to intrinsic dimensionality as shown in the figure 4.

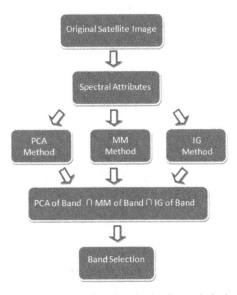

Fig. 4. Overall of our band selection technical

In the PCA-MIG, we divide step into three parts and combine the results by intersection. In the following sections, the clustering results over the original and the resulting reduced data have been compared.

3   Result

In this section, the author present and compare the experimental results obtained by applying each of the techniques, and then evaluating its effectiveness in clustering.

3.1   Experimental Setup

3.1.1   Hyperspectral Data

For this paper, hyperspectral test data was obtained from the SMMS imaging. For experiments we focus on a collection of data taken in June 12, 2010 in the northern part of Amnat charoen province, Thailand. These data contain 65,535 instances by 115 bands.

3.1.2 Unsupervised Classification Method

The experiments performed in this paper use the simple K-mean from the WEKA software package [7]. The simple K-Mean is the common unsupervised classification method used with remote sensing data. The effectiveness of the K-Mean depends on reasonably accurate estimation of the k cluster for each spectral class.

3.2   Experimental Results

In this paper, the PCA-MIG is performed using the two techniques described above. While performing the PCA, we have performed 3 experiments as shown in the figure 5.

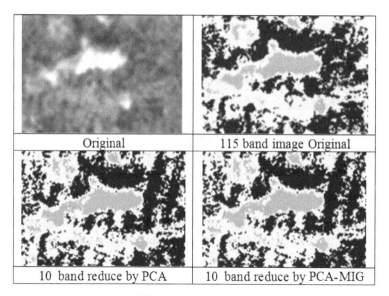

| Original | 115 band image Original |
| 10 band reduce by PCA | 10 band reduce by PCA-MIG |

Fig. 5. Result imagery

Table 1. Show that the percentage of each class in original image and 10 band reduced by PCA-MIG is similar to each other

	4 Cluster 115 band Original image	4 Cluster 10 band reduce by PCA	4 Cluster 10 band reduce by PCA-MIG
Cluster1	22677 (34%)	22677 (35%)	22677 (34%)
Cluster2	1974 (3%)	1398 (2%)	2093 (3%)
Cluster3	20294 (30%)	20302(31%)	20185 (30%)
Cluster4	20590 (31%)	21158(32%)	20580 (31%)
All cluster	100%	100%	100%

From the experiment, it was found that clustering using PCA combined with maxima-minima functional gives the nearest result with 115 band original image and can reduces numbers of attribute from 115 down to 10 attributes. These selected attributes were used as input for clustering algorithms. Table 1 shows the percent clustering obtained for various classes for an example experiment. It can be seen that there are some similarities between the original image clustering and the PCA-MIG techniques. Because features from maxima and minima functional can use preserved features that separate different object classes. After processing, both the representation information and class specific information are included.

Table 2. Show that the percentage of each class in original image and 7 attribute reduced by PCA-MIG is similar to each other

	7 Cluster 36 attributes Original image	7 Cluster 7 attributes reduce by PCA	7 Cluster 7 attributes reduce by PCA-MIG
Cluster1	489 (8%)	675 (10%)	608 (9%)
Cluster2	1299 (20%)	1343 (22%)	1362 (21%)
Cluster3	927 (14%)	760 (12%)	742(12%)
Cluster4	572 (9%)	598 (9%)	586 (9%)
Cluster5	1315 (20%)	1504 (23%)	1504 (23%)
Cluster6	1043 (16%)	627 (10%)	881 (14%)
Cluster7	790 (12%)	928 (14%)	752 (12%)
All cluster	100%	100%	100%

Another experiment was tested a proposed band selection on the statlog (landsat satellite) data set from UCI databases [8]. The database consists of the multispectral values of pixels in 3x3 neighborhoods in the satellite imagery, and the classification associated with the central pixel in each neighborhood. One frame of Landsat MSS imagery consists of four digital images of the same scene in different spectral bands. Two of these are in the visible region (corresponding approximately to green and red regions of the visible spectrum) and two are in the (near) infra-red. Each pixel is an 8 bit binary word, with 0 corresponding to black and 255 to white. The spatial resolution of a pixel is about 80m x 80m. These data contain 6435 instances. Each instance consists of 36 attributes. The proposed process was implemented on java environment, and tested on CPU 2.80 GHz Intel(R) Core 2 Duo processor with 1 GB of RAM. From table 2, it can be noticed that the differences in clustering between original image and PCA-MIG techniques are very closed as shown in Table 2.

4   Conclusions

Hyperspectral imagery software becomes widely used. Although hyperspectral imagery provides abundant information about bands, their high dimensionality also substantially increases the computational burden. Dimensionality reduction offers one approach to hyperspectral imagery analysis. In this paper, we present a band selection method using principal components analysis (PCA) and maxima-minima functional for hyperspectral imagery such as small multi-mission satellite (SMMS). We tested the proposed process on satellite image data such as small multi-mission satellite (hyper spectral) for unsupervised classification.   We compared this classification results between original images and PCA-MIG by clustering.   The experimental results show that the differences in clustering between original image and PCA-MIG techniques are very closed.   Because features from maxima and minima functional can use preserved features that separate different object classes and, both the representation information and class specific information are included. The outcome of this research will be used in further steps for analysis tools in hyperspectral imagery processing.

Acknowledgments. The authors would like to thank Graduate School, Kasetsart University. The financial support from Thailand Graduate Institute of Science and Technology (TGIST) is gratefully acknowledged. The scholar student ID is TG-22-11-53-005D and the grant number is TGIST 01-53-005.

References

1. Richards, J.A.: Remote Sensing Digital Image Analysis: An introduction. Springer, Heidelberg (1986)
2. Small Multi-Mission Satellite (SMMS) Data Available,
   http://smms.ee.ku.ac.th/index.php
3. Agarwal, A., El-Ghazawi, T., El-Askary, H., Le-Moigne, J.: Efficient Hierarchical-PCA Dimension Reduction for Hyperspectral Imagery. In: 2007 IEEE International Symposium on Signal Processing and Information Technology, December 15-18, pp. 353–356 (2007)
4. Kaewpijit, S., Le-Moige, J., El-Ghazawi, T.: Hyperspectral Imagery Dimension Reduction Using Pricipal Component Analysis on the HIVE. In: Science Data Processing Workshop. NASA Goddard Space Flight Center (February 2002)
5. Koonsanit, K., Jaruskulchai, C.: Band Selection for Hyperspectral Image Using Principal Components Analysis and Maxima-Minima Functional. In: Theeramunkong, T., Kunifuji, S., Sornlertlamvanich, V., Nattee, C. (eds.) KICSS 2010. LNCS, vol. 6746, pp. 103–112. Springer, Heidelberg (2011)
6. Cheng, X., Chen, Y.R., Tao, Y., Wang, C.Y., Kim, M.S., Lefcourt, A.M.: A novel integrated PCA and FLD method on hyperspectral image feature extraction for cucumber chilling damage inspection. ASAE Transactions 47(4), 1313–1320 (2004)
7. Kirkby, R., Frank, E.: Weka Explorer User Guide. University of Waikato, New Zealand (2005)
8. Frank, A., Asuncion, A.: UCI Machine Learning Repository. University of California, School of Information and Computer Science, Irvine, CA (2010),
   http://archive.ics.uci.edu/ml/support/Statlog
9. Zhao, Z., et al.: Advancing Feature Selection Research,
   http://featureselection.asu.edu/
   featureselection_techreport.pdf (retrieved: September 2, 2010)
10. Jackson, J.E.: A User Guide to Principal Components. John Wiley and Sons, New York (1991)
11. Jolliffe, I.T.: Principal Component Analysis. Springer (1986)
12. Cover, T.M., Thomas, J.A.: Information Gain. In: Elements of Information Theory. Wiley (1991)

NestedCube: Towards Online Analytical Processing on Information-Enhanced Multidimensional Network

Jing Zhang[1,2], Xiaoguang Hong[1,2,*], Zhaohui Peng[1,2], and Qingzhong Li[1,2]

[1] School of Computer Science and Technology, Shandong University
[2] Shandong Provincial Key Laboratory of Software Engineering
zhangjing123551@163.com, {hxg,pzh,lqz}@sdu.edu.cn

Abstract. The boom of web 2.0 applications has given rise to an ever increasing amount of networks. These networks can be modeled as information-enhanced multidimensional networks in which nodes and edges both have multidimensional attributes. We consider extending data warehousing and OLAP technology toward such new multidimensional network. So Nested Cube, a new data warehousing model, which can support OLAP queries on information-enhanced multidimensional network is proposed in this paper. On the basis of Nested Cube, bidirectional two-ply OLAP query is introduced to mine information deeply. Finally we apply the model on real data set DBLP and the experiment results show that Nested Cube is a powerful and effective tool for decision support on information-enhanced multidimensional network.

Keywords: OLAP, nested cube, data cube, information network.

1   Introduction

Data warehousing is used widely in business market for analyzing large amounts of data, aimed at enabling the analyst to make better and faster decisions. As an important technology of data warehouse, Online Analytical Processing (OLAP) technology allows us to analyze the data from different perspectives and multiple granulates [1,2]. OLAP is mainly supported through data cube which has proved a powerful model [4].

Information networks are ubiquitous due to the popular use of web, blogs, and various kinds of online databases. Examples of information networks are: co-author network, social networks such as Facebook and Renren. The information hidden in the networks can reflect the relationship between people or groups. So it is essential for us to mine the network data deeply.

The information network can be modeled as a graph with vertices representing entities and edges depicting relationship between entities [3]. The traditional OLAP techniques are useful for analyzing and mining structured data, but they face challenges in processing networks data.

The information network can be modeled as a multidimensional network and graph cube which can support data warehousing and OLAP on multidimensional network is

---

* Corresponding author.

Z. Bao et al. (Eds.): WAIM 2012 Workshops, LNCS 7419, pp. 128–139, 2012.

proposed in [3]. In multidimensional network, vertices are associated with multidimensional attributes and edges just stand for simple relationship. In reality, there is abundant semantic information between entities. For example, in co-author network, vertices stand for authors and edges stand for collaboration relationship between any two authors. The description of relationship is not detailed and too general and we can describe the relationship vividly by the papers they complete together. These papers have multidimensional attributes such as title, conference, date, content, and so on.

In this paper, we propose information-enhanced multidimensional network to model information network and on that basis a new data model, nested cube, is proposed to support OLAP. The contributions of our work can be summarized as follows:

1. We model the information network as the information-enhanced multidimensional network in which not only vertices are associated with multidimensional attributes, but also the edges have multidimensional attributes. In this way semantic information between entities can be represented clearly.
2. We propose a new data model, nested cube, to extend multidimensional analysis on new multidimensional network. The outer layer is a graph cube and the measure of cuboid query is a network which contains a traditional data cube. The attributes of vertices constitute the dimensions of graph cube and the attributes of edges constitute dimensions of data cube. The data cube is nested in the graph cube.
3. On the basis of nested cube, we formulate the bidirectional two-ply OLAP query. Traditional OLAP operations are performed in one cube. But in this paper we can perform OLAP query on outer graph cube first and obtain a measure network based on which the second-layer OLAP query on data cube which nested in the measure can be formulated. Likewise, we can perform OLAP query from the inner data cube to the outer graph cube.
4. Cube is generated dynamically. The second-layer OLAP query is based on the measure of the first-layer OLAP query. So the cube is generated dynamically after the first-layer OLAP query.
5. We evaluate our method on real data sets and the experimental results demonstrate the effectiveness of Nested Cube.

The rest of this paper is organized as follows. Section 2 gives definitions and descriptions of information-enhanced multidimensional network and nested cube model. In Section 3 bidirectional two-ply OLAP query is given in detail. Section 4 describes the experiment results. Section 5 introduces related work. Section 6 concludes the paper and gives our future work.

2   Nested Cube Model

Many information networks can be modeled as an information-enhanced multi-dimensional network, we define it as follows.

DEFINITION 1. [INFORMATION-ENHANCED MULTIDIMENSIONAL NET-WORK] Information-enhanced multidimensional network, M, is a graph denoted as N = (V , E , T , A , B), where V is a set of vertices, E is a set of edges and A = {A_1, A_2, . . ., A_n} is a set of n vertex-specific attributes, i.e., $\forall u \in V$, there is tuple A(u) of u , denoted as A(u) = (A_1(u), A_2(u), . . . , A_n(u)), where A_i(u) is the value of u on i-th attribute, $1 \leq i \leq n$. T is a set of tuples and B = {B_1, B_2, . . . , B_m} is a set of m attributes of T, i.e., $\forall e \in E$, there exist corresponding several tuples $T_s \subseteq T$. $\forall t \in T$, t can be denoted as B(t) = (B_1(t), B_2(t),......, B_m(t)) , where B_j(t) is the value of a tuple t on j-th attribute, $1 \leq j \leq m$.

We can illustrate the definition of information-enhanced multidimensional network by an example.

EXAMPLE 1. Figure 1 shows a simple social network. There are ten vertices and sixteen edges in the graph, as shown in Figure 1(a). Every vertex stands for a person that contains several multidimensional attributes including ID, Name, Nation, Gender, Profession, Degree, Age, as shown in Figure 1(b). These attributes can describe the properties of these persons. Every person can share information with his/her friends. So every edge contains several pieces of information. The relationship between edge and information are shown in Figure 1(c). Multidimensional attributes which can describe the properties of shared information including Date, Category, Topic are shown in Figure 1(d). For example, Jim and Lucy share information about war in politics in 2006.

Example 1 models a social network by information-enhanced multidimensional network, which not only presents the friendship between people but also enriches the semantic of friendship by describing the shared information between people. There can be more than one information between two people, i.e., every edge in multidimensional network can contain more than one piece of information.

DEFINITION 2. [VERTEX GROUPING] Given an information-enhanced multi-dimensional network M = {V, E, T, A, B} and a possible aggregation A' = {A_1', A_2', . . ., A_n'}, where A_i' equals A_i or *. V'={g_1, g_2, . . . , g_t}, $t \geq 1$ is called a vertex grouping of M, if and only if: 1. $g_1 \cup g_2 \cup ... \cup g_t = v$ 2. $\forall g_i, g_j \in G$, $g_i \cap g_j = \varnothing$ 3. $\forall u_i, u_j \in g_i$, A'(u_i)=A'(u_j).

DEFINITION 3. [EDGE MERGING]Given an information-enhanced multi-dimensional network M = {V, E, T, A, B} and a possible aggregation A' = {A_1', A_2', . . ., A_n'}, where A_i' equals A_i or *. A vertex grouping of M, G={g_1, g_2, . . . , g_t}, $t \geq 1$ can be obtained. E'={(g_i, g_j) | $\exists (u_s, u_t) \in E$ and $u_s \in g_i \wedge u_t \in g_j$} is a new edge set. $\forall$ (g_i, g_j)$\in$ E' is the merging of all edges between nodes in g_i and nodes in g_j. $\forall e' \in E'$, there exist corresponding several tuples $T_{s'i} \subseteq T$, $T_{s'i}=T_{s1} \cup T_{s2} \cup ... \cup T_{sn}$, T_{si} is the tuples corresponding to each e which is merged to form e'. T'=$T_{s'1} \cap T_{s'2} \cap T_{s'3}... \cap T_{s'p}$, p is the number of E'.

DEFINITION 4. [MEASURE NETWORK] Given an information-enhanced multidimensional network M = {V, E, T, A, B} and a possible aggregation A' = {A_1', A_2', . . . , A_n'}, where A_i' equals A_i or *. The measure network, A', is also an multidimensional network M' = (V', E', T', A, B). V', E', T' can be obtained by vertex grouping, edge merging respectively.

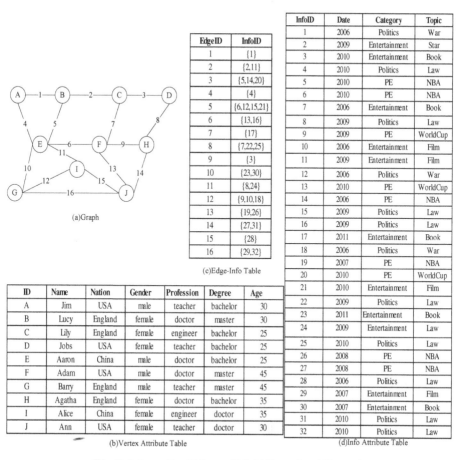

Fig. 1. Information-Enhanced Multidimensional Network

DEFINITION 5. [NESTED CUBE] Nested cube is that a data cube embeds in a graph cube. It is based on information-enhanced multidimensional network M (V, E, T, A, B). The attributes of vertices constitute the dimensions of graph cube. And the attributes of T constitute the dimensions of data cube. The graph cube can be obtained by computing all possible aggregations of A. In the measure network of a aggregation of A', M' = { V', E', T', A, B}, T' is a set of tuples which stand for entities which have multidimensional attributes, and they satisfy the condition of data cube. The data cube embeds in the measure of graph cube, so the nested cube is obtained.

EXAMPLE 2. Figure 2 presents a nested cube lattice, each cuboid of which is generated from the information-enhanced multidimensional network shown in figure 1. The lattice which is the lower part of nested cube is a lattice of graph cube, and the lattice which is the higher part of nested cube is a lattice of data cube.

As shown in figure 2, given a aggregation A'(nation, *, *, *, *), we can obtain a measure network which indicates that how the people of USA, England and China share information between each other and what are the details of the shared information. The shared information between people in China and England are

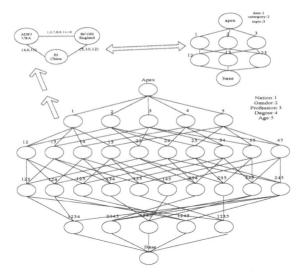

Fig. 2. Nested Cube

organized by many tuples which has multidimensional attributes and a data cube is generated. Given another aggregation B' (*, category, *), we can obtain a numeric measure in this data cube.

3   OLAP on Nested Cube

We propose bidirectional two-ply OLAP query to analyze nested cube. This kind of OLAP query includes From-Node-To-Edge operation and From-Edge-To-Node operation. In order to make it convenient to understand the definition of the new OLAP operation, we use example 1 to describe it. Consider some typical OLAP-style queries that might be asked on an information-enhanced multidimensional network:

1. How do people that belong to different nations share information and what are the number and category of these shared information?
2. How do people that belong to different nations and genders share information and what is the temporal distribution of these shared information?
3. What is the network structure among people in different professions who share a certain class of information with each other?

These queries clearly involve some kind of aggregations on information-enhanced multidimensional network. For the first and second query, the first part answer are two measure networks corresponding to the cuboid (nation, *, *, *, *) and (nation, profession, *, *, *) in outer graph cube, and the second part of answer are measures corresponding to the cuboid (*, catergory, *) and cuboid (date, *, *) in inner data cube. The third query firstly analyzes the shared information on edges and then analyzes the nodes which stand for people. In the following sections, we will formulate and address From-Node-To-Edge operation and From-Edge-To-Node operation separately.

3.1   From-Node -to-Edge Operation

In order to obtain the result of first query, we should group the nodes on attribute nation, i.e., the nodes which have the same value on attribute nation are grouped together. Then by edge merging, we can get the shared information between any two nodes. Finally selected shared information is aggregated on category. The result is shown in figure 3.

Figure 3(a) shows the nodes are grouped into three parts. And the nodes belonging to the same nation are grouped together. (1, 3, 7, 8, 9, 14, 16) are edge IDs between people in USA and that in England by edge merging. Figure 3(b) indicates the shared information between people in USA and that in England. And the result of aggregated measure on category is shown in Figure 3(c).

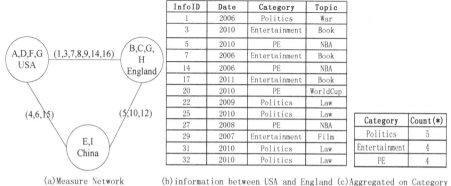

InfoID	Date	Category	Topic
1	2006	Politics	War
3	2010	Entertainment	Book
5	2010	PE	NBA
7	2006	Entertainment	Book
14	2006	PE	NBA
17	2011	Entertainment	Book
20	2010	PE	WorldCup
22	2009	Politics	Law
25	2010	Politics	Law
27	2008	PE	NBA
29	2007	Entertainment	Film
31	2010	Politics	Law
32	2010	Politics	Law

Category	Count(*)
Politics	5
Entertainment	4
PE	4

(a)Measure Network   (b)information between USA and England (c)Aggregated on Category

Fig. 3. Result of Query 1

As the name suggests, from-node-to-edge operation is to analyze nodes first and analyze edges secondly, i.e., firstly OLAP on graph cube and then OLAP on data cube which is generated dynamically in the first step.

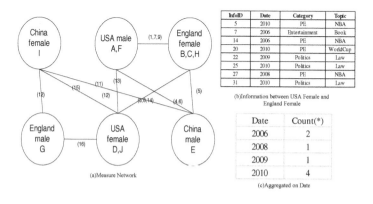

InfoID	Date	Category	Topic
5	2010	PE	NBA
7	2006	Entertainment	Book
14	2006	PE	NBA
20	2010	PE	WorldCup
22	2009	Politics	Law
25	2010	Politics	Law
27	2008	PE	NBA
31	2010	Politics	Law

(b)Information between USA Female and England Female

Date	Count(*)
2006	2
2008	1
2009	1
2010	4

(c)Aggregated on Date

(a)Measure Network

Fig. 4. Result of Query 2

There are not essential differences between query 1 and query 2. In order to illustrate the meaning of from-node-to-edge operation, we give the result of query 2. Figure 4 shows the result of query 2. The nodes are grouped into six parts. And the nodes belonging to the same nation and gender are grouped together. (3, 8, 14) are edge IDs between people in USA female and that in England female by edge merging. Figure 4(b) indicates the shared information between them. And the result of aggregated measure on date is shown in Figure 4(c). Algorithm1 outlines a baseline algorithm to address from-node-to-edge operation in detail. In Algorithm 1, we create a container: MapOne. MapOne contains a mapping from the values of nodes on an attribute combination to the nodes in original network. A' (x) stands for the value of x on aggregation A'.

Algorithm 1: From-Node-To-Edge Operation
  First Input: An Information-Enhanced Multidimensional Network (V, E, T,
              A, B), An aggregation A'
First Output: the Measure Network (V', E', T', A, B)
  Begin:

```
      MapOne (values, nodes) = NULL
      For each u in V do
        If A'(u) not in MapOne.values
          Put A'(u) in MapOne.values
       Endif
       Put u in V'
      Endfor
      Return V'
      For each e= (u, v) in E do
        u' = MapOne(A'(u)) v' = MapOne(A'(v))
       put t' corresponding to e'= (u', v') in T'
        If e' not in E'
         put e' in E'
       Endif
      Endfor
      Return E', T'
```

  End
  Second Input: Measure Network (V', E', T', A, B) and a aggregation B'
  Finally Output: Numeric measure on B'
  Begin:

```
      Array value = NULL
      For each tuple in Ts corresponding to every e'
          If (B'(tuple) not in value)
             Put B'(tuple) in value
          Endif
       Endfor
      For each value
          Put aggregation function on value
          Return measure
      Endfor
```

  End

3.2   From-Edge -to-Node Operation

In order to get the result of query 3, we should analyze what categories there are in all the shared information and their quantity distributions. Then according to a certain type of information, people who share this kind of information are selected and analyzed on attribute profession. Figure (5) shows the result of query 3.

According to the result, people pay more attention to political information, so we want to know the network structure among people in different professions who share political information. Figure5 (b) shows the structure which indicates that the teacher and doctor shared most political information among all profession combinations. Of course, you can select any kind of information such as PE.

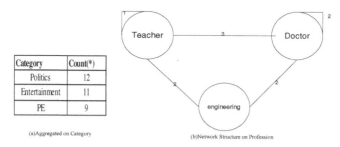

Category	Count(*)
Politics	12
Entertainment	11
PE	9

(a)Aggregated on Category

(b)Network Structure on Profession

Fig. 5. Result of Query 3

As the name suggests, from-edge-to-node operation is to analyze information on edges first and analyze nodes secondly, i.e., firstly OLAP on data cube and then OLAP on graph cube which is generated dynamically in the first step. Algorithm2 outlines a baseline algorithm to address from-edge-to-node operation in detail.

Algorithm 2: From-Edge-To-Node Operation

First Input: An Information-Enhanced Multidimensional Network M (V, E, T,
A, B), An aggregation B'
First Output: Numeric measures on B'

```
Begin:
      Array Value = NULL
      MapOne (values, nodes) = NULL
      For each tuple in T
         If (B'(tuple) not in Value)
         Put B'(tuple) in Value
         Endif
      Endfor
      For each value in Value
         Put aggregation function on value
         Return measures
       Endfor
End
```

Second Input: the same information-enhanced multidimensional network M (V, E,
T, A, B) as above, a aggregation A' and Value
Finally Output: network structure (V", E", E".numbers)
Begin:

```
For every value in Value
    Find every t' in T' satisfy B(t') = value
    Select E' corresponding to T'
    Select V' corresponding to E'
    For every u in V' do
        If A'(u) not in MapOne.values
            Put A'(u) in MapOne.values
        Endif
        Put u in V''
    Endfor
    For every e'= (u', v') in E' do
    u'' = MapOne(A'(u')) v'' = MapOne(A'(v'))
     If e''=(u'', v'') not in E''
       put e'' in E''
       e''.number=0;
    EndIf
    else e''.numbers++
    Endfor
Endfor
Return V'', E'', E''.numbers
```
End

4   Experiments

In this section, we present the major experimental results. We examine a real data set
DBLP from effectiveness perspectives. All the algorithms and experiments are
implemented in Java and tested on a Windows PC with Intel dual-core processor
2.33GHz and 2G of RAM.

4.1   DBLP Data Set

We download the DBLP bibliography data in March, 2012 and further extract a subset
of publication information from fifty conferences in six different research areas:
database (DB), data mining (DM), artificial intelligence (AI), information retrieval(IR),
hardware(HW) and media(Media). In our co-author network, there are 49220 papers
and 57514 authors. For each author, there are three dimensions of information: Name,
Productivity and Researchtime. Productivity can be obtained by the amount of papers
the author published. We discretize the publication number of an author into four
different buckets. If publication number is between 1 and 10, we consider the
productivity of the author as poor. If publication number is between 11and 40, we
consider the productivity of the author as fair. If publication number is between 40 and

100, the productivity of the author is good. If publication number is more than 100, the productivity of the author is excellent. Researchtime can be obtained by taking the date of publishing the first paper from the date of publishing the last one. Similarly, researchtime of an author can be discretized into three buckets. If researchtime is between 1 and 10, researchtime can be expressed as short. If it is between 11 and 20, it is considered as middle. If it is between 21 and 34, it is long. For each paper, there are four dimensions of information: Title, Date, Conference and Area.

4.2   Effectiveness Evaluation

We first evaluate the effectiveness of Nested Cube as a powerful decision support method in the co-author network and present some interesting findings by addressing bidirectional two-ply OLAP query on the network. In the experiments, we are interested in the characteristics of papers between researchers from different perspectives. Upon the nested cube built on the co-author network, we first issue a From-Node-To-Edge operation (*, productivity, */*, *, *, area) and the result is shown in Figure 6. From the figure 6(a), we can conclude that the poor community cooperates a lot with the poor community and the fair community, while the cooperations between excellent and excellent are not frequent. Figure 6(b) displays paper distribution between poor and poor on area. The papers about HW occupy a large proportion and the papers about AI come second.

Next we formulate a From-Edge-To-Node operation (*, *, *, area/*, *, researchtime) and result is shown in figure 7. Figure 7(a) indicates that the proportion of papers about AI is large in all the papers. And we secondly mine the cooperation pattern on researchtime of authors who publish papers about AI. Figure 7(b) shows that short community and short community cooperate AI paper most.

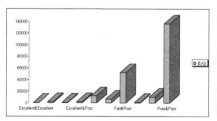 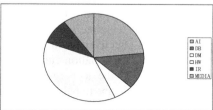

Fig. 6. (a) Co-authorship between Different Productivity (b) Paper Distribution on Area

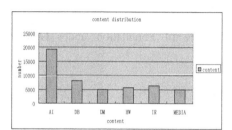 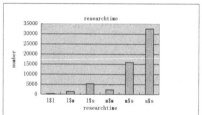

Fig. 7. (a) Paper Distribution on Area (b)Cooperation Distribute on Researchtime

5   Related Work

As an important notion in data analysis, OLAP technology has drawn a lot of attention from many research communities [5, 6, 7]. Traditional OLAP cubes studies [8, 9] focus on numeric measures. However, in recent years many different demands emerge and we have some other forms of data to process. So many researchers are willing to extend OLAP technology to new emerging data in different application domains, such as sequences [10], spatial data [11], text [12, 13] and mobile data [14]. This paper aims to apply OLAP technology to information network. In [3], information network was modeled as a graph with vertices which have multidimensional attributes representing entities and edges depicting relationship between entities and some OLAP operations on that graph were proposed. But in our paper, information network is modeled as a graph whose vertices and edges both have multidimensional attributes. Nested Cube allows us to mine networks data deeply by performing bidirectional two-ply OLAP query on two cubes.

[15] aims to provide OLAP functionalities on graphs which is different from those of Nested Cube. Because these graphs contain attributes themselves which can support I-OLAP. The vertices of graphs contain attributes which can support T-OLAP. These OLAP operations focus on aggregatting a set of graphs into a summary graph. The input of Nested Cube is a single graph rather than a set of graphs.

6   Conclusion and Future Work

In this paper, we propose an information-enhanced multidimensional network which can better model the networks data. Then we address the problem of supporting warehousing and OLAP technology for information-enhanced multidimensional networks. This paper studies this exact problem by first proposing a new data warehousing model, Nested Cube. We formulated bidirectional two-ply OLAP query on Nested Cube to mine information deeply. Our experimental results have demonstrated the effectiveness of Nested Cube as the first method for warehousing and OLAP toward information-enhanced multidimensional networks. However, this is merely the tip of the iceberg. Out next work will consider the materialization algorithms of nested cube to balance the storage space and response time.

Acknowledgment. This work was supported by the National Key Technologies R&D Program (GrantNo. 2009BAH44B02), the National Natural Science Foundation of China (Grant No.61003051), the Natural Science Foundation of Shandong Province of China under GrantNo. ZR2010FM033 and Shandong Distinguished Middle-aged and Young Scientist Encouragement and Reward Foundation under GrantNo. BS2009DX040.

References

1. Agarwal, S., Agrawal, R., Deshpande, P., Gupta, A., Naughton, J.F., Ramakrishnan, R., Sarawagi, S.: On the Computation of Multidimensional Aggregates. In: 22th International Conference on Very Large Data Bases, pp. 506–521. Morgan Kaufmann, Mumbai (1996)

2. Chaudhuri, S., Dayal, U.: An Overview of Data Warehousing and OLAP Technology. In: ACM SIGMOD International Conference on Management of Data, pp. 65–74. ACM Press, Tucson (1997)

3. Zhao, P., Li, X., Xin, D., Han, J.: Graph Cube: On Warehousing and OLAP Multidimensional Networks. In: ACM SIGMOD International Conference on Management of Data, pp. 853–864. ACM Press, Athens (2011)

4. Gray, J., Chaudhuri, S., Bosworth, A., Layman, A., Reichart, D., Venkatrao, M., Pellow, F., Pirahesh, H.: Data Cube: A Relational Aggregation Operator Generalizing Group-By, Cross-Tab, and Sub-Totals. In: Data Mining and Knowledge Discovery, vol. 1, pp. 29–53. Kluwer Academic, Montreal (1997)

5. Chen, B.C., Chen, L., Lin, Y., Ramakrishnan, R.: Prediction Cubes. In: 31th International Conference on Very Large Data Bases, pp. 982–993. ACM Press, Trondheim (2005)

6. Chen, Y., Dong, G., Han, J., Wah, B.W., Wang, J.: Multi-dimensional Regression Analysis of Time-Series Data Streams. In: 28th International Conference on Very Large Data Bases, pp. 323–334. Morgan Kaufmann, Hong Kong (2002)

7. Tian, Y., Hankins, R.A., Patel, J.M.: Efficient Aggregation for Graph Summarization. In: ACM SIGMOD International Conference on Management of Data, pp. 567–580. ACM Press, Vancouver (2008)

8. Beyer, K.S., Ramakrishnan, R.: Bottom-up Computation of Sparse and Iceberg Cubes. In: ACM SIGMOD International Conference on Management of Data, pp. 359–370. ACM Press, Philadelphia (1999)

9. Zhao, Y., Deshpande, P., Naughton, J.F.: An array-based Algorithm for Simultaneous Multidimensional Aggregates. In: ACM SIGMOD International Conference on Management of Data, pp. 159–170. ACM Press, Tucson (1997)

10. Lo, E., Kao, B., Ho, W.S., Lee, S.D., Chui, C.K., Cheung, D.W.: OLAP on sequence data. In: ACM SIGMOD International Conference on Management of Data, pp. 649–660. ACM, Vancouver (2008)

11. Han, J.: Olap, Spatial. In: Encyclopedia of GIS, pp. 809–812. Springer, New York (2008)

12. Zhang, D., Zhai, C., Han, J.: Topic Cube: Topic Modeling for OLAP on Multidimensional Text Databases. In: 9th SIAM International Conference on Data Mining, pp. 1123–1134. SIAM Press, Sparks (2009)

13. Lin, C.X., Ding, B., Han, J., Zhu, F., Zhao, B.: Text Cube: Computing IR Measures for Multidimensional Text Database Analysis. In: 8th IEEE International Conference on Data Mining, pp. 905–910. IEEE Computer Society, Pisa (2008)

14. Li, J., Zhou, H., Wang, W.: Gradual Cube: Customize Profile on Mobile OLAP. In: 6th IEEE International Conference on Data Mining, pp. 943–947. IEEE Computer Society, Hong Kong (2006)

15. Chen, C., Yan, X., Zhu, F., Han, J., Yu, P.S.: Graph OLAP: Towards Online Analytical Processing on Graphs. In: 8th IEEE International Conference on Data Mining, pp. 103–112. IEEE Computer Society, Pisa (2008)

MRFM: An Efficient Approach to Spatial Join Aggregate

Yi Liu, Luo Chen, Ning Jing, and Wei Xiong

College of Electronic Science and Engineering, National
University of Defense Technology, Changsha 410073, China
liu.yi.nudt@gmail.com

Abstract. Spatial join aggregate(SJA) is a commonly used but time-consuming operation in spatial database. Since it involves both the spatial join and the aggregate operation, performing SJA is a challenging task especially facing the deluge of spatial data. A popular model nowadays for massive data processing is the shared-nothing cluster using MapReduce. Thus, to explore SJA in MapReduce, a **Map-Reduce-Filter-Merge**(MRFM) algorithm is proposed.Map step divides the total SJA task into disjoint sets, then Reduce step aggregate each set individually, a Filter operation will filter those aggregate results of single assignment spatial objects.Finally, Merge step further aggregate the partial results of multiple assignment spatial objects using an efficient merge algorithm. Extensive experiments in large real spatial data have demonstrated the efficiency, effectiveness and scalability of the proposed methods.

Keywords: MapReduce, Spatial Join Aggregate, Spatial Database.

1 Introduction

Exponential increase in the amount of spatial data make the spatial query processing more challenging. Spatial join aggregate (SJA) is an important and frequently used operation in spatial databases. Since both the spatial join and the aggregate operation are expensive, especially on large spatial data sets, SJA is a costly operation. SJA retrieve the summarized information from the result of spatial join that the user are more interested in. SJA has been applied to many scenarios(e.g.,spatial statistical analysis, data warehouses,etc). For example, in VLSI design, a circuit layer consists of numerous wires, represented as rectangles. When two layers are placed together, the intersection between the wires of different layers may cause electro-magnetic interference. A SJA can be applied to retrieve the wires and intersecting number from the other layer. The result indicates the positions where the topology of the circuit can be improved to reduce interference.Similarly, the spatial semi-join aggregate returns the objects of one data set and the intersection or containment counts with with the other.

SJA has some similarity with spatial range aggregate (RA) queries. Several techniques[1-3] have been proposed for the efficient processing RA in spatial databases. However, the application of these techniques to SJA would incur large

Z. Bao et al. (Eds.): WAIM 2012 Workshops, LNCS 7419, pp. 140–150, 2012.

computational overhead, because the processing of a spatial semi-join aggregate on two data sets R and S requires $|R|$ RA queries, while a full join aggregate requires $|R|+|S|$ RA queries.

In this work, we investigates the problem of executing SJA for large-scale spatial data sets. MapReduce is used to execute SJA operation, since MapReduce is a popular model for large-scale data processing. We first propose the baseline approach using two MapReduce stages. The first stage performs the partial aggregate spatial join and the second stage perform the further merge operation for the results generated from the first stage. However,due to the unnecessary aggregate operation for single assignment objects, the baseline approach does not scale well for large result data. In light of this limitation, we introduce a Map-Reduce-Filter-Merge(MRFM) algorithm,Map-Reduce operation is similar to the baseline algorithm, while Filter step filter those aggregate results for single assignment objects, which will considerably decrease the computing complexity and data size performed in Merge phase. Thus an efficient sequential merge algorithm can be applied to execute the small scale merge operation.

2   Background

2.1   Spatial Join Aggregate

Given two data sets R and S, SJA retrieves the objects in data set R or S with the intersection or containment counts from the other data set. In the execution stage of SJA, if R is regarded as outer relation and S inner relation, then the form of a SJA query Q is: "select op(expression(t_i)) from R, S where $R.geometry$ intersection $S.geometry$". where op represents aggregate functions, t_i is the join tuple of $R \bowtie S$.

SJA: The spatial join aggregate SJA(R, S) of R and S is:

$$\text{SJA}(R,\ S)=\{(t,\text{op}(t,S(R)))|\ \text{for all}\ t \in R(S)\ \}$$

Semi-SJA: The spatial semi-join aggregate SJA(R, S) of R and S is denoted as Semi-SJA(R, S) and expressed as:

$$\text{Semi-SJA}(R,\ S)=\{(t,\text{op}(t,S))|\ \text{for all}\ t \in R\ \}$$

Aggregation functions are divided into three classes [4]: distributive, algebraic and holistic. Distributive aggregates (e.g., *count,max,min,sum*) can be computed by partitioning the input into disjoint sets, aggregating each set individually and then obtaining the final result by further aggregating the partial results. Algebraic aggregates can be expressed as a function of distributive aggregates: *average*, for example, is defined as *sum/count*. Holistic aggregates (e.g., *median*), on the other hand, cannot be computed by dividing the input into parts. In this paper, we only consider the spatial join *count* queries on spatial data sets,but the solutions apply to any distributive,or algebraic(not holistic) aggregates with straightforward. For the spatial join *count* queries,op(t,S) is:

$$\text{op}(t, S)=\{(t,|S'|)|\ S' \subseteq S\ \&\&\ \forall s \in S',\ t\ \text{intersection}\ s\ \text{is}\ true\ \}$$

Namely, the output of SJA is a list of the form $(t, count)$, where t represents a spatial object and $count$ is the size of its intersect with or contains objects (from the other data set).

2.2 MapReduce Basics

A MapReduce[5] program typically consists of a pair of user-defined map and $reduce$ functions. The map function is invoked for every record in the input data sets and applies user-defined logic to every record to produce a list of intermediate key/value pairs. The MapReduce framework aggregates the values with the same intermediate key and produces a partitioned and sorted set of intermediate results. The reduce function fetches sorted data, applies user-defined logic to those values, and produces a list of output values. The signatures of map and reduce are:

$$map(k1, v1) \rightarrow list(k2, v2)$$
$$reduce(k2, list(v2)) \rightarrow list(k3, v3).$$

3 Baseline Method

The straightforward method for SJA in MapReduce consists of two MapReduce stages executed in sequence.Firstly,partial aggregate spatial join stage.Secondly, merge stage. A further merge operation aggregate the partial results.

3.1 Stage 1: Partial Aggregation Spatial Join

The goal of this stage is to partition R and S into spatial disjoin buckets, then compute the partial spatial join aggregate results in each bucket. In this work, we adopts a tile-based partitioning method which splits join region regularly into N_T same-size tiles(N_T is far larger than the Reducer number), The spatial objects in R and S are hashed into the tiles with which extent is overlapped.

Lemma 1. Given a join region T, the partition, denoted by $\pi(T)$, is $\pi(T) = \{T_i|i = 0...N_T - 1\}$. T_i is a tile of $\pi(T)$.then $\pi(T)$ meet the following conditions:

(1) $\cup_{i=0}^{N_T-1} T_i = T$;
(2) $i \neq j \Longleftrightarrow T_i \cap T_j = \phi$.

Lemma 2. Let $\pi(T)$ be a partition of T, R_i and S_i is the tuple sets associated with T_i. If the join results of R_i and S_i in different partition has been avoided. The following holds:

$$\text{SJA}(R, S)=\bigcup_{i=0}^{N_T-1}\text{SJA}(R_i, S_i).$$

proof. From lemma 1, the following is true:

(1) $R = \bigcup_{i=0}^{N_T-1} R_i$;

(2) $S = \bigcup_{i=0}^{N_T-1} S_i$;

(3) $i \neq j \Longleftrightarrow R_i \cap S_j = \phi$.

then, according to assignment law:

$$\text{SJA}(R,S) = \text{SJA}(\bigcup_{i=0}^{N_T-1} R_i, \bigcup_{i=0}^{N_T-1} S_i) = \bigcup_{i=0}^{N_T-1} \text{SJA}(R_i, S_i).$$

Lemma 2 indicates that SJA(R,S) is divided into disjoint sub tasks $\text{SJA}(R_i, S_i)$ $(i = 0...N_T - 1)$. Let r be the reducer number, N_T tasks will be mapped into the r reducers using a round robin scheme.

Map Phase. In map phase, the input to this phase are the relations of R and S. To identify the heterogeneous data sources, the first task of map phase is to homogenize the data sources. In this work, tag "r" is used to tag objects from R. In addition, the map phase hash the spatial objects to the tiles it overlays. Algorithm 1 shows the process.

Algorithm 1: Partition spatial objects.

Input: Spatial relation R and S

Output: A set of (*tileid, object*) pairs

1.**class** Mapper

2.   **method** Map(objectid *oid*,object *o*)

       //step 1 Homogenization,insert a data-source tag for *o*

3.   *o*=homogenize(*o*)

       //step 2 compute the tiles *o.MBR* overlays

4. $TempTileList = \text{GetTiles}(o.MBR)$

       //step 3 map *o* to tiles it overlays

5.   **foreach** t in $TempCellList$ **do**

6.   Emit(tileid t, object o)

Partition Function. Let t be the output key at the end of Map phase. t meets:$0 \leq t < N_T$. The partition function f is defined as:$f(t) = t \bmod r$.

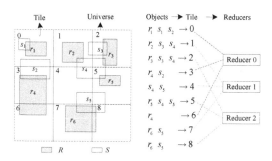

Fig. 1. The example spatial partition

For example, consider Fig.1 where the join region is divided into 9 tiles, r is set to 3. The object is first mapped to the tiles overlays, then tiles are mapped to reducers using a round robin scheme. Thus tiles 0,3 and 6 are mapped to reducer 0, tiles 1,4 and 7 are mapped to reducer 1, and tiles 2,5 and 8 are mapped to reducer 2.

Reduce Phase. The input to reduce phase is the tile number tid as the key and associated spatial objects R_i and S_i as the values. Reduce phase execute the partial aggregate spatial join. Algorithm 2 shows the process.

Algorithm 2: Forward-Sweep SJA with Duplicate Avoidance.

Input: key:tid, values: R_i, S_i

Output: A set of $(object, count)$ pairs

1.**class** Reducer
2.   **method** Reduce(tileid ti,objects R_i, S_i)
3. $interLst \leftarrow$Hashtable(tuple t,count c)
4.   sorts R_i and S_i in x axis order
5.   **while** $(R_i \neq \varnothing \; \delta\delta \; S_i \neq \varnothing)$
6.   Let r=Head(R_i) and s=Head(S_i)
7.   **if** $r.x_l < s.x_l$ **then**
8. R_i.remove(r)
9.   let $count \leftarrow 0$
10.   **while**($s \neq \varnothing \; \delta\delta \; r.x_h > s.x_l$)
11.   **if** r intersect s is true **then** // using duplication avoidance
12. $count \leftarrow count + 1$
13.   updateInterLst($interLst,s,1$)
14. $s=S_i$.nextElement()
15.   End **while**
16. $count \leftarrow count+interLst$.get($r$)
17.   output(key r,value $count$)
18. $interLst$.remove(r)
19.   **else**
20.   let $count \leftarrow 0$
21.   **while**($r \neq \varnothing \; \delta\delta \; s.x_h > r.x_l$)
22.   **if** s intersect r is true **then** // using duplication avoidance
23. $count \leftarrow count + 1$
24.   updateInterLst($interLst,r,1$)
25. $r=R_i$.nextElement()
26.   End **while**
27. $count \leftarrow count+interLst$.get($s$)
28.   output(key s,value $count$)
29. $interLst$.remove(s)
30.   End **while**

Prior to processing any input key-value pairs, line 3 initialize a hashtable for holding spatial object and its temp aggregate value. Line 4 sorts R_i and S_i in x axis order. Then line 5-30 scans forward in the sorted lists R_i and S_i to search the objects that will intersect the current object in the future. In each scan forward operation for the current object $r(s)$, line 12,23 count the intersection number for $r(s)$,while for $s(r)$, algorithm just update the aggregate value 1 to $interLst$, since $r(s)$ is removed from $R_i(S_i)$.Before $r(s)$ is removed in line 18,29, the map function output the results. To avoid generating duplicates among different tiles,algorithm 2 uses the reference points technique proposed in [6] to avoid duplicates.

3.2   Stage 2: Merge Phase

The results obtained from the first stage are the partial aggregate results in each tile. The object of this stage is to further aggregate the partial results.

The input key to the map function is the spatial object and the value is the partial aggregation number. The map function just emit them unchangedly. In the shuffle phase,the partition function re-partitions them according to the key,which represent a spatial object.In the reduce phase, the input key represents a spatial object and the values is a series of spatial aggregation results from all partitions. The reduce function just sum those aggregation results and form the final result.

4   Map-Reduce-Filter-Merge Algorithm

In baseline algorithm, when each record o appears only in one tile or is replicated into one bucket, the result obtained in the first stage is the final aggregate result for o. The baseline algorithm do not take this situation into account.

4.1   Filter Optimization for Single Assignment Object

The basic idea of filter optimization is to detach the aggregate result of SA objects from the total result. The filter is twofold.

Tile Level Filter. If the object is only replicated to one tile, we insert an additional *assignment* tag into the attributes of the object in the map phase of the first round. But for the tuples assigned multiple, we do not change it. In this work, tag 1 is used as the single tile assignment tag. In the above example,for the tuples assigned to tile only once(e.g., r_1,s_1),the output key/value will be:

$$r_1 \longrightarrow (0,r_1 \ r \ 1)$$
$$s_1 \longrightarrow (0,s_1 \ 1)$$

Reducer Level Filter. If the object is mapped to tile list tl, for $\forall t \in tl, f(t) \equiv j$. Then tl are all mapped to reducer j. Hence the aggregate value obtained in reducer j will be the final result. For the tuples only replicated to one reducer, we use tag 2 as the single reducer assignment tag.

Algorithm 4 shows the filter process in reduce phase of the first round. The difference from algorithm 3 is that each reducer will initialize a hashtable H_{SR} to hold the single reducer assignment records and its aggregate value. At the end of the reduce function, if key contains *single tile assignment tag*, since it can be identified as the final result,the key and value will be directly written into the result sets in line 8. If key contains *single reducer assignment tag*, we update it to H_{SR} in line 10. While the other key and value pairs will be transferred to the next stage. After all tasks replicated to this reducer have been completed, the reducer invoke the close method to emit the aggregate result for single reducer assignment objects.

Algorithm 4: Filter operation in Reducer.
Input: Input: key:tid, values: R_i, S_i
Output: A set of $(object, count)$ pairs

1.**class** Reducer
2.   **method** Initialize
3. H_{SR} ← new Hashtable(tuple o, count c)
4.   **method** Reduce(tileid ti,objects R_i, S_i)
5.   Algorithm 3 line 3-30,but line 17 and 28
          are revised to line 6-12
6.   if($r(s)$ contains single tile tag)
8.   Send $r(s)$ and *count* to result sets
9.   if ($r(s)$ contains reducer tile tag)
10. H_{SR}.update($r(s)$,*count*))
11.   if ($r(s)$ is MA objects)
12.   output(key $r(s)$,value *count*)
13.   **method** Close
14.   for((tuple o, count c)∈ H_{SR})
15.   Send o and c to result sets

4.2   Merge Optimization

The goal of this phase is to further aggregate the partial results from different reducers for MA spatial objects.

Algorithm 5: Sequential merge algorithm.
Input: Partial Aggregate Result List ($PARL$)
Output: Final Merged Aggregate Result(FR)

---

1. FL ← new Hashtable(Object o, List L)
//Map each record into FL
2. foreach result r in $PARL$
3.   (o,*count*) ← Head (r)
4.   while ((o,*count*)≠ ∅)
5.   updateFL(FL,o,*count*)
6.   (o,*count*)← r.nextElement
//sum the aggregate values for each object
7.sum=0
8.**foreach**((Object o, List l) in FL) **do**
9.   sum ← count(l)
10.   FR.add(object o, count *sum*)

Two algorithms can be used to solve this problem. When the size of the results generated from the first round is still very high, the proposed algorithm in Section 3.2 is a good choice. However, the computation intensity and data size of results have been greatly decreased after the filter operation in most situations. A more efficient sequential algorithm is better alternative. Algorithm 5 shows the process. Algorithm 5 is inspired by the idea of MapReduce, since the the data is distributed in different partitions. Algorithm 5 first initialize a hashtable FL in line 1, where key is used to hold objects and value is a list to hold corresponding partial aggregate results.Line 2-6 just map each record from different reducers into FL. Function $updateFL$ update the object and its aggregate result in FL. Then line 7-9 summarize the partial aggregate values and add the final output to the result set FR.

5   Performance Evaluation

In this section, we conduct extensive experiments to evaluate the practical performance of our proposed MRFM algorithm under MapReduce framework,moreover we compare it with the baseline algorithm.

The evaluations are performed on a 32 nodes IBM x3650 cluster and Apache Hadoop version 0.20.1 is used for our experiments. Three spatial datasets from TIGER/Line project are taken as the testbed. They are the road network, hydrography and census data of U.S. (except Hawaii,Alaska). The detailed information of the datasets are listed in Table 1.

Table 1.   Datasets used in the experiments

Dataset	Objects	Size(GB)	Description
R	29,692,784	7.3	the road network information in U.S
H	7,066,849	2.7	the hydrography information in U.S
C	8,137,053	3.9	the census block information in U.S

Two experiments are performed to evaluate the performance of our proposed algorithms. The first one is $R \times H$ spatial intersection join aggregate (SIJA), the second one is $C \times H$ spatial contain semi-join aggregate (semi-SCJA).The join region is partitioned into $2^k \times 2^k$ tiles. k varied from 7 to 10. For each pair c and h with semi-SCJA, if $c.MBR$ contains $h.MBR$ is true, we judge that c contains h is true due to the extra complexity of exact spatial contain operation.

5.1   Comparison with Baseline Algorithm

First, we test MRFM algorithm with the baseline algorithm on 32-nodes environment by using different tile size. The reducer number on cluster is fixed at 64. Fig.2 shows that under different tile size, the performance of the partial aggregate join phase in two experiments for the two algorithms is very approximate. thus, the filter operation do not damage the performance of the first stage. Furthermore, the performance of merge phase in MRFM is more efficient than that in baseline algorithm. Hence, the filter optimization yields certainly performance benefit.

Another observation is that for the two algorithms, there exists an optimal tile size. In the latter experiments, we chose tile size $2^8 \times 2^8$ for the two experiments.

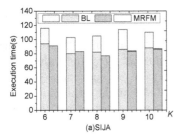 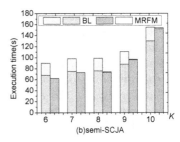

Fig. 2. Comparison with baseline algorithm

To test the effects of filter operation in MRFM, a simple MapReduce program is written to calculate the ratio of the result number obtained from the two stages. Fig.3 shows the results. When tile size varied from $2^6 \times 2^6$ to $2^{10} \times 2^{10}$, the percent by result number for filtered results increased from 98.5% to 78.05% for SIJA and from 93.34% to 34.15% for semi-SCJA, which indicates that the filter operation can decrease indubitably the computing intensity in Merge phase considerably under a appropriate tile partition.

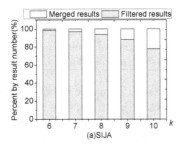 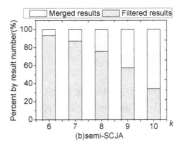

Fig. 3. Filtered results analysis

5.2   Effects of Reducer Number and Node Number

Finally, we test the proposed algorithm by using different reducer number on different nodes number cluster.

Fig.4 shows the effects of node number and reducer number on MRFM performance. Obviously, the performance of MRFM has direct relationship with node number and the reducer number. Firstly, with the increasing of node number, the performance of MRFM for the two experiments improves obviously. It should be noted that the efficiency improvement is not as significant as that when node number is larger than 16 for both SIJA and semi-SCJA. The reason is the that the reducer number in 16-nodes is sufficient to process SIJA and semi-SCJA for the given spatial data sets. Secondly, for a certain node number, when reducer number $r \leq 8N$, performance of MRFM improve with the increase of r. But when $r > 8N$, performance of MRFM recedes with the increase of r. That is because 8 reducer tasks could be executed simultaneously at every nodes. Since each node has 8 cores. When $r > 8N$, the reducer could not be executed simultaneously in one cycle.

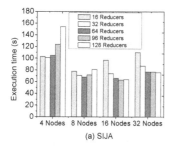
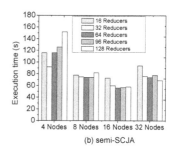

(a) SIJA   (b) semi-SCJA

Fig. 4. Effects of reducer number

6   Related Work

The MapReduce paradigm was initially presented by Google in [5]. Since then, it has gained a lot of attentions in academic [8,9,10]areas. A comparison of the MapReduce paradigm with parallel DBMS has been done in [8]. an in-depth study of the performance of MapReduce has been investigated in [9].In [10] the authors proposed extending the interface with a new function called merge in order to facilitate joins.

In the spatial database domain, an increasing number of spatial processing applications are implemented with MapReduce with the help of its open-source implementation such as Apache Hadoop[11]. The problem of road alignment by matching satellite imagery and digital line maps with MapReduce was studied in [12]. This work concentrates on the complexity of the problem, which is a rather challenging task. [13] presented their experiences in applying the MapReduce model to solve two important spatial problems: the bulk-construction of R-trees and the quality computation of aerial images. [14] proposed the first spatial join algorithm with MapReduce on small scale clusters. To the best of our knowledge, there is no previous work on parallel processing spatial join aggregate with MapReduce on cluster.

7   Conclusion

In this paper we studied the problem of answering spatial join aggregate queries under the MapReduce framework. We first proposed a baseline approach. The baseline approach did not consider the single assignment of objects, which aggregate result can be obtained in the first Map-Reduce stage. The filter algorithm, which is the core of the whole approach, has been tested that can minimize the workload in merge phase.

An interesting future research topic is how to combine the spatial index with MapReduce to answer complex spatial queries on large spatial data sets, Since MapReduce process large data using a brute-force approach. Any progress in that direction would be of great value as more and more spatial data are being acquired around the world and integrated into large-scale databases.

Acknowledgments. This work is supported by National High-Tech Research and Development Plan of China(Nos.2011AA120306)and the National Natural Science Foundation of China(Nos.61070035).

References

[1] Tao, Y., Papadias, D.: Range aggregate processing in spatial databases. IEEE Transactions on Knowledge and Data Engineering 16(12), 1555–1570 (2004)

[2] Jurgens, M., Lenz, H.: The Ra*-Tree: An Improved R-Tree with Materialized Data for Supporting Range Queries on OLAP-Data. In: Proc. DEXA Workshop (1998)

[3] Papadias, D., Kalnis, P., Zhang, J., Tao, Y.: Efficient OLAP Operations in Spatial Data Warehouses. In: Jensen, C.S., Schneider, M., Seeger, B., Tsotras, V.J. (eds.) SSTD 2001. LNCS, vol. 2121, p. 443. Springer, Heidelberg (2001)

[4] Gray, J., Bosworth, A., Layman, A., Pirahesh, H.: Data Cube: A Relational Aggregation Operator Generalizing Group-By, Cross- Tabs and Subtotals. In: Proc. Intl. Conf. Data Eng. (1996)

[5] Dean, J., Ghemawat, S.: MapReduce: simplified data processing on large clusters. Commun. ACM 51(1), 107–113 (2008)

[6] Dittrich, J.P., Seeger, B.: Data redundancy and duplicate detection in spatial join processing. In: ICDE, pp. 535–546 (2000)

[7] UC Bureau, Census 2010 Tiger/Line data (2010)

[8] Pavlo, A., Paulson, E., et al.: A comparison of approaches to large-scale data analysis. In: SIGMOD Conference, pp. 165–178 (2009)

[9] Jiang, D., Ooi, B.C., Shie, L., Wu, S.: The Performance of MapReduce: An Indepth Study. In: VLDB 2010 (2010)

[10] Yang, H., Dasdan, A., et al.: Map-Reduce-Merge: simplified relational data processing on large clusters. In: SIGMOD Conference, pp. 1029–1040 (2007)

[11] White, T.: Hadoop: The Definitive Guide. Yahoo! Press, Sebastopol (2009)

[12] Wu, X., Carceroni, R., et al.: Automatic alignment of large-scale aerial rasters to road-maps, Geographic Information Systems. In: Proceedings of the 15th Annual ACM International Symposium on Advances in Geographic Information Systems, Article No. 17 (2007)

[13] Cary, A., Sun, Z., Hristidis, V., Rishe, N.: Experiences on Processing Spatial Data with MapReduce. In: Winslett, M. (ed.) SSDBM 2009. LNCS, vol. 5566, pp. 302–319. Springer, Heidelberg (2009)

[14] Zhang, S., Han, J., Lin, Z., et al.: SJMR: Parallelizing Spatial Join with MapReduce on Clusters. In: SSDBM Conference (2009)

A Distributed Inverted Indexing Scheme
for Large-Scale RDF Data

Xu Li, Xin Wang[*], Hong Shi, Zhaohua Sheng, and Zhiyong Feng

School of Computer Science and Technology, Tianjin University, Tianjin, China
lixu86@yahoo.com.cn, {wangx,serena,zyfeng}@tju.edu.cn,
shengzhaoli24353@sina.com

Abstract. With the development of the Linked Data project, enormous RDF data have been published on the Web. A scalable system is required to provide an efficient retrieval for large-scale RDF data. This paper presents a distributed inverted indexing scheme for large-scale RDF data. A scalable inverted index is built using the underlying data structure of Cassandra which is a distributed key-value storage system. We optimize the indexing scheme with the characteristics of RDF data model to effectively support the fast keyword search. The loading, encoding and indexing procedures are implemented for RDF data simultaneously using the MapReduce framework. The experimental results show that our indexing scheme can effectively support keyword retrieval for large-scale RDF data.

Keywords: RDF, large-scale, inverted index, distributed indexing, MapReduce.

1 Introduction

Since the first release of the Linked Data [2] project in October 2007, Linked Data has been growing rapidly. Due to its further development, more and more data on the Web are available. By September 2011, The Linked Open Data has included 295 data sets, 31 billion RDF triples, and 504 million RDF links.

However, how to efficiently retrieve the piece of data in billions of triples has become a crucial problem. Unlike traditional Web documents, the RDF data have special data structures. Since the current theoretical models in Information Retrieval are not designed for RDF data, users may get irrelevant results if they just input simple keywords in a traditional Web retrieval system. The transfer from document retrieval to data retrieval has raised a new challenge for Web retrieval systems. A highly efficient RDF data retrieval model is the major problem in the adoption of the Semantic Web.

On the other hand, how to store, manage, and use these enormous amounts of data more efficiently has also become an important issue. The traditional database system appears to be inadequate in the context of current massive data processing, which

---

[*] Corresponding author.

Z. Bao et al. (Eds.): WAIM 2012 Workshops, LNCS 7419, pp. 151–161, 2012.
© Springer-Verlag Berlin Heidelberg 2012

requires a great deal of computing resource. A single server or computer has become unsatisfied for the current requirements.

In this paper, we study in how to establish an efficient retrieval model for the large-scale RDF data. In order to achieve large-scale data processing, we use the MapReduce framework to load RDF data, execute the dictionary encoding and build indexes through two successive MapReduce jobs. The whole process only takes one operation without any other external support. In order to provide efficient query services, three structured indexes and an inverted index are created using the underlying data structure provided by Cassandra. Also, based on the characteristics of RDF data, we optimize the structure of the inverted index to provide users with more accurate query services when the inverted index is integrated with other indexes.

This paper is organized as follows: in Section 2, we review the related work in the retrieval of RDF data. A theoretical model of inverted indexing and query processing is presented in Section 3. Based on the definitions in the previous section, in Section 4, we describe in details the specific method that processes data with MapReduce. In Section 5, we take the DBpedia dataset as an example to carry out a set of experiments and present the performance results. Finally, in Section 6, we conclude the paper and outlook the future work.

2   Related Work

With the rapid development of the Semantic Web, applications about RDF data require a reliable and efficient data management system. Based on the RDBMS [10] a great number of improved RDF storage systems have already emerged such as [8][9]. These systems have achieved the management of large-scale RDF data and the support of complex SPARQL queries. But SPARQL queries require the use of the accurate query language, which is not convenient for common users. In addition, SPARQL queries are also limited by the format of data and cannot support different data sources. There are also some other approaches [5][6][7] concerning on keyword-based retrieval of RDF data. In those works, theories of Information Retrieval are introduced to the Semantic Web, and models of the respective index and rank algorithm are proposed respectively. Although keyword search can provide users with good query experiences, current indexing models are not appropriate to the structure of the RDF data model. The K-Search [4] and Semplore [3] are among the first work that combines the keyword retrieval and RDF structure. The SWSE [11] project has proposed a highly efficient indexing model based on the entity retrieval. However, in their system, RDF data are only treated as ordinary Web documents, and they do not take full advantage of the RDF graph structure to optimize the index structure.

In response to the challenges posed by the massive data storage and analysis, CumulusRDF [12] and Jingwei [13] system have done some exploratory work in the distributed management of large-scale RDF data. The Jingwei system takes the distributed key-value storage cluster as the RDF underlying storage system, designs a storage and indexing scheme specifically for the features of the RDF data model, and creates *SPO*, *POS*, and *OSP* structured indexes to effectively support the fast execution of triple pattern queries. However, the data processing of the Jingwei system has low execution efficiency since it does not use the distributed computing technology. Based on part of the theory of the Jingwei system, our work does some improvements to make it support the keyword search and to increase the data processing efficiency.

3   Distributed Indexing Scheme

3.1   Basic Definitions

Before building the inverted index, we first give the definition of the document. The document is the basic unit to create an inverted index. It can be a separate document, or part of a document, or an entirety containing both documents and the relevant content. Considering RDF data in particular, the semantic information of the data should be fully considered in the document definition. RDF data are different from the traditional unstructured document data. RDF documents should be a series of data collections that have specific inherently links and are able to express the same common information. Detailed definitions of these concepts are described as follows.

Definition 1. *(RDF Triple) Assume that there are two disjoint infinite sets U and L, where U is the set of RDF URIs and L is the set of RDF literals. An RDF triple is a tuple of the form*

$$(s, p, o) \in U \times U \times U \cup L$$

In this tuple, s is called the subject, p the predicate, and o the object.

Definition 2. *(RDF Graph) An RDF graph is a finite set of RDF triples. Let G be an RDF graph.*
1. A subgraph of G is a subset of G.
2. The universe of T, denoted by univ(G), is the set of elements of $U \cup L$ that occur in the triples of T.
3. The size of G, denoted by $|G|$, is the number of triples it contains.
4. subj(G) denotes the set of all subjects in G.
5. pred(G) denotes the set of all predicates in G.
6. obj(G) denotes the set of all objects in G.

Definition 3. *(RDF Document) Given an RDF graph G and a subject $s \in subj(G)$, a subject document for s is a subgraph D_s of G where $subj(D_s) = \{s\}$.*

The RDF document is defined as a collection of all triples with the same subject in Definition 3. Our document definition adds certain semantic information to the RDF document. There is the one-to-one correspondence between each subject and entity in RDF data. A triple is a description of the entity, and triples with the same subject are a collection to describe the same entity. The RDF document constructed by these triples contains all the relative information of the entity.

3.2   Index Structure

This section describes the basic theoretical model of the inverted index for RDF data and the improved scheme. The work in this paper is partly based on the Jingwei system, as pointed in related work. The Jingwei system has designed three structured indexes (*SPO*, *POS*, and *OSP*) which are not only the RDF data storage repository, but also provide the index table needed for triple pattern queries.

3.2.1   Basic Inverted Index

The inverted index is essentially a list of pairs consisting of a document and a word, i.e., each term and its corresponding document IDs are a format of one-to-many mappings. Cassandra provides three nested key-value pairs, i.e., {key→super column}, {super column→column name} and {column name→value}. The three nested key-value pairs may provide 4-layered storage, namely, the row key (*RK*), the super column (*SC*), the column name (*CN*) and the column value (*CV*). The feasibility of using Cassandra to create an index has been confirmed through experiments in [13]. We further extend this theory using the underlying data structure provided by Cassandra to create an inverted index. The structure of the basic inverted index is shown in Fig. 1. In Cassandra, the key is used to store a term, while the column name is used to store a document ID, whereas column value is null. We define the subject ID as a document ID in Section 4.

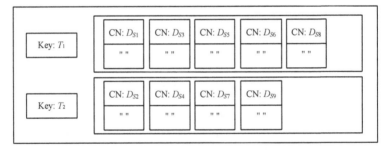

Fig. 1. The Structure of Basic Inverted Index

The above structure solves the problem that the inverted index occupies too much storage space. As the scale of RDF data in Linked Data increases rapidly, creating an inverted index certainly takes up large amounts of storage space, and reliable retrieval performance cannot be guaranteed. Cassandra has high horizontal scalability, and therefore is ideal for storing large-scale data. Thus, our method for creating the inverted index in the distributed situation can effectively tolerate the storage pressure caused by large indexes.

3.2.2   Improved Inverted Index

Based on the inverted index created using the structure showed in previous subsection, given an arbitrary word, all document IDs can be obtained. However, there are still some disadvantages. This index structure only realizes the basic functions of the inverted index with limited information provided. The system can only provide document IDs but not the triples which contain terms. Also, the particularity of the RDF data model is not taken into consideration in this solution. So we further improve this index structure to make it more suitable for indexing RDF data. As shown in Fig. 2(a), the improved inverted index makes use of the super column in Cassandra. The correspondence between the contents stored in Cassandra and the inverted index structure is listed in Table 1.

Table 1. The description of the inverted index

Cassandra	Inverted index
RK	The term
SC	The id of the document containing the term
CN	The id of the RDF resource containing the term in this document
CV	The location information of the resources. 0 indicates that the resource is a subject, 1 a predicate, 2 an object, and 3 a literal

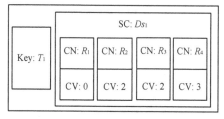 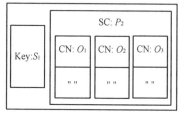

(a) Improved Inverted Index   (b) Index Structure of *SPO*

Fig. 2. Index Structures

For example, $\{RK: T_1\}$ means that the term of the posting list is T_1; $\{SC: Ds_1\}$ means that the document ID of the term is Ds_1; $\{CN: R_2\}$ indicates that the resource ID of the term is R_2; $\{CV: 2\}$ indicates that resource R_2 appears as an object in the document Ds_1 and is non-literal. All these expressions together give the meaning that the T_1 appears in the resource R_2 of document Ds_1 and R_2 is the object of non-literal.

3.3   Query Model

Based on the improved inverted index, we call the structured indexes according to the value in the column value to execute the nested key mapping query. Table 2 lists the available indexes corresponding to the 4 kinds of different situations. The *SPO* index structure given by the Jingwei system is shown in Fig. 2(b).

Table 2. Available index in different situations

Column value	Triple pattern query	Available index
0	$(S, ?X, ?Y)$	*SPO*
1	$(S, P, ?Y)$	*SPO*
2	$(S, ?X, O)$	*OSP*
3	$(S, ?X, O)$	*OSP*

The document IDs and resource IDs of the term are firstly fetch from the inverted index. Then available structured index is selected according to the column value. As a result, all the triples where this term occurs are able to be precisely found. The combination of the keyword search and the triple pattern query can provide users with comprehensive query services.

4   Indexing with MapReduce

Here we show how to do dictionary encoding, meanwhile, create the structured indexes and the inverted index by using two successive MapReduce jobs. The first job assigns a unique code to every resource, and the second job creates the indexes and rebuilds the triples.

4.1   First Job: Encoding Resources

In order to parse RDF data conveniently, the input of first job is compressed files in the N-Triples format and the mapper receives a tuple in the form of *<Null, Triple>* where the triple is coded in N-Triples format.

First, each triple is assigned a unique id in the map process. The number can be divided into many different ranges by the unique mapper task id. For example, the first mapper task can offer the number 0 to 50; the second mapper task can offer the number 51 to 100, and so on. In this way we avoid the risk that the triples processed by different mapper tasks are assigned the same id. Each triple generates three outputs which are composed of two parts: key and value. The key is one of the subject, predicate, and object. The value is a tuple in the form of *<Code, String>* where the code consists of the triple ID and the position in the triple and the string is the subject in the triple.

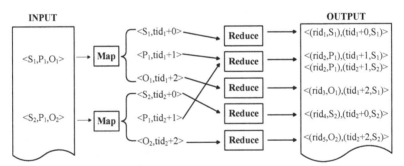

Fig. 3. First Job: Encoding Resources

During the reduce phase, we assign a unique id to every resource and store the mapping relationships of the resource and its id. The resource ID is generated in a similar way to the triple id. In the meantime, the program creates a column family named *Resource2Id* in Cassandra, then inserts the resource as a row key and inserts the resource ID as a column value. The *Id2Resource* index is built similarly. In this way, we compress the RDF data so as to reduce effectively the storage space. The output of first job consists of the resource ID, the triple ID, the resource URI, and the subject URI in the triple. The implementation procedure is shown in Fig.3 and Algorithm 1.

Algorithm 1. First job

1: **map** (*key*, *value*):
2:        *counter* ← *counter* + 1
3:        *triple_id* ← *task_id* in first 3 bytes + *counter*
4:        *resources* [] ← split(*value*)
5:        **EMIT** { *resources*[0], (*triple_id* + *subject*, *resources*[0]) }
6:        **EMIT** { *resources*[1], (*triple_id* + *predicate*, *resources*[0]) }
7:        **EMIT** { *resources*[2], (*triple_id* + *object*, *resources*[0]) }

1: **reduce** (*key*, **iterator** *values*):
2:        *counter* ← *counter* + 1
3:        *resource_id* ← *task_id* in first 3 bytes + *counter*
4:        **create** column family *Resource2Id* and *Id2Resource*
5:        **set** *Resource2Id*{*RK*←*key*, *CN*←"id", *CV*←*resource_id*}
6:        **set** *Id2Resource*{*RK*←*resource_id*, *CN*←"resource", *CV*←*key*}
7:        **for** (*value* in *values*)
8:                **EMIT**{ (*resource_id*, *key*), (*triple_id* + *position*, *triple_subject*) }

4.2   Second Job: Rebuilding the Triples and Creating Indexes

The second job is more sophisticated to implement than the previous one. The mapper reads the files generated by the first job, and emits a tuple where the key is set as the triple ID and the value is set as the resource ID plus the position of the resource. In the meantime, by parsing the resource URI in the key we extract terms to complete the creation of the inverted index. The program creates a super column family named *InvertedIndex* in Cassandra, then inserts the term as a row key, the document ID as a super column name, the resource ID as a column name and the position as a column value. Here, we take the subject ID in the triple as the document ID which can be gained by using the *Resource2Id* index.

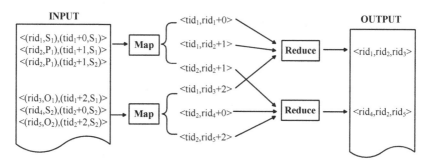

Fig. 4. Second Job: Building the Indexes

In the reduce phase, resources will be regrouped based on the triple ID and triples will be rebuilt in accordance with the resource ID and the position. After that, we adopt a similar approach to create three structured indexes: *SPO*, *POS* and *OSP*. For example, the program creates a super column family named *SPO* in Cassandra, and then inserts the subject as a row key, the predicate as a super column name and the object as a column value. *OSP* and *POS* indexes are built similarly. The implementation procedure is shown in Fig.4 and Algorithm 2.

Algorithm 2. Second job

1: **map** (*key, value*):
2:     **EMIT**(*triple_id, resource_id + position*)
3:     *term* ← generate *term* from *resource*
4:     *doc_id* ← *subject_id*
5:     **create** super column family *InvertedIndex*
6:     **set** *InvertedIndex* { *RK*←*term*, *SC*←*doc_id*, *CN*← *resource_id*, *CV*← *position* }

1: **reduce** (*key*, **iterator** *values*):
2:     **create** super column family *SPO,POS,OSP*
3:     **for** (*value* in *values*)
4:         **if** *position* = *subject* **do**
5:             *subject* ← *resource_id*
6:         **if** *position* = *predicate* **do**
7:             *predicate* ← *resource_id*
8:         **if** *position* = *object* **do**
9:             *object* ← *resource_id*
10:        **set** *SPO*{ *RK* ← *subject*, *SC* ← *predicate*, *CN* ← *object*, *CN* ← *null* }
11:        **set** *POS*{ *RK* ← *predicate*, *SC* ← *object*, *CN* ← *subject*, *CN* ← *null* }
12:        **set** *OSP*{ *RK* ← *object*, *SC* ← *subject*, *CN* ← *predicate*, *CN* ← *null* }
13:            **EMIT**{ *null, triple*(*subject, predicate, object*) }

---

5   Experiments

In this section, we take the DBpedia 3.6 dataset as an example to compare the basic inverted index with the improved inverted index.

5.1   Data Preprocessing

Before indexes are created, we must first preprocess the data. The purpose of this process is to remove the unimportant data and extract terms. The RDF data which is different from the traditional document data contains a great number of meaningless tokens. They occur repeatedly, and have not actual contents, so there is no need to create the inverted index for them. At the same time, the size of indexes will be effectively reduced after removing them. Another important task of data preprocessing is to extract terms from the RDF data. The analyzer provided by Lucene can obtain terms from text documents, and remove symbols and stop words at the same time. The analyzer, however, is only able to extract the terms composed by a single word but not a phrase. We observe that specific noun phrases are connected by "_" through the analysis of the DBpedia dataset. Therefore, we use this characteristic to get phrase terms by parsing characters, without the help of other tools.

5.2   Performance Evaluation

The experimental results in [13] demonstrate that using the underlying data structure provided by Cassandra to create indexes is feasible. The read performance of Cassandra is fully able to meet user's query requirements, and we have proved that the system is enough stable through a series of stress testing. The emphasis of this paper is the improvement of the retrieval effect and user experience, so here we only show the execution time of data processing and query efficiency through a series of experiments. The comparisons of query results are given in section 5.3.

For the evaluation of the program, we have used a 4-nodes cluster. Each node equipped with two dual core processors with 4GB memory and a local 7200 RPM SATA disk. The nodes are interconnected through Gigabit Ethernet. The programs are written in Java 1.6 and based on the open source project Apache Lucene. The version of Hadoop is 0.20 and the version of Cassandra is 1.0. The operating system is a 64-bit Ubuntu 10.04 server. The experimental data is the instances of the DBpedia 3.6 dataset (13.8 million triples).

Experiment 1: The execution time of data processing. Execute the program under the same experimental conditions. Table 4 shows the execution time and the average value.

Table 3. Execution time of data processing

Serial number	First job (s)	Second job (s)	Execution time (s)
1	484	5558	6042
2	474	5497	5971
3	482	5516	5998
Average	480	5524	6004

Experiment 2: Select randomly 5 vocabularies from the Web as test cases. Table 5 shows these query results and execution time.

Table 4. Query results and execution time

Serial number	Query	Results	Execution time (s)
1	information retrieval	4	0.096
2	index	464	0.112
3	Lincoln	3587	0.174
4	cloud computing	30	0.097
5	American	17363	0.286

From the Table 4 we notice that the second job spends more time than the first job. The results show that most of the time in the process of data processing is used to create three structured indexes and the inverted index, and the loading and encoding process consume little time. The main reason of this phenomenon is that DBpedia instances contain such enormous amounts of information that large numbers of terms are generated. Table 5 shows that the query execution time affected by the number of results, and it needs to spend more time when to get more results. However, the query execution time in milliseconds is sufficient to meet user's requirements.

5.3   Demonstration Scenarios

Table 6 demonstrates the query results of the basic inverted index and the improved inverted index, and compares their differences. The query case is taken from the fifth query in Table 5.

Table 5. Demonstration of query results

Indexing model	Results
Basic	S: <http://dbpedia.org/resource/Indiana> P: <http://xmlns.com/foaf/0.1/name> O: <"State of Indiana"@en> … S: <http://dbpedia.org/resource/Indiana> P: <http://dbpedia.org/ontology/language> O: <http://dbpedia.org/resource/Southern_**American**_English> …
Improved	S: <http://dbpedia.org/resource/Indiana> P: <http://dbpedia.org/ontology/language> O: <http://dbpedia.org/resource/Southern_**American**_English> … S: <http://dbpedia.org/resource/Indiana> P: <http://dbpedia.org/ontology/language> O: <http://dbpedia.org/resource/Inland_Northern_**American**_English> …

The above results sufficiently confirm that the improved inverted index can accurately return all matching triples. But the basic inverted index can only provide the RDF documents which contain the query terms, and users need to browse the entire document to obtain the data that they need.

6   Conclusions and Future Work

In this paper, we propose a distributed indexing scheme for large-scale RDF data, i.e., creating the inverted index using the underlying data structure of Cassandra. This scheme has a high degree of flexibility and ensures high efficiency. At the same time, it is easy to improve the indexing structure depending on the query requirements. Distributed indexes also solve the problem of the large space requirement brought by large-scale data. Processing data using MapReduce greatly improves the efficiency of index building, and the work is highly integrated. The improved inverted index that is based on the features of the RDF data model can provide precise query services, and the advantages are shown by our experimental results. In the future, we will continue to optimize the index structures to effectively support the ranking of RDF data, which will lay the foundation of the further development of the Semantic Web search engine.

Acknowledgements. This work was supported by the National Science Foundation of China (No. 61100049, 61070202) and the Seed Foundation of Tianjin University (No.60302022).

References

1. Chang, F., Dean, J., Ghemawat, S., Hsieh, W.C., Wallach, D.A., Burrows, M., Chandra, T., Fikes, A., Gruber, R.E.: Bigtable: A Distributed Storage System for Structured Data. In: Proc. of OSDI, pp. 205–218 (2006)
2. Bizer, C., Heath, T., Berners-Lee, T.: Linked data - the story so far. International Journal on Semantic Web and Information Systems 5(3), 1–22 (2009)
3. Wang, H., Liu, Q., Penin, T., Fu, L., Zhang, L., Tran, T., Yu, Y., Pan, Y.: Semplore: A scalable IR approach to search the Web of Data. Web Semantics: Science, Services and Agents on the World Wide Web 7(3), 177–188 (2009)
4. Bhagdev, R., Chapman, S., Ciravegna, F., Lanfranchi, V., Petrelli, D.: Hybrid Search: Effectively Combining Keywords and Semantic Searches. In: Bechhofer, S., Hauswirth, M., Hoffmann, J., Koubarakis, M. (eds.) ESWC 2008. LNCS, vol. 5021, pp. 554–568. Springer, Heidelberg (2008)
5. Cheng, G., Ge, W., Qu, Y.: FALCONS: Searching and browsing entities on the semantic web. In: Proceedings of the World Wide Web Conference (2008)
6. Ding, L., Pan, R., Finin, T.W., Joshi, A., Peng, Y., Kolari, P.: Finding and Ranking Knowledge on the Semantic Web. In: Gil, Y., Motta, E., Benjamins, V.R., Musen, M.A. (eds.) ISWC 2005. LNCS, vol. 3729, pp. 156–170. Springer, Heidelberg (2005)
7. Guha, R., McCool, R., Miller, E.: Semantic search. In: Proceedings of the 12th International Conference on World Wide Web, pp. 700–709 (2003)
8. Weiss, C., Karras, P., Bernstein, A.: Hexastore – sextuple indexing for semantic web data management. Proceedings of the VLDB Endowment 1(1), 1008–1019 (2008)
9. Harth, A., Umbrich, J., Hogan, A., Decker, S.: YARS2: A Federated Repository for Querying Graph Structured Data from the Web. In: Aberer, K., Choi, K.-S., Noy, N., Allemang, D., Lee, K.-I., Nixon, L.J.B., Golbeck, J., Mika, P., Maynard, D., Mizoguchi, R., Schreiber, G., Cudré-Mauroux, P. (eds.) ASWC 2007 and ISWC 2007. LNCS, vol. 4825, pp. 211–224. Springer, Heidelberg (2007)
10. Beckett, D., Grant, J.: Semantic Web Scalability and Storage: Mapping Semantic Web Data with RDBMSes. In: SWAD-Europe Deliverable, W3C (January 2003)
11. Hogan, A., Harth, A., Umbrich, J., Kinsella, S., Polleres, A., Decker, S.: Searching and browsing Linked Data with SWSE: the Semantic Web Search Engine. J. Web Sem. 9(4), 365–401 (2011)
12. Ladwig, G., Harth, A.: CumulusRDF: Linked Data Management on Nested Key-Value Stores. In: SSWS (2011)
13. Wang, X., Jiang, L., Shi, H., Feng, Z., Du, P.: Jingwei+: A Distributed Large-Scale RDF Data Server. In: Sheng, Q.Z., Wang, G., Jensen, C.S., Xu, G. (eds.) APWeb 2012. LNCS, vol. 7235, pp. 779–783. Springer, Heidelberg (2012)

MSMapper: An Adaptive Split Assignment Scheme for MapReduce*

Wei Pan, Zhanhuai Li, Qun Chen, Shanglian Peng, Bo Suo, and Jian Xu

School of Computer Science, Northwestern Polytechnical University,
Xi'an 710072, China
panwei1002@nwpu.edu.cn

Abstract. MapReduce as a popular platform has been extensively used for solving data-intensive applications. A number of tuning parameters can be applied to improve the performance of MapReduce. Among these parameters, the number of map tasks (mappers) driven by the number of logical input splits has a dramatic effect on the performance. However, subject to one-to-one correspondence between mappers and splits, the tradeoff between mapper-level parallelism and mapper startup costs must be carefully evaluated based on the input size and the split size. Meanwhile, the manual parameter configuration is lack of flexibility to meet the performance requirements of different jobs. In this paper, an adaptive split assignment scheme is proposed to decouple the number of mappers from the number of splits. We introduce the MSMapper(Multi-Split Mapper), a modified self-tuning mapper in which multiple splits can be assigned to one mapper. And with aid of inter-MSMapper communication, we reveal the potential that map tasks can be constructed without dependence on the number of splits, while the modified MapReduce architecture can sustain fine-grained load balancing and fault tolerance, as well as coarse-grained task startup overhead. We built our prototype on top of the Hadoop MapReduce realization, and present a comprehensive evaluation that shows the benefits of the MSMapper in common scenarios where split sizing problems arise. The results show that the modified version can improve the performance by a factor of 2.5.

1 Introduction

MapReduce is one of the most representative distributed parallel computing paradigm initially proposed by Google for data-intensive processing on large commodity clusters. To support distributed and parallel applications, two types of upper layer primitives, Map and Reduce, are proposed by MapReduce to abstract complex operations. With this abstraction, an enormous amount of nontrivial details, such as load partitioning and assignment, task scheduling, fault-tolerant,

* This work is sponsored by the National Natural Science Foundation of China (Nos. 61033007,60970070), the National High Technology Research and Development Program (863 Program) of China (No. 2012AA011004), and the National Basic Research Program (973 Program) of China (No. 2012CB316203), NWPU basic research foundation (Nos. JC20110227,JC20110225,JC201261).

Z. Bao et al. (Eds.): WAIM 2012 Workshops, LNCS 7419, pp. 162–172, 2012.

and load balancing can be handled by the framework. Among these details, load partitioning and assignment is the most crucial link by which the parallelism of one MapReduce job will be determined. MapReduce divides the input of a job into equal-size disjoint logical partitions called splits based on parameters related to split size. And then MapReduce framework assigns one map task per split, which is known as one-to-one assignment, each map task will invoke user-defined map function for each record in the split without communication with other map tasks. In view of parallel computing, this is a classic domain decomposition technique which is ideally suited for data-intensive applications. This illustrates how easily MapReduce system scales as the amount of data is increased.

At the same time, we also have noticed the strong connection between the performance of MapReduce job and the number of map tasks, as well as the number of splits. For a given load of one job, less number of map tasks with larger split size cannot fully exploit the available resources, and the quality of the load balancing and fault tolerance will decline. Conversely, although having better load balancing and fault tolerance, more number of map tasks with fine-grained splits will incur the excessive overhead of splits management and mapper start-up. Thus the tradeoff between mapper-level parallelism and mapper start-up costs must be appropriately considered based on the input size and the split size. In this point, the MapReduce user or administrator should carefully fine-tune the system to get better performance. But it is virtually difficult to find a "one-size-fit-all" parameter to meet the performance requirements of different jobs in dynamic runtime environment, especially for parameters related to split size. Based on these observations, we believe that one important improvement for the MapReduce framework is to decouple the number of splits from the number of mappers. Our solution to this problem is a new type of map task: MSMapper (Section 3.1), which can dynamically obtain multiple splits to amortize mapper startup costs without sacrificing fine-grained load balancingand fault tolerance (Section 3.3). One-to-many correspondence of map tasks and splits can be used to replace one-to-one correspondence by using MSMapper.

In this paper, we make the following contributions: (1)We introduce an adaptive split assignment scheme, where more than one splits can be assigned to one enhanced map task, named MSMapper, and a loose mapping model and adaptive scheme can be achieved. (2)The concept of adaptive-split (Section 3.1) was proposed as input to MSMapper, which can expand dynamically to accommodate multiple normal splits. There are two constructing strategies (Section 3.3) for adaptive-split: cumulative-incremental strategy and preset-balanced strategy, which are provided by using newly added inter-MSMapper communication mechanism (Section 3.4). (3)We implemented the proposed approaches on Hadoop 0.20.2 and evaluated our prototype on common scenarios where split sizing problems arise (Section 4). The results show that our adaptive techniques significantly improve the progress of job as more splits are involved, due to the elimination of mapper start-up cost.

In addition, we introduce necessary background on Hadoop MapReduce in (Section 2), survey related work in (Section 5); and conclude in (Section 6).

2   The Relation between Split Size and Job Performance

Picking the appropriate split size for a MapReduce job can dramatically change the performance of Hadoop. In current implementation, the number of mappers is determined through split-size-related parameters rather than the direct configuration. More details about these parameters can be seen in Table 1.

Table 1. Symbols and parameters used in Hadoop analysis

Symbol	Parameter Name&Description	Default Value
D_S	Input data size of the MapReduce Job	\
N	Number of slavers in cluster	\
B_R	dfs.replication: HDFS block replication factor	3
B_S	dfs.block.size: HDFS block size	64MB
S_{min}	mapred.min.split.size: The smallest valid size in bytes for a split.	1
S_{max}	mapred.max.split.size: The largest valid size in bytes for a split.	Long.MAXVALUE
Mn_s	mapred.tasktracker.map.tasks.maximum: ①	2
Rn	mapred.reduce.tasks: The number of reduce tasks per job	1
T_J	The total runming time for MapReduce job	\
T_R	The running time for reducer	\
T_O	The startup cost of mapper	\
T_P	The processing cost of mapper	\

Hadoop uses input formatter (so-called InputFormats) to divide the input data of size D_S into Sn logical splits of size S_S, which is determined by B_S, S_{min} and S_{max}. There exists a relationship among them as following: $S_S = \max(S_{min}, \min(S_{max}, B_S))$, if condition $(S_{min} < B_S < S_{max})$ is met, the relation $S_S = B_S$ can be deduced, that is to say input splits are created one per block by default. Based on this assumption, we can also see $Sn = D_S/S_S = D_S/B_S$, With current split assignment scheme and split-based job decomposition approach, the maximum number of parallel mappers Mn is theoretically equal to Sn. But in the runtime, the actual number of parallel mappers is limited to the available map slots which is related to N and Mn_s (①Mn_s: The maximum number of mappers that could be run on a slaver at one time), so the number of mappers executed simultaneously is $N \times Mn_S$. This implies that many waves of mappers are often necessary to larger jobs. And then our analysis below mainly focuses on the effects of these parameters on startup cost and CPU cost. We summarize our main result in the following proposition.

Proposition 1. *Given the workload description D_S and the hardware description N, the startup cost and CPU cost in a Hadoop job is:*

$$T_J = \frac{D_S}{B_S} \cdot T_O + Wn \cdot T_P + \lambda_{[Rn>0]} \cdot T_R \qquad (1)$$

where Wn is defined to be:

$$Wn = \frac{D_S}{B_S \times N \times Mn_S} \qquad (2)$$

Analysis: Given the workload D_S and block size B_S, we have $\frac{D_S}{B_S}$ map tasks in total, and $N \times Mn_S$ represents maximum load capacity of cluster per wave. So, (2) is the number of waves involved one job. Moreover, $\lambda_{[.]}$ is an indicator function, for map-only job, $\lambda_{[.]} = 0$. On the whole, the overhead of MapReduce job processing mainly consists of two parts:CPU-cost and IO-cost. Our adaptive technique presented in this paper focuses on reducing the CPU-cost, and only when *combiner* can be used, the IO-cost might be reduced. so we mainly analyze the possibility of improving performance of job from the angle of CPU-cost. And further, from (1) we can see that the total execution time of each job has three main parts: task startup time, mapper execution time and reducer execution time. With a given D_S, obviously, the overhead of mapper creation will dominate the total job execution time if B_S are too small. However picking the appropriate B_S for different jobs which have different input size could be very tough. Mostly Hadoop experts tune MapReduce jobs based on rules of thumb, but [1] argued that rules of thumb tend to suboptimal in dynamical environment. Therefore, adaptive technology should be applied to optimize the split assignment scheme by replacing manual configuration.

3   Adaptive Split Assignment Scheme

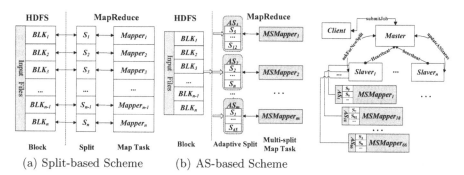

Fig. 1. Job Decomposition Approach

Fig. 2. Task Assignment Framework

3.1   Adaptive-Split and MSMapper

As shown in Figure 1(a), input data "Input Files" are divided into physical blocks $\{BLK_1, BLK_2, ..., BLK_n\}$ according to preset parameter (B_S) and loaded into HDFS before job processing. When the job is submitted, MapReduce will utilize customized input formatter to partition the input data into logical splits $\{S_1, S_2, ..., S_n\}$. To simplify the analysis, this paper assumes equal size between a logical split and its corresponding physical block. And with current split assignment scheme, it also can be seen from Figure 1(a) that each map task $Mapper_i$ corresponds to a split S_i. As can be concluded from theoretical analysis of section 2, the static optimization techniques which tries to manually set the split(block) size cannot satisfy optimal performance requirements of different jobs in the same cluster.

Figure 1(b) illustrates the two core components of our proposed adaptive split assignment scheme: adaptive-split(AS) and Multi-Split Mapper(MSMapper). Each AS is composed of some normal splits as shown in figure. AS classifies normal split S_i according to corresponding block locations by following the principle of same location precedence, which satisfies the condition: $AS_1 \cup AS_2 \cup ... \cup AS_m = \{S_1, S_2...S_n\}$. Compared with normal split, AS can be considered as a aggregation load. AS is called *adaptive* because the number of splits within it is not definite, which can be adjusted according to capability of node, requirements of load balancing and fault tolerance during job execution period.

The MSMapper differs from common mapper is in that MSMapper takes AS as its input, through which MSMapper could handle more than one normal split while only pay for task startup overhead once. As shown in Figure 1(a) and 1(b), compare with mapper, few MSMappers would be required to fulfil same job($m \ll n$). And although AS is the high-level input of MSMapper, the low-level inputs of MSMapper are still normal splits. So MSMapper does not change the original interface of the map function.

3.2   MSMapper Mechanism

The primary mechanisms of normal mapper are retained by MSMapper, while certain changes have been made in split assignment scheme and inter-MSMapper communication. The main differences can be seen as following: (1)During job initialization, the number of mappers is no longer directly driven by the number of splits. The modified system first constructs AS and maintain them into some newly created data structure (see details in 3.3), and then master schedules MSMappers driven by ASs on slavers with data locality. (2)During task execution in map side, MSMapper could dynamically take multiple normal splits in its life span by using AS. But in low-level terms, note that MSMapper is still running on normal splits, which guarantees the existing applications can be run in the new system without alteration. So MSMapper exhibits great downward compatibility and agility. (3)Moreover, MSMapper will need to make several additional key decisions: output-timing and stop-timing. For details see Section 3.3. (4)MSMapper expands existing communication mechanism in Hadoop by adding some real-time communication interfaces and enhanced old protocol.

With the aid of AS and MSMapper we successfully decouple one-to-one correspondence between splits and mappers. These techniques also helps system to

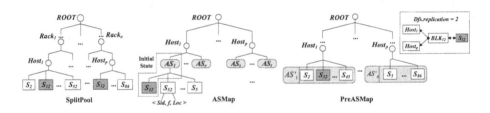

Fig. 3. The core data structures for adaptive-split

control task startup cost and make use of a series of supporting technologies such as construction of AS, dynamic load shedding and migration to keep fine-grained load balancing, and more details are given in the following sections.

3.3   Construction Strategy for Adaptive-Split

Construction of AS is one of the core techniques used to dynamically adjust the size of AS according to the current capacity of nodes and fault tolerance. Two kinds of AS construction strategies are implemented and presented in the following sections respectively.

Cumulative-Incremental Adaptive-Split Construction Strategy. First, we introduce two main-memory data structures SplitPool and ASMap maintained in master. SplitPool is used to keep the dependency relationship between normal splits and nodes as a global view, which is created at job initialization time. As shown in Figure 3, SplitPool uses tree-like data structure to abstract network topology of the cluster and register normal split (S_i) into the physical host node ($Host_i$) which each contains a list of local splits, such as $\{S_2, S_{12}, ..., S_{52}\}$ for $Host_1$. The tree-like data structure can effectively reflect network distances between two nodes. Then using SplitPool, Master assigns the nearest normal splits to AS, after that AS will be taken by MSMapper invoked by host as input according to the distance, such as $MSMapper_1$ run on $Host_1$ takes AS_1 as its input. And all these mapping relations should be dynamically maintained by ASMap which is regarded as a snapshot of the current status of AS construction. The changes of ASMap's status will take place in new request for split, load balancing and fault tolerance.

The strategy is called cumulative-incremental, that is because when slaver requests new MSMapper from master, a new AS is initialized with only one normal split. During MSMapper execution, master will dynamically allocates new splits to AS when it receives a new split request from same MSMapper with new communication interface *askForNewSplit* where previous normal split has been processed. So the AS size will gradual increase as MSMapper running, and MSMapper can tackle multiple splits within AS at once without the startup overhead of multiple tasks. It should be pointed out that all these background operations are totally transparent to end users, without sacrificing the ease of use.

Preset-Balanced Adaptive-Split Construction Strategy. Besides SplitPool and ASMap, the preset-balanced strategy also includes a special main-memory data structure named PreASMap, which records the mapping between pre-assigned AS(pAS) and normal splits. Compared with AS, pAS has contained multiple normal splits after initialization, such as AS_1' of PreASMap as shown in Figure 3, contains multiple normal splits $\{S_2, S_{12}, ...S_{45}\}$ after initialization. PreASMap is created by the master when the job is initiated, which also uses tree-like structure to organize the pAS. After initialization, within PreASMap each host node corresponds to one pAS, and the normal splits within pAS are also organized according to network distance for optimization.

It's worth pointing out, when the replication factor (B_R) for HDFS block is greater than 1, there may exist redundancy between pASs in PreASMap, and such redundancy also can be seen in SplitPool. Namely, given the universal set of job's splits $\{S_1, S_2, ..., S_n\}$, and two sets $A = \{S_i | S_i \in Host_l\}$ and $B = \{S_j | S_j \in Host_p\}$, if $A \cup B \neq \emptyset$, it means that there is redundancy between the split lists of host nodes, as well as pASs. But different from SplitPool, PreASMap needs additional preprocessing to remove this redundancy. For instance, in Figure 3, when B_R=2, the HDFS block BLK_{12} corresponding to split S_{12} has a copy both in $Host_l$ and $Host_p$ respectively. So in SplitPool, S_{12} is registered into both node $Host_l$ and $Host_p$. However, when PreASMap is created, S_{12} is only registered into only one pAS, such as AS_1', after redundancy processing.

The strategy is called preset-balanced, that is because, in initial *preset* phase, to guarantee normal splits evenly distributed over pASs, we use round-robin algorithm to allocate normal splits and ensure each split is allocated only once even if there are more than one copies on different nodes. And in *balanced* phase, master could adjust the content of pAS according to the current capacity of nodes and fault tolerance. That is, some splits should be removed from pAS taken by MSMapper which might invoked by straggler. After load shedding, these loads will be transferred to some highly capable nodes to ensure load balancing over cluster. Same as cumulative-incremental strategy, these operations are also transparent to end users.

Dynamic Load Balancing. Two above-mentioned adaptive strategies use different balance methods to ensure the fine-grained dynamic load balancing on the new MSMapper-based MR system. The cumulative-incremental strategy adopts the typical task-pool mechanism, where master maintains a to-do task (split) list and slaver applies for new task according to their own capacity after completing previous task. This mechanism can make sure each idle node keeps busy and, further, the dynamic load balancing of the whole cluster. For the preset-balanced strategy, where load balancing has two implications: first, in pAS initial phase, master assumes that all nodes are isomorphic and have same capacity without prior knowledge. Thus, the normal splits will be evenly distributes among pASs by round-robin algorithm. In this phase, loads to be allocated are in a static equilibrium. Second, if there does exists stragglers in cluster, master can re-balance all loads among nodes based on the execution progress reported by MSMapper with new communication interface:*updateASStatus*. In this way, the whole cluster could achieve dynamic load balancing.

Fault Tolerance. Unlike mapper, MSMapper can continually process multiple splits, and special concern should be given to it when the task or the node fails. There are two important decisions needed to be made after a normal split has been processed by MSMapper.

One is whether to store the intermediate results to the local storage now, or waiting for the aggregation result of more splits. For the immediate output pattern, shuffle can start as soon as possible, which can reduce the length of the parallel pipeline with the help of the overlap of MSMapper executing and network transportation. And better fault tolerance can be achieved based on eager spill mode. For the delayed output pattern, the aggregation of more intermediate results makes *combiner* gets more benefits, which can further reduce the intermediate results to save the expensive network IO cost. It is a good choice for aggregation operations which fit for *combiner*.

The other decision should be made is whether to choose another normal split to continue the process or to stop execution of MSMapper. Theoretically, MSMapper could execute continuously till all splits exhausted, on this occasion, other jobs might be starved from not being able to get a free slot for a long time. we designed an stop indicator for MSMapper, called MSMapper-TTL(Time To Live) based on the quantity of normal splits consumed, continuous running time and the amount of output. If a threshold is exceeded during MSMapper execution, MSMapper will terminate and spill all the rest of outputs to disk.

And for recovery from failures, the location information of intermediate results produced by MSMapper will also be synchronized to master with *updateASStatus*. As shown in Fig. 3, the splits in ASMap contains a state tuple, $< Sid, Loc, Status >$. *Sid* is the identification of splits. *Loc* is the location information of intermediate results. *Status* contains several execution states, such as running, done, and etc. Master, with the help of information maintained in ASMap, can use the location information of the intermediate result produced by completed splits to realize fine-grained fault tolerance.

3.4 Inter-MSMapper Communication

In order to support the construction of AS and provide load balancing and fault tolerance based on AS, we need to modify the original communication protocol based on RPC in Hadoop, such as *heartbeat* and *ping*. We also need to add several new protocols to support MSMapper based on AS more effectively. The modified MR system, with the help of message-passing mechanism designed in [2], adds a real-time communication protocol for MSMapper - *askForNewSplit* and *updateASStatus* as shown in Fig.3. Further details of implementation can be referred in [2].

4 Experimental Evaluation

4.1 Evaluated Systems and Setup

For evaluation, we used a cluster of 17 nodes with 4 Core 2.66GHz CPU, 4GB memory and Ubuntu Linux 9.10 each, one as master and the rest as slavers. The implementation of MSMapper is based on Hadoop 0.20.2 which is the baseline to all the evaluated systems. The mainly parameters associated with experiment are listed in Table 1, without specification, the default values are used.

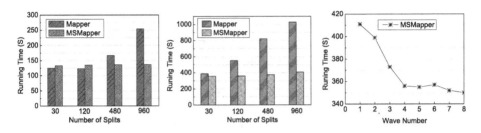

Fig. 4. Map-only Job   **Fig. 5.** WordCount Job   **Fig. 6.** Multi-wave effect

4.2 Map-Only Job with MSMapper

Figure 4 shows the performance of running MSMapper on cluster with 15 slavers using the dataset generated by TeraGen which is build-in MR program in Hadoop. Each record in the dataset is characterized by fixed format with equal-size, which also is key/value pair. The characteristics make sure that each map task could handle the almost equal numbers of key/value pairs, therefore, further to ensure that each map task has roughly the same processing time, with minimal impact on the input data skew. Experiment analyzed the performance gap between normal mapper and MSMapper by running a map-only job which contain a inverse operation on the value of each record. It can avoid a great network and local disk IO.

With $Mn_S = 2, N = 15$ and the volume of TeraGen dataset is 7.68GB, Fig. 4 shows the total running time of the map-only job on 30 map task slots, by taking different values of B_S: 256MB, 64MB, 16MB and 8MB, that is, the number of splits can vary from 30 to 960. For normal mapper, job could finish with 1, 4, 16 and 32 waves of mappers. But for MSMapper, job can finish in only one wave in any case by using AS whose size can dynamically vary according to the size of B_S. And the cost model for MSMapper-based job is abstract as following: $T'_J = (N \times Mn_S) \cdot T'_O + 1 \cdot T'_P$. Compared to Proposition (1), for MSMapper, task startup cost is a constant with regard to D_S and B_S, and T'_O slightly larger than T_O coming with the cost of data structure maintenance,such as ASMap. And we can also deduce that $1 \cdot T'_P \approx Wn \cdot T_P$. Obviously, for map-only job $T_R = 0$. Therefore, MSMapper-based job doesn't incur the overhead of starting multiple map tasks.

4.3 MapReduce Job with MSMapper

In order to investigate how the MSMapper improve the performance of normal MapReduce job, an example is the *WordCount* which is a commonly used benchmark for MR systems, The following settings is used for evaluation. The *Wikipedia* corpus file was used as *WordCount*'s dataset ,and in order to guarantee that every slaver has sufficient workload, we expand the dataset to 45GB with simple full-copy method, which can boost the effectiveness of *combiner*. With $Mn_S = 4, N = 16$, Fig.5 shows that, *WordCount* is able to achieve more impressive benefits than map-only job by using MSMapper. Node that, *combiner* is just an optimization to minimize the amount of intermediate data that has to written to disk, shuffle on network directly into the reducer. So for aggregation operations which is apt for *combiner*, more mapper output might be aggregated in MSMapper and more benefit can be achieved.

4.4   MapReduce Job with Multiple Waves of MSMappers

To explore the performance improvement achieved by wave factor, we show in Fig.6 the diversity of running times by using overlapped MSMapper. In above two experiments, we manually set a single wave of MSMappers to handle all workload. However, Section 3.3 analysis convinces us that more waves may actually provide more better performance by adjusting the number of MSMappers dynamically according to MSMapper-TTL. As shown in Fig.6 though more waves of MSMappers will introduced extra startup overhead, the overlap between MSMappers and shuffle can not only amortize these startup cost, but also offer optimal performance without introducing other overheads. Four or five waves is the rule of thumb which is predicted from our experience during evaluation.

5   Related Work

There are a number of ongoing research projects aiming to better understand the MapReduce, some are addressing performance issues from the system's implementation model point of view. Hadoop++ [3] is an improved version of Hadoop that incorporates support for index-scans and co-partitioning by using non-intrusive approach. To solve the issue that lacks built-in support for iterative programs, [4] [5] and [2] extend MapReduce with new programming models to serve these applications, and also in [2], inter-task communication is introduced to improve performance and expressive ability of Hadoop MapReduce, as well as the main support for our adaptive technologies presented in this paper. In [6],the authors propose HOP in which intermediate data is pipelined between tasks or jobs, and some applications characterized by pipelining behaviors such as, online aggregation and continuous queries, can benefit from HOP. The study in [7] employs a purely hash-based framework which replace sort-merge implementation to enable incremental processing and fast in-memory processing.

Some studies focus on the optimization opportunities presented by configuration parameters and semantics of the MR program. In [8], the authors make a case for techniques to automate the setting of MapReduce job configuration parameters. Starfish [1] is a self-tuning system for analytics on large datasets, which can provide good performance automatically according to user needs and system workloads. And in the combined performance models of Hadoop [9], cost-based optimization [10], concise statistical summaries [11], starfish can handle the significant interactions arising among choices made at different levels. Manimal [12] uses a static analysis-style mechanism for detecting optimizable code in MapReduce programs automatically. And Our work fills a void by providing a adaptive split assignment scheme to get good performance automatically without use intervention.

6 Conclusion

This paper demonstrated that MapReduce framework with adaptive split assignment scheme and inter-task communication can result in significant performance benefits. Our empirical and theoretical analysis showed that one-to-one correspondence between splits and mappers poses a critical tradeoff. We introduce MSMapper and AS, with various AS construction strategies to enable both fine-grained load balancing/fault tolerance and coarse-grained task startup overhead. Evaluation of our MSMapper-based Hadoop prototype showed that our new adaptive technologies can significantly improve the progress of map-only job, as well as normal MapReduce job.

References

1. Herodotou, H., Lim, H., Luo, G., Borisov, N., Dong, L., Cetin, F.B., Babu, S.: Starfish: A self-tuning system for big data analytics. In: CIDR 2011, Asilomar, CA, United states, pp. 261–272 (2011)
2. Wei, P., Li, Z., Sai, W., Qun, C.: Evaluating large graph processing in mapreduce based on message passing. Chinese Journal of Computers 10, 1768–1784 (2011)
3. Dittrich, J., Quian-Ruiz, J.A., Jindal, A., Kargin, Y., Setty, V., Schad, J.: Hadoop++: Making a yellow elephant run like a cheetah (without it even noticing). PVLDB 3(1), 518–529 (2010)
4. Bu, Y., Howe, B., Balazinska, M., Ernst, M.D.: Haloop: Efficient iterative data processing on large clusters. PVLDB 3(1), 285–296 (2010)
5. Ekanayake, J., Li, H., Zhang, B., Gunarathne, T., Bae, S.H., Qiu, J., Fox, G.: Twister: a runtime for iterative mapreduce. In: HPDC 2010, pp. 810–818. ACM (2010)
6. Condie, T., Conway, N., Alvaro, P., Hellerstein, J.M.: Mapreduce online. In: NSDI 2010, pp. 21–21 (2010)
7. Li, B., Mazur, E., Diao, Y., McGregor, A., Shenoy, P.J.: A platform for scalable one-pass analytics using mapreduce. In: SIGMOD 2011, pp. 985–996 (2011)
8. Babu, S.: Towards automatic optimization of mapreduce programs. In: SOCC 2010, Indianapolis, Indiana, USA, pp. 137–142 (2010)
9. Herodotou, H., Babu, S.: Hadoop performance models. Technical report (2010)
10. Herodotou, H., Babu, S.: Profiling, what-if analysis, and cost-based optimization of mapreduce programs. PVLDB 4(11), 1111–1122 (2011)
11. Herodotou, H., Dong, F., Babu, S.: No one (cluster) size fits all: automatic cluster sizing for data-intensive analytics. In: Proceedings of the 2nd ACM Symposium on Cloud Computing, SOCC 2011, pp. 18:1–18:14. ACM, New York (2011)
12. Jahani, E., Cafarella, M.J., Ré, C.: Automatic optimization for mapreduce programs. PVLDB 4(6), 385–396 (2011)

Driving Environment Reconstruction and Analysis System on Multi-sensor Network

Chunyu Zhang[1], Yong Su[1], Jiyang Chen[2], and Wen Wang[1,3]

[1] Research Institute of Highway, M.O.T, Beijing CHENGDA Traffic Technology CO. LTD,
P.R. China, Beijing 100088, No.8 Xitucheng Rd. Haidian District
chunyuzhang0320@hotmail.com, yongsu@163.com, wen.wang@rioh.cn
[2] Highway Administration of Liaoning Province, P.R. China, Shenyang Province Shisan Rd.
Heping District
chenjiyang486@sina.com
[3] BeiJing University of Aeronautics & Astronautics, School of Economics and Business
Management, China, Beijing 100191, No.37 Xueyuan Rd. Haidian District

Abstract. We construct a driving environment reconstruction and analysis system based on multi-sensors network onboard and some functional subsystem as well. With the data acquired, processed and stored, the real comprehensive driving environment, which includes vehicle dynamic state information, traffic environment information and driving behavior information, can be established accurately and provide what had happened in and around the vehicle. Besides, this system can also provide the researchers with additional and important information, for example traffic sign, moving object and driver gaze information. Practical results show this system is a very powerful technical framework to deep incident analysis and a quantitative evaluation measure to the effect of passive and active safety technologies, which can promote and formulate vehicle safety measures or reduce serious injuries and disabilities in addition to the reduction of fatalities and injuries in general.

Keywords: driving environment reconstruction, multi-sensors.

1   Introduction

With the development of economy and the growth of vehicle quantity, fatal road accidents have been decreasing, however the overall accident and injury counts keep increasing. Traffic safety becomes a research hot topic and increasingly arouses the attention of scholars and administrators. It is necessary to analyze contributing factors leading to the traffic accident and find the solution to reduce them [1]. According to statistics of American academic, ITS-based traffic safety technology is a very useful and powerful technical means to solve traffic safety problem.

Although in recent years many ITS-based methods have been taken, for example many new technologies adopted in cars, advanced traffic information service system and etc., there is still lack of effective technical platform to collect comprehensive driving environment data, concerning driver characteristics performance, driver behavior, driving environment, vehicle maneuver, which is essential to understand critical incidents, near-crashes, pre-crash causal and contributing factors. In addition,

Z. Bao et al. (Eds.): WAIM 2012 Workshops, LNCS 7419, pp. 173–180, 2012.

many passive and active safety technologies have been introduced in the market over the past several years, especially the fast development of passive safety technologies. The variety of automobile assessment tests have been done before those technologies commercialized on the market, to promote the safety technologies by quantitative evaluation, we need to further the improvements in accident analysis and establish performance evaluation techniques. On the basis of these factors, the establishment of the driving environment reconstruction and analysis system is expected. Moreover, with the rapid advancement of electronic information technology and the development of new technologies, the establishment of quantitative methods for measuring the effect of safety technologies is possible.

In this paper the research effort is to construct driving environment perception and reconstruction system, which can record not only vehicle dynamics, but also driver behavior and traffic environment information. A primary goal was to provide vital exposure and pre-crash data necessary for understanding causes of crashes, supporting the development and refinement of crash avoidance countermeasures, and estimating the potential of these countermeasures to reduce crashes and their consequences.

2   Previous Work

In recent years, drive recorders have been designed to record the vehicle dynamics behavior at a certain frequency during driving condition, such as speed, acceleration, etc. it has been installed in commercial vehicle in many countries, for example, in china, Japan and American, etc. The application of drive recorders has an important role on combating fatigue driving, speeding, other traffic violations, vehicle safety, traffic accident analysis and identification [5], [6]. However, currently equipped drive recorders on commercial vehicle emphasize on recording only the vehicle running status information and neglect the driver behavior information and driver response delay, in fact, driver behavior characteristics should be the main factor to traffic safety[3], [10], [11]. To obtain safe driving, a driver should be in the center of the safety analysis [10]. Therefore drive recorders' function on deep incident analysis and driver behavior analysis is limited.

Some driving environment reconstruction and analysis systems is expected to construct to analyze the cause of traffic incident and such kind of problems related to traffic safety, such kind of similar research work has already been done by American National Highway Traffic Safety Administration, Virginia Tech Transportation Institute and Japan Automobile Research Institute‚ Ministry of Land, Infrastructure and Transport.

Iowa University establishes public center and focus on the human factors & vehicle safety. The researchers carry out their research on helping teens learn to drive more safely using video cameras in their cars.

Japanese Nissan System equipped with radar, GPS, vehicle status sensors and 6 channel video sensors which monitoring front and rear view of environmental information, driver's facial expression, driver's limb movements. Researchers try to find cognitive risk level of different ages and different personalities.

American Virginia System consists of multi-sensors: radar detects obstacle, GPS records vehicle position, four channel video sensors respectively monitors and records driver's facial expression, driver limb movements, front of and rear environmental information, vehicle sensors collect dynamic vehicle running status information, such as steering wheel angle, brake pedal pressure and so on. With the comprehensive data acquired, researchers can assess the danger of driver's operating movement quantitatively and deduce the conclusion that traffic accident is mainly due to driver's distraction [7], [8], [9].

At present there is no such kind of system yet. In this paper, we describe our driving environment reconstruction and analysis system, compared with similar system in American and Japan, our system has a noticeable characteristic, it not only can record the comprehensive information of driver, vehicle and environment, but also can analyze driver behavior, judging if the driver is in distraction. Our system can be a very powerful technical framework to deep incident analysis and a quantitative evaluation measure to the effect of passive, active safety technologies, which can promote and formulate vehicle safety measures or reduce serious injuries and disabilities in addition to the reduction of fatalities and injuries in general.

3   System Framework

Many existing driver recorder is applied to record vehicle dynamics, however, much research work neglects the preview of a driver behavior and driver response delay. Moreover, the behavior of high-speed vehicles differs greatly from other robots. On the basis of these factors, we proposed an integrated interactive driving environment reconstruction and safety analysis system framework, which is designed to record vehicle dynamics, traffic environment, driver behavior information at a certain frequency and analyze driver behavior as well.

The system consists of several major components and subsystems that were installed on vehicle: onboard sensor network module, sensor fusion module, driver behavior analysis module, and traffic sign recognition subsystem, vehicle tracking subsystem, GPS positioning subsystem and etc. The noticeable characteristic of our system lies in it is not essential for data collection, but also provide the researchers with additional and important information, such as driver gaze, road traffic sign position. These subsystems include traffic sign recognition subsystem that can inform the driver what kind of speed limit sign is ahead, vehicle tracking subsystem detects and tracks vehicle ahead, GPS positioning subsystem collects information on vehicle position for vehicle localization.

The specific framework of the reconstruction system is shown as below (see Fig.1). In this framework, we consider a driving environment reconstruction and analysis system as a vehicle-driver-environment interactive closed-loop system. We concentrate on not only the current driving situation, but also the essence worthy of mining under the collecting data, such as the driver response delay and so on.

Onboard sensor network in Fig.1 includes all sensors equipped on different part of vehicle, which provides real-time information about drivers, traffic environment simultaneously. Each of the sensing subsystems in the car was independent, so that any failures that occurred were constrained to a single sensor type. In general, the external sensors capture object appearance, range, and road surface condition outside a vehicle, and the interior sensors collect vehicle state information, such as speed, acceleration, and steering angle etc.

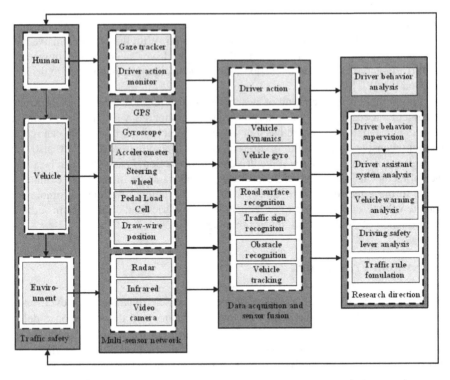

Fig. 1. Function of driving environment reconstruction and analysis system

Traffic environment sensors include radar, infrared and video cameras, which were designed to mainly percept the situation information around running vehicle, record, detect and recognize vehicle, obstacles, pedestrians on road, traffic sign, road surface condition, weather condition and so on.

Driver behavior sensors are composed of video cameras, which were designed to mainly monitor and judge driver's action by recording facial expression, upper limb movement, limb movement and tracking gaze movement of eye.

Vehicle state sensors consist of GPS, Gyroscope, Accelerometer, steering wheel, Pedal load cell, draw-wire position, which were designed to record vehicle position, velocity, acceleration, direction angle, yawing angle of running vehicle.

Data acquisition and sensor fusion module consist of some subsystem, which is a very important part of the framework. Sensor fusion module realizes acquisition, processing, fusion and storage of sensor information of different kinds, which model the driving environment together, moreover, tracking moving object, recognizing traffic sign and road surface condition, monitoring driver's action and tracking driver's gaze as well. In addition, multi-signal processing is done on a fast moving vehicle, the behavior of high-speed vehicles differs greatly from other robots, so it is very essential to process signal at very high speed in accordance with certain high efficient algorithms and special mechanism of processing, storage technique.

Driver's behavior analysis module discerns the driver's action, fatigue and distraction situation, at present with our integrated system we can deduce driver attention situation by judging if the driver's gaze is in accordance with the object moving on the road in the video.

In the above framework, we also list some potential research direction which can be done by our system, for example quantitatively evaluation measures of safety technologies of car, supervision of driver's action in emergency and etc.

4   Hardware Architecture

According to functional requirements of system, the onboard hardware system should have the following characteristics: easy to equipped sensors, high data processing speed, ability to anti-vibration, ability to operate under a relatively poor circumstance and so on. To satisfy the above condition, we selected the following sensors and set its corresponding interface, see Fig. 2.

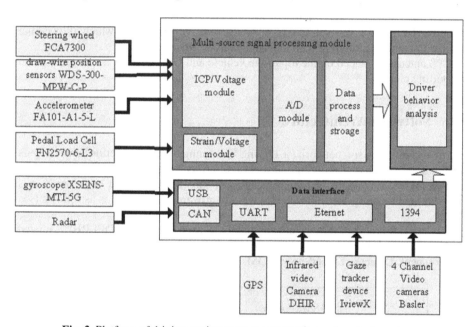

Fig. 2. Platform of driving environment reconstruction and analysis system

Some kinds of the vehicles status sensors can't be connected to the computer directly, so signal conditioning modules and data conversion module are used in conjunction. The gyroscopes, radar and GPS can connected with the computer by the USB, CAN, and serial port directly. Driving environment and driver's operation information are gathered mainly by multi-channel video sensors which connect with computer through Ethernet port and 1394.

Pentium-based computer integrated with human-computer interface acquires and stores data from multi-sensor network distributed around the vehicle, which is mounted on the supporting plate special designed on the vehicle. Data storage was achieved via the system's hard drive, which was large enough to store driving environment data for several days without downloading data.

The video subsystem was particularly important as it provided digital and continuous scenes of the happenings in and around the vehicle. This subsystem included six camera views monitoring the driver's face and driver upper limbs, and lower limbs, the forward view, the rear view.

What is worth mentioning is the selection of sensors, especially the video cameras. There are must be a balance between image quality and horizontal visual angle. In general, when the horizontal visual angle becomes wider, the image quality becomes poorer on the other hand, consequently driver behavior, recognition based on video and detailed accident analysis becomes difficult. Therefore we have to look for the optimal image specification.

The forward-view infrared camera was mounted in front of center mirror. This location did not occlude any of the driver's normal field of view.

Accelerometer sensor, which obtains longitudinal and lateral kinematic information, is put on the middle of the vehicle.

Two radar antennas were mounted behind license plates on the front and rear of the vehicle. The location was selected to be as unobtrusive as possible which allowed the vehicle instrumentation to remain inconspicuous to other drivers.

GPS antenna is placed on the top of the vehicle and GPS receiver is mounted on the supporting plate.

5   Software Architecture

Based on multi-sensor network, a driving environment reconstruction and analysis system is constructed. Taking into account the further function extension and also the post-secondary development requirements, software design should be adopted in a modular, open, configurable structure. Overall software architecture is as follows (see Fig. 3).

Development tools of this system are mainly VC++6.0/2005 and DELPHI 7, related programming technologies include image processing, data acquisition, driver development, DirectX, DirectShow, COM, DLL and so on.

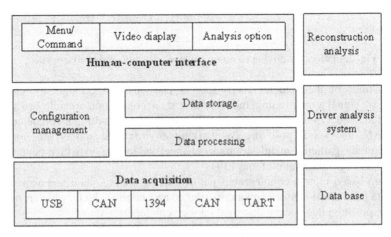

Fig. 3. System software architecture

6   Environmet Reconstruction and Analysis System

Driving environment reconstruction and analysis system is arranged as follow (see Fig. 4) including the following functions:

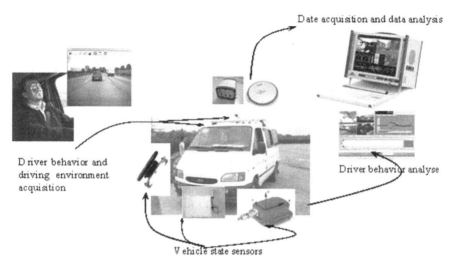

Fig. 4. Driving environment reconstruction and analysis system

- 6-channel video data acquisition and storage in real-time, including front and rear view of the vehicle road environment, driver facial expression, eye and gaze movement, upper limb movement and lower limb movement information.
- 2-channel Radar signal acquisition and storage, detecting the object in front and rear of environment.
- Steering wheel sensor, accelerometer sensor, GPS, gyroscope, draw-wire position sensors and pedal Load sensor implement the vehicle state signal acquisition and storage.
- Driver's gaze device analyze driver's gaze in real-time and, judge if the driver is in distraction.

7   Conclusion

In this paper, we propose to establish driving environment reconstruction and analysis system based on multi-sensor network. This system can automatically collect and store vehicle estate, traffic environment and driving behavior simultaneously. Using the data acquired by this system, deep incident analysis, quantitative evaluation measure to the effect of passive, active safety technologies, and evaluation of risky driving behavior and crash risk, calculation of relative risk of engaging in secondary tasks and etc. can be analyzed. This construction of system can facilitate this process and have a positive effect on traffic safety research.

Acknowldgement. This project "application research of information exchange and safety service based on android embedded platform" is supported by "basic scientific research operation cost of state-leveled public welfare scientific research courtyard". Thanks to the hard work to all participates and suggestion of the reviewer.

References

1. Parker, M.J., Zegeer, C.V.: Traffic conflict techniques for safety and operations. 2008 Observers Manual. Publication. No. FHWA-IP-88-027. U.S. Department of Transportation (January 1989)
2. Lin, Q., Cheng, B., Lai, J., et al.: A new method for analysis of traffic conflict by using video drive recorders. In: Proceedings of ITS Conference, Beijing (2007)
3. So, K., Noboru, K., Toshihiro, A., et al.: Analysis o f mechanism of rear end collision accident by drive data recorders. In: Proceedings JSAE Annual Congress; Klauer, S.G., Dingus, T.A., Neale, V.L., et al.: Impact on Driver Inattention on Near Crash/Crash Risk: An Analysis Using the 100 Car Naturalistic, Driving Study Data (Report No. DOT HS 810 594). National Highway Traffic Safety Administration, Washington, DC (2006)
4. Klauer, S.G., Dingus, T.A., Neale, V.L., et al.: Impact on Driver Inattention on Near Crash/Crash Risk: An Analysis Using the 100 Car Naturalistic Driving Study Data (Report No. DOT HS 810 594). National Highway Traffic Safety Administration, Washington, DC (2006)
5. Noboru, K., So, K., Tsuyoshi, K., et al.: Analysis o f rear end collision near miss and its occurrence by drive data recorder. In: Proceedings JSAE Annual Congress (2006)
6. Tetsuya, N., Kenichi, Y., Horishi, N., et al.: Development and validation of a drive recorder for automobile accidents. In: Proceedings JSAE Annual Congress (2000)
7. Hanowski, R.J., Blanco, M., Nakata, A., Hickman, J.S., Schaudt, W.A., Fumero, M.C., Olson, R.L., Jermeland, J., Greening, M., Holbrook, G.T., Knipling, R.R., Madison, P.: The drowsy driver warning system field operational test, data collection methods final report. Report No. DOT HS 810 035. National Highway Traffic Safety Administration, Washington, DC (2008)
8. Hanowski, R.J., Olson, R.L., Hickman, J.S., Bocanegra, J.: Driver distraction in commercial vehicle operations. In: First International Conference on Driver Distraction and Inattention in Gothenburg, September 28-29 (2009)
9. Feng, L.Q., Jia, F.R., Bo, C., et al.: Analysis o f causes of rear end conflicts using naturalistic driving data collected by video drive recorders. In: Proceedings SAE Annual Congress, New York (2008)
10. Suetoml, T., Kido, K.: Driver behavior under a collision warning system-a driving simulator study. SAE paper (2005)
11. Sayer, J.R., Devonshire, J.M., Flanagan, C.A.: Naturalistic driving performance during secondary tasks. In: Proceedings of the Fourth International Driving Symposium on Human Factors in Driver Assessment, Training and Vehicle Design (2007)

LuSH: A Generic High-Dimensional Index Framework

Zhou Yu, Jian Shao, and Fei Wu

College of Computer Science, Zhejiang University,
Hangzhou, China, 310027
{yuz,jshao,wufei}@zju.edu.cn

Abstract. Fast similarity retrieval for high-dimensional unstructured data is becoming significantly important. In high-dimensional space, traditional tree-based index is incompetent comparing with hashing methods. As a state-of-the-art hashing approach, Spectral Hashing (SH) aims at designing compact binary codes for high-dimensional vectors so that the similarity structure of original vector space can be preserved in the code space. We propose a generic high-dimensional index framework named LuSH in this paper, which means Lucene based SH. It uses SH as high-dimensional index and Lucene, the well-known open source inverted index, as underlying index file. To speedup retrieval efficiency, two improvement strategies are proposed. Experiments on large scale datasets containing up to 10 million data show significant performance of our LuSH framework.

Keywords: High-dimensional Index, Spectral Hashing, Lucene.

1   Introduction

With the rapid development of Internet, increasing amounts of unstructured data (e.g. image, video, music) are produced at every moment. How to well organize these large scale data and execute efficient retrieval is an urgent research issue. Unlike structured data, it's hard to compare similarity between two unstructured data. Taking image data as an example, the common method is extracting typical visual features (e.g. color histogram, Tamura texture[1], CEDD[2], SIFT[3], etc.) from data and then similarity of original data is converted to distance comparing in feature space (e.g., Euclidean distance). Unfortunately, to maintain the characteristics of original data, the extracted features are usually high-dimensional vectors. Hence, building an index structure to provide efficient retrieval in high-dimensional feature space is of crucial importance.

The trending method of high-dimensional index is Spectral Hashing (SH)[13]. It's an improvement of Semantic Hashing[4], which aims at designing compact binary codes for high-dimensional vectors so that the similarity structure of original vectors space can be preserved in the code space, then the similarity retrieval is executed based on these codes. In this way, the feature-extracted unstructured data object can be represented as a "small" code. Assume all data are hashed to binary codes

Z. Bao et al. (Eds.): WAIM 2012 Workshops, LNCS 7419, pp. 181–191, 2012.

(hamming code), given a query code of same metric, the Near Neighbors (NN) retrieval scenario is as follows: linear scan all codes in the database, sort hamming distance to query code and return the codes within a small distance r. Since the codes are all in low-dimensional hamming space, the retrieval is relatively fast.

Although the SH algorithm is a state-of-the-art index in high-dimension, there are still some problems to apply it in a system-level application. Firstly, linear scan based NN retrieval is infeasible when the scale of database reaches tens of millions and the concurrent access user number reaches hundreds or more. Secondly, how to fuse SH with underlying index storage should be taken into account, which involves cache, disk access, and index storage optimization.

In this paper, we propose an index framework based on SH. Since the underlying index storage is complex, the open source inverted index, Lucene[5] is applied to fuse with SH, taking over the management of index storage. To speedup retrieval, multi-probe retrieval and ID mapping strategies are proposed. We name our framework "LuSH", which means a Lucene based SH index.

The rest of the paper is organized as follows. In Section 2, we review the related work on semantic hashing related algorithm and Lucene index. In Section 3, we present our LuSH index framework and describe our improved algorithm in retrieval. In Section 4, we demonstrate the performance of our framework on large scale datasets. In Section 5, we conclude the paper and end up with future work.

2   Related Work

2.1   Semantic Hashing

Traditional tree-based multi-dimensional index such as R-Tree[6] and KD-Tree[7] are not made for high-dimensional data. They are best fit for the data less than ten dimensions. As the dimension increases, the efficiency of these algorithms decreases rapidly. This is called the "curse of dimensionality"[8]. It has been proved that if the scale of database N and dimension d don't satisfy $N \gg 2^d$, the efficiency of the index algorithm is worse than linear scan in the entire database.

Semantic Hashing[1] is proposed in this context. It aims at embedding high-dimensional data into a low-dimensional space, while remaining the semantic similarity structure of data in the original space. Semantic hashing is a concept, many algorithms are proposed according to this.

One of the most well-known hashing methods is Locality Sensitive Hashing (LSH)[9,10]. Hash code is generated by multiple projections to random hyperplanes. The projection is locality sensitive that if two high-dimensional points is similar, they have a large probability to share the same code. Due to random projection, LSH suffers from redundancy of hash code length and hash tables.

Some machine learning algorithm is proposed to make up the drawback of LSH. Restrict Boltzman Machine (RBM)[11], as a deep learning model algorithm, provides a new thought of hashing, and gives better performance compared with LSH. A simpler machine learning method Boosting SSC[12], uses adaBoost to train a set of learners and gets final hash code from these composite weak learners. These machine learning hashing algorithms achieve a more precise result than the random projected LSH, however, high complexity restricts their practical applicability.

Spectral Hashing[13] is proposed by Weiss in 2009, which has been demonstrated to outperform other hashing methods above.

2.2 Spectral Hashing

Spectral Hashing can be seen as an extension of Spectral Cluster[19]. The concrete formulation is as follows:

$$\min \sum_{i,j} W_{ij} \left\| y_i - y_j \right\|^2$$

$$\text{s.t.} \quad y_i \in \{-1,1\}^K, \quad \Sigma_i y_i = 0, \quad \frac{1}{n} \Sigma_i y_i y_i^T = \mathbf{I} \tag{1}$$

Where W_{ij} is the similarity matrix of sample i and j. y_i denotes the binary hashing code of sample i and K the number of code length. This formulation carry through the motivation of semantic hashing, similarity relationship of samples in the original high-dimensional space is preserved in low-dimensional hamming space.

Solution of equation 1 is NP hard. Spectral relaxing by removing the constraint $y_i \in \{-1,1\}^K$, equation 1 is transformed to a spectral analysis problem, whose solutions are simply extracting k eigenvectors from the Laplacian matrix of W with minimal eigenvalues.

After the equation is solved, the following procedure is to learn a hash function with the solved eigenvectors and eigenvalues for novel input queries (out of sample problem), whose solutions are eigenfunctions of the weighted Laplace-Beltrami operators defined on manifolds[14,15,16]. Specifically for a case of uniform distribution on $[a, b]$, the eigenfunction of 1D-Laplacian is as follows.

$$\Phi_k(x) = \sin\left(\frac{\pi}{2} + \frac{k\pi}{b-a} x\right) \tag{2}$$

The final hash function $f(x) = sgn(\Phi_1(x)\Phi_2(x), \dots \Phi_K(x))$

2.3 Lucene Index

Lucene is an open source inverted index supported by the Apache Software Foundation. Due to its high efficiency, it's widely used in full text search engines, database management systems, file system, etc.

The purpose of inverted index is to allow efficient full text retrieval. Traditional index used in database system, e.g. B-Tree[17], does not support the keywords based full text retrieval, thus, inverted index is proposed. It not only stores the document itself, but also the inverted table mapping from keyword to document. In Lucene, for storage concern, the document and inverted table are associated using an internal id (see Fig.1).

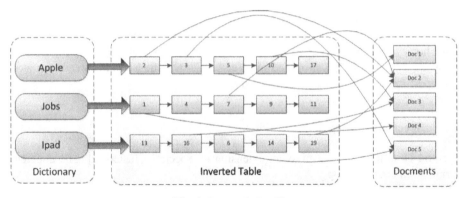

Fig. 1. Lucene index file

3   LuSH Framework

In this section, we describe the overall architecture of our LuSH. For better explanation, we use Content based Image Retrieval System (CBIR) as a typical application (see Fig.3). The framework can be partition into three modules: Training, Index and Retrieval (shown in Fig.3 with brown dot lines), we detail the three modules individually in the following subsections.

3.1   Training Module

The purpose of the module is to learn a hash function using a training set. We prepare a portion of samples (assume n samples) which represent the entire database. Generally, random sampling strategy is used. Then, we extract feature of these samples to get a matrix $X \in R^{n \times d}$ and solve the objective function in section 2.2, and get K minimal eigenvalues and their corresponding eigenvectors. The hash function can be seen as the projections of an input data to the K eigenvectors and binarize the each bit at a threshold.

In consideration of storage optimization, the code length K we choose is the multiple of 8 (size of a byte). In practice, 32 bits (size of an integer) or 64 bits (size of a long integer) is preferred.

3.2   Indexing Module

When the hash function $f(x)$ is obtained, the next step is to use this function to encode all the data objects in the entire database. The same visual feature as we used in the training step is extracted for each item in the dataset. Assume the database has N data objects, every object has its unique identifier (UID), and we gain N hamming codes of which the code length is K. We then add the N hamming codes and their corresponding UID into index.

We design a generalized index interface for our index structure, which is for future extension. Lucene is an implementation of the interface just like a plug-in. The generic index interface we design contains functions as follows:

- **Initialize:** load the index file and trained hash function
- **Add:** input the document containing hash code and corresponding UID, return a boolean value of whether it's successfully added into index
- **Delete:** input the document containing hash code and corresponding UID, return a boolean value of whether it's successfully deleted from index
- **Update:** input the old and new documents, execute delete operation for the old document at first, then add the new document
- **Commit:** commit the operation up to now and makes the updates to index retrievable

In Lucene, document is the basic index unit. We design a document consist of an hash code and its UID. The original data and feature vectors are optionally stored depending on application requirements. When executing an add (delete or update) operation to the index file, each hash code is a keyword in the dictionary and internal ids are appended to it if these documents share the same hash code. The other information is stored in the document structure with the internal id as its entry.

3.3   Retrieval Module

After the indexing procedure for the whole database is completed, we can execute efficient NN retrieval using the index structure. Similar with the index building procedure, for any query object, we first extract features and then use the hash function again to obtain a hash code. To achieve efficient NN retrieval using the hash code, two strategies are proposed.

3.3.1   Multi-probe Retrieval

When we try to execute NN retrieval for a given hash code, linear scanning in the database is used in traditional SH. It's not efficient enough when database is large scale. Since in most applications, the NN we expect to fetch is a small portion of the entire database (e.g. hamming distance within 2), it's easy to enumerate all the possible NN we need, which we called a probing sequence. Then we use the probing sequence to execute exact query (find the same hash code) in the database. Each exact code retrieval problem can be simply solved in $O(logn)$ time complexity with a good index structure (binary search). Furthermore, each probe is independent from each other, so retrieval for each probe can be highly parallel computed. This is a remarkable improvement compared with the $O(n)$ linear scan method when n is of very large scale.

The multi-probe sequence Δ for a query code is generated using Algorithm 1. For example, given a probe radius R, the sequence Δ has $\sum_{r=0}^{R} C(K,r)$ elements as shown in Fig.2, where $C(K,r) = \frac{(K(K-1)...(K-r+1))}{r!}$. Since every probe in Δ is sorted with hamming distance from the original query code, the fetched results are also in ascending ordered by the distance from query code.

Algorithm 1. *Generate Multi-Probe Sequence*

Input: query points hash code q, code length of hash code K, probe radius R
Output: Multi-probe sequence Δ

$\Delta \leftarrow \emptyset$
for $r = 0$ to R
    Set of probes p with distance r from q are defined as: $\Omega_r = \{ p \mid dist(q,p) = r \}$
    $\Delta \leftarrow \Delta \cup \Omega_r$
end for

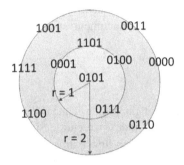

Fig. 2. An example of generating a multi-probe sequence with probing radius within 2. The query code is "0101" in the center of the circle. The sequence has 1+4+6 = 11 elements.

3.3.2   ID Mapping

As demonstrated in section 2.3, in the index structure of Lucene, internal ids are stored in inverted table. The query hash code is first located in the dictionary, then, internal ids linked to the hash code are fetched. After this, a second fetch step using the internal ids is executed to get the UID in the documents.

The second fetch step is redundant as the whole document is taken out. However, in our LuSH framework, we only need the UID. To speed up retrieval, we add an extra structure to store the mapping from internal id to UID in memory. Consider that the mapping maybe to too large to store in memory, cache strategy is used.

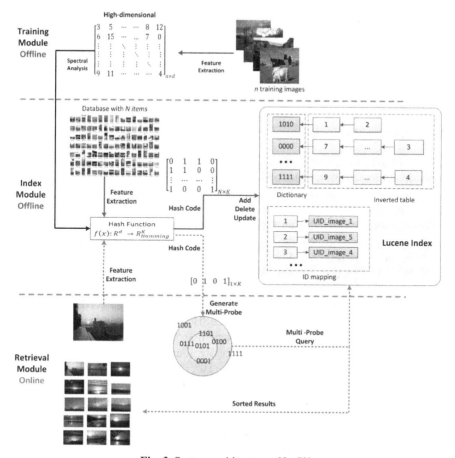

Fig. 3. System architecture of LuSH

4 Experimental Results

In this section, we evaluate the performance of our LuSH framework on a typical application of CBIR system and test the accuracy and efficiency on large datasets (up to 10 million).

4.1 DataSet and Evaluation Criterion

To evaluate the performance of LuSH, we conduct experiments on two image datasets as follows.

- Flickr Images: an image dataset crawled by ourselves, which contains 100,000 images of varies of species.
- Sogou Images[18]: a huge scale image dataset provided by Sogou Lab. It contains up to 10 million images with categories of people, animals, building, scenery, etc.

To evaluate the performance of LuSH, four criterions are concerned.

- precision of the retrieval result
- Index storage size
- Index building time
- query response time

4.2   Experimental Setup

To apply LuSH in a CBIR system, the first thing is to extract visual features from images. In our experiment, Color and Edge Directivity Descriptor (CEDD)[2] is used.

The two datasets have different emphases in our experiments. The Flickr Images dataset is relatively tiny; it's suitable to evaluate precision of the retrieval result since the computation for ground truth for a very large dataset is time costing. The Sogou Images dataset is used to test the performance of our index structure, i.e. index size, build time and query response time.

For the Flickr Images dataset, we evaluate the precision of our LuSH. 10,000 images are random selected from the database as the training set and another 100 images as query set. The ground truth for each query is the top 1% NN in Euclidean distance and the NN we retrieval using our LuSH are within hamming distance 2. To show the correlation of precision and code length K, K ranges from 8 to 64.

For the Sogou Images dataset, we generate four subsets in different size with random sampling, i.e. sizes of the subsets are 10^4, 10^5, 10^6, 10^7. We build index for the four subsets using a multi-thread strategy (20 threads). When index building procedure is completed, we analyze the size of index file and the index building time, then use 100 random samples to execute query in these indexes and evaluate average query response time. Since the scale of dataset is large, code length K is set 64 and radius for NN is within 2 to achieve high precision.

4.3   Experimental Results Analysis

The precision result of LuSH on Flickr Image dataset is shown in Fig.4. As the code length increase, the precision increase simultaneously, accompanying with the time cost for computing. For the scale of 100,000 images, $K = 40$ is a relatively ideal trade-off.

The size of index files and index build time on Sogou Images dataset are demonstrated in Table 1, the size of index file, index build time are linear increase with the size of dataset, which is in our expectation.

Focus on the largest subset with $N = 10$ million, the size of index file is about 1GB, which is 10^{-5} of the original images and 10^{-2} of the extracted feature vectors. This is an impressive compression ratio. The corresponding indexing time is less than one hour, since it's an offline step, it satisfies most of the system-level application's requirement.

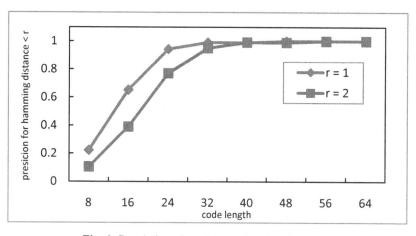

Fig. 4. Correlation of precision and hash code length

Table 1. Size of index file and index build time

Size of dataset (*N*)	Size of Index File(*MB*)	Index Building Time(*s*)
10,000	1.5	15
100,000	13	63
1,000,000	124	446
10,000,000	1151	3353

We then evaluate the average query response time. Since linear scan is not applicable to this large scale datasets. We don't compare our multi-probe strategy with it. We compare the standard Lucene's "quadric" fetch method with our ID mapping strategy and show result in Table.2. It is obvious that our ID mapping strategy gets better results than the standard method in all of the four different sizes of subsets.

Table 2. Average query response time

Size of dataset (*N*)	Average query response time(*ms*)	
	Standard	ID mapping
10,000	8.97	**7.15**
100,000	30.08	**25.2**
1,000,000	201	**180.7**
10,000,000	1960	**1840**

5   Conclusion and Future Work

In this paper, we propose a generic high-dimensional index framework: LuSH, which fuses the Spectral Hashing with Lucene index. To speedup retrieval efficiency, two strategies are proposed:

- Multi-probe retrieval strategy is used instead of linear search in SH.
- An extra mapping from Lucene's internal id to UID is added in Lucene index structure to avoid redundant fetch operation.

We use CBIR system as a typical application to demonstrate the architecture of LuSH and then do experiment on large scale image datasets (up to 10 million) to show the performance of LuSH framework.

In the future, we consider extending LuSH to distributed index environment.

Acknowledgment. This work is supported by the HEGAOJI Project under grant No.2010ZX01042-002-003-001, the National Natural Science Foundation of China No. 61105074, 61103099, the China Postdoctoral Science Foundation, and the Fundamental Research Funds for the Central Universities.

References

1. Tamura, H., Mori, S., Yamawaki, T.: Textural features corresponding to visual perception. IEEE Transactions on Systems, Man, and Cybernetics 8(6), 460–472 (1978)
2. Chatzichristofis, S.A., Boutalis, Y.S.: CEDD: Color and Edge Directivity Descriptor: A Compact Descriptor for Image Indexing and Retrieval. In: Gasteratos, A., Vincze, M., Tsotsos, J.K. (eds.) ICVS 2008. LNCS, vol. 5008, pp. 312–322. Springer, Heidelberg (2008)
3. Lowe, D.G.: Distinctive image features from scale-invariant keypoints. International Journal of Computer Vision (IJCV) 60(2), 91–110 (2004)
4. Salakhutdinov, R., Hinton, G.: Semantic hashing. International Journal of Approximate Reasoning 50(7), 969–978 (2009)
5. Apache Lucene Project, http://lucene.apache.org/
6. Guattman, A.: R-Tree: A Dynamic Index Structure for Spatial Searching. In: ACM SIGMOD Int. Conf. on Management of Data, Boston, pp. 47–57 (1984)
7. Bentley, J.L.: Multidimensional binary search Trees used for associative searching. Communications of the ACM 18(9), 509–517 (1975)
8. Bellman, R.: Adaptive control processes: a guided tour. Princeton University Press (1961)
9. Indyk, P., Motwani, R.: Approximate nearest neighbors: towards removing the curse of dimensionality. In: Proceedings of the Thirtieth Annual ACM Symposium on Theory of Computing, pp. 604–613. ACM, New York (1998)
10. Andoni, A., Indyk, P.: Near-optimal hashing algorithms for approximate nearest neighbor in high dimensions. In: FOCS, pp. 459–468 (2006)
11. Salakhutdinov, R., Hinton, G.: Learning a nonlinear embedding by preserving class neighbourhood structure. In: AI and Statistics, p. 5 (2007)
12. Shakhnarovich, G., Viola, P., Darrell, T.: Fast pose estimation with parameter sensitive hashing. In: ICCV, pp. 750–757 (2003)

13. Weiss, Y., Torralba, A., Fergus, R.: Spectral hashing. In: Advances in Neural Information Processing Systems, pp. 1753–1760 (2009)
14. Bengio, Y., Delalleau, O., Roux, N., et al.: Learning eigenfunctions links spectral embedding and kernel PCA. Neural Computation 16(10), 2197–2219 (2004)
15. Coifman, R., Lafon, S., Lee, A., et al.: Geometric diffusions as a tool for harmonic analysis and structure definition of data: Diffusion maps. Proc. of the National Academy of Sciences of the United States of America 102(21), 7426 (2005)
16. Belkin, M., Niyogi, P.: Towards a theoretical foundation for Laplacian-based manifold methods. Journal of Computer and System Sciences 74(8), 1289–1308 (2008)
17. Comer, D.: Ubiquitous B-tree. ACM Computing Surveys (CSUR) 11(2), 121–137 (1979)
18. SogouP2.0, http://www.sogou.com/labs/dl/p2.html
19. Ng, A.Y., Jordan, M.I., Weiss, Y.: On Spectral Clustering: Analysis and an algorithm. Journal of Advances in Neural Information Processing Systems 2, 849–856 (2002)

Improving Text Search on Hybrid Data [*]

Huaijie Zhu, Xiaochun Yang, Bin Wang, and Yue Wang

College of Information Science and Engineering,
Northeastern University, Liaoning, 110819, China
zhuhjneu@gmail.com, yangxc@mail.neu.edu.cn

Abstract. In our real life, there is much hybrid data which contains not only unstructured data but also structured data. In general, the majority techniques of text search on hybrid data are only focused on unstructured data (text) ignoring the structured data. So this may lead a bad ranking of the searching results. In this paper, we describe a new method about improving text search using structured data. Our contributions are summarized as follows: (i) We build the uniform problem model; (ii) Ours is the first approach adopting the mutual information of feature words to qualify the relevance (similarity) between two texts; and (iii) We utilize several rules to consider the structured data to improve text search and build our approach. Finally, experimental results show the relevance function and our approach guarantees the search results with high recall, top-k precision, Mean Average Precision and good search performance, respectively.

1 Introduction

In our real life, the majority of data exists in a hybrid form. This data in a hybrid form consists of part of unstructured data besides the typical structured data. The unstructured data is mainly in text form. However, there are still a lot of issues to be resolved in text search on the hybrid data. For example, in large IT companies, a variety of computer problems happen every day. These problem records have two kinds of attributes: structured attributes (like category, time and solvers) and unstructured attributes (like description) as shown in Table 1. When searching the relevant problems, text search without considering structured data will lead bad ranking results to users so that it can not cater to users' information need. And the hybrid data is also shown in online shopping. Take the taobao for example, the structured data contains the shop name, clothing name, price and so on. In addition, there are some unstructured data such as the description of commodities and the evaluation to commodities. Also, the text search can not help users search the desirous ranking commodities.

[*] The work is partially supported by the National Natural Science Foundation of China (Nos. 60973018), the National Natural Science Foundation of China (No. 60973020), the Doctoral Fund of Ministry of Education of China (No. 20110042110028) and the Fundamental Research Funds for the Central Universities (Nos. N110804002, N110404015).

Z. Bao et al. (Eds.): WAIM 2012 Workshops, LNCS 7419, pp. 192–203, 2012.

Table 1. An example of computer problems and solutions in an IT company

id	...	description	solution	solvers
r_1	...	User's [LAN] account is locked.	Verified user and reset a new account.	E and F
r_2	..	User needs to have [LAN] password reset.	Verified User and reset password.	A
r_3	..	User's [LAN] account is locked.	Verified user and unlocked account.	C and B [1]
r_4	...	User needed mapped drives from his old PC added to his new Laptop	Setup mapped drives on users new Laptop.	C, D, and E
r_5	..	User needs to have [LAN] password reset	Null	B, C, and D
r_6	..	User's [LAN] account is locked.	Verified user and unlocked account.	B
r_7	..	User's [LAN] account is locked.	Verified user and reset a new account	E, H, and G

[1] C cooperated with B to finish the problem.

Previous work on text search does not consider text search on hybrid data. And using these previous approaches on hybrid data makes the ranking results unsatisfactory. Table 1 shows an example about some computer problem data of a company. In the table, each row is a problem record and each record consists of a lot of attributes in the columns. "description" is the matching attribute on which we do the text search. But the users also care about "solution" and "solvers" attributes. Consider a query of the problem about "LAN account is locked." The IT company wants to get the relevant problem results. If we use the text search without considering the structured data, there are some problems for the results. In detail, if we use cosine as the similarity function, we return the results as follows: records r_1, r_3, r_6, and r_7. The results have some problems:

(1) These results have missed record r_2 because of a low similarity score for the query text and the description of record r_2. In fact, in the computer problems, we think that "LAN account is locked." is relevant to "User needs to have [LAN] password reset";

(2) For records r_3 and r_6, we can observe that the solvers of record r_6 is subset of that of record r_3 and their descriptions are similar or same. Obviously, we would like to choose solvers B to help us solve our problem instead of B and C together because the company would like to apply the less persons to address the problem. In some sense, the record r_3 is useless because of the existence of record r_6;

(3) For records r_1, r_6 and r_7, we will recommend solvers B independently prior to {E, F} and {E, H and G} because the company would like to apply the less persons to address the problem.

These problems cause the ranking of the results unreasonable. In fact, the company would like to get the results as shown in Table 2. Each solution plan contains the problem description and the corresponding solution or solvers. We can see that the solution is diverse or the solvers groups are feasible for the relevant problems.

To address this problem, we would like to have a method for measuring the relevance between the records and query text that captures more of relevance of texts and considering structured data to help improve text search. In order to

Table 2. Solution plans

description	solution or solvers
User's [LAN] account is locked	Verified user and unlocked account.
User needs to have [LAN] password reset	Verified User and reset password.
User's [LAN] account is locked.	Verified user and reset a new account
User's [LAN] account is locked	B
User needs to have [LAN] password reset	A
User's [LAN] account is locked	E and F
User's [LAN] account is locked	E, H and G

get this objective, we construct our relevance function by improving the mutual information and dig some rules using the additional structured information. And we dig some rules for the structured data. Our approach combines the relevance function and the rules to filter some duplicate results and to gain the good ranking results. In other words, we also improve the search performance by structured data.

In our work, there are many challenges. First of all, when using mutual information, how to compute a suitable score to the relevance of two phases like "account is locked" and "password reset"? In the computer problems, "account is locked" is closely related to "password reset." Secondly, for the structured data, how to consider it in our approach to help improve text search? At last, how to search the results from huge data set quickly?

In the remaining of this paper, we give the related work in Section 2, we describe the problem definition in Section 3, we present a Mutual Information based approach, called MI in Section 4 and an improved approach, called MI-S, by adopting structured data in Section 5. The experimental results are given in Section 6. Finally, we make a conclusion.

2   Related Work

With the development of information, there is a lot of related work about text search(retrieval). Some approaches are aimed at constructing the accurate and reasonable similarity function, while others are focused on improvement or expansion on query object (texts, keywords, documents etc) such as [1] whose idea is that complex word forms are segmented into relevant subwords and expansion of query keywords based on sematic relationship [2] and the context [3]. The similarity functions of current text retrieval include two categories: the first is retrieval model based on statistic such as vector space model (traditional methods like cosine coefficient [4]) and mined association rule between words for measuring the similarity of two documents [5], and the other is retrieval model based on semantic analysis (or LSA) [6], such as the kernel of [7] is sematic meaning based on the case grammar and sematic relationship. However, the ideas of these approaches are either computing the similarity of single two text sets based on the shard feature words or considering some natural language like semantic analysis. So these ideas have ignored mutual information of feature words in special context. What is more, there is no good taking the corresponding structured data into account to improve the text search.

Also, keyword search on structured and unstructured Data is related to our work. [8] is a tutorial about an overview of the state-of-the-art in supporting keyword search on structured data, outline the problem space in this area, introducing representative techniques that address different aspects of the problem, and discussing further challenges and promising directions for future work.

Different from these works mentioned above, our work is considering how to improve text search using structured data. It cares about not only the matching object which is text but also the structured data which is shown up with text. Also, we use the mutual information of words to build relevance function.

3   Problem Definition

Given a collection of problem records R, a query text q about the problem, each record r in R consists of several attributes.

Problem Statement. *Give a similarity or relevance function. Find and rank the results from the collection of problem records according to function scores between the results and q.*

For the sake of simplicity, Table 3 displays notations and their descriptions used in the rest part of the paper.

Table 3. Notations used in this Paper

Notations	Description
$W(t)$	The word set of a text t.
$|W(t)|$	The number of words in $W(t)$.
$freq(w_1, \ldots, w_k)$ $(1 \leq k)$	The occurrence frequency of words $w_1, \ldots, w_k$.

4   A Mutual Information Based Approach (MI Approach)

For the sake of addressing our problem and returning ranking results to users, our basic approach to solve the proposed problem is computing the mutual information scores between the query text and the value of matching attribute in all the records. We call this approach mutual information based approach (MS approach for short). In the following subsection, we give our relevance function improving mutual information of feature words in special context.

4.1   Mutual Information Score

Our relevance function denotes the relevance (similarity) between query text and the value of A_M (matching attribute) in the problem. We adopt mutual information to build the relevance function. It considers not only the same words but also related words.

As is shown in Equation 1, $MI(x, y)$ denotes the mutual information of two random variables x and y. Mutual information is a quantity that measures the mutual dependence of the two random variables. $p(x, y)$ is the joint probability of x and y. $p(x)$ and $p(y)$ are the probabilities of x and y respectively. And in this paper, we take the frequency of words as probability. That is, $p(x)$ is the frequency of x and so is $p(y)$, $p(x, y)$ is the occurrence frequency of x and y. we make some improvement for Equation 1 and construct our mutual function score.

$$MI(x, y) = \log(\frac{p(x, y)}{p(x) * p(y)}). \tag{1}$$

We show mutual information between two words and mutual information between a query p and a text t in the following subsections.

Mutual Information between Two Words. Given two words w_1 and w_2, we use $c(w_1, w_2)$ to measure their relevance degree (see Equation 2). We use $freq(w_1)$ to express the frequency of w_1, $freq(w_1, w_2)$ to express the occurrence frequency of both w_1 and w_2. Obviously, $freq(w_1, w_2)$ can not be greater than $freq(w_1)$ (or $freq(w_2)$).

$$c(w_1, w_2) = \begin{cases} \frac{freq(w_1, w_2)}{freq(w_1) - freq(w_1, w_2)}, & if freq(w_1) \neq freq(w_1, w_2), \\ 1, & if freq(w_1) = freq(w_1, w_2). \end{cases} \tag{2}$$

Based on $c(w_1, w_2)$ and $c(w_2, w_1)$, we can get the mutual information between two words in Equation 3. We can gain that the mutual information of the same two words is 1. So we set the max value of mutual information between two words is 1 and min value is 0.

$$MI(w_1, w_2) = \begin{cases} 1, & if \log(c(w_1, w_2) + c(w_2, w_1)) > 1, \\ 0, & if \log(c(w_1, w_2) + c(w_2, w_1)) < 0, \\ \log(c(w_1, w_2) + c(w_2, w_1)), & otherwise. \end{cases} \tag{3}$$

Mutual information between a Query p and a Text t. We use $MI(p, t)$ to denote the mutual information score between p and t (see Equation 4).

$$MI(p, t) = |W(p) \cap W(t)| + \frac{(\sum_{w_j \in W(P)} \sum_{w_i \in W(t) - W(p)} MI(w_i, w_j))}{|W(p)|}. \tag{4}$$

Definition 1. *Give two text t_1 and t_2, and a threshold δ. We say t_1 is relevant to t_2 if $MI(t_1, t_2) \geq \delta$.*

5   Improved MI Approach Using Structured Data (MI-S Approach)

Using MI approach, we may ignore some structured data and get some redundant or useless results. This can lead bad ranking results. To observe this disadvantage, consider the following simple example.

Example 1. For the data in Table 1, consider a query "LAN account is locked." If we rank relevant records using MS approach, we get the ranking results as follows: records r_1, r_3, r_6, r_7, r_2, and r_5. Obviously, these ranking results are not optimal. (1) For record r_3 and r_6, we can observe that the solvers set of record 6 is subset of that of record r_3 and their descriptions are similar or same. Obviously, we would like to choose solvers B to help us solve our problem instead of B and C together because the company would like to apply the less persons to address the problem. In some sense, the record r_3 is useless because of the existence of record r_6; (2) The solvers set of record r_5 is also subset of that of record r_6. Moreover, its solution is null. (3) For the ranking of results, we will recommend solvers B independently prior to {E, F} because of the less persons. So we need to consider structured data to improve the ranking results.

To overcome this advantage, we should make the best use of structured data to improve the MS approach. However, the structured data may contain many kinds of attributes. In fact, we need not consider all the attributes to cater to the different users' information need. Accordingly, we only take some useful attributes into account. In order to address various kinds of problems and make the most of useful attributes, we firstly build a uniform problem model.

5.1   A Hybrid Data Model

As all the kinds of problems have the common attributes that we care about to deal with our problems, we create a uniform problem model. For the sake of searching our desired problems, the problem model is about the common attributes that we all would like to care about.

Each problem is in the form of (A_M, A_T, A_P, A_U) where A_M (called matching attribute) is the description of the problem, A_T (called target attribute) is the solution to the problem, A_P (called target set attribute) is the person set who solve the problem, and A_U is useless of addressing the problem. According this uniform model, we choose the corresponding attributes to approach problems in different domain. For example, in the example shown in Table 1, we can see that "description" is matching attribute, "solution" is target attribute, "solvers" is target set attribute and other attributes are useless attributes.

Table 4. Uniform problem model

A_M	A_T	A_P	A_U
matching attribute	target attribute	target set attribute	useless attribute

For the matching attribute A_M, we also use the mutual information to capture its relevance. We give rules for target attribute A_T and target set attribute A_P in following subsections.

5.2   Observations and Rules

For the sake of making the most of the target attribute A_T and target set attribute A_P in the problem model to improve the text search, we use the following two observations. And based on these observations, we make some definitions and rules.

Observation 1. *Coming back to Table 1, for attribute description and solution, the similar or same solution can be for the same description, but the similar description can have different solutions. If users search the similar solutions which are already in our candidate solutions, most of users do not want to see many similar solutions in the result. That means, the solutions in the results should be diverse. So based on this idea, we only choose only one of the similar solutions.*

Given two records r_1 and r_2. Let A_M denote matching attributes, A_T denote target attributes and A_P denote target set attributes. $r_1[A]$ denotes the value of attribute A of record r_1.

Rule 1. *We choose one of $r_1[A_M, A_T]$ and $r_2[A_M, A_T]$ as the result, if $r_1[A_T]$ is relevant to $r_2[A_T]$.*

For example, in Table 1, suppose $r_3[description, solution]$ is already in the result set. Since $r_6[solution]$ is similar (same) to $r_3[solution]$, we need not to put $r_6[description, solution]$ in the result set.

Rule 2. *We choose both $r_1[A_M, A_T]$ and $r_2[A_M, A_T]$ as the result, if*
(i) $r_1[A_M]$ is relevant to $r_2[A_M]$, and
(ii) $r_1[A_T]$ is irrelevant to $r_2[A_T]$.

For example, in Table 1, suppose $r_1[description, solution]$ is already in the result set. Since $r_3[solution]$ is not relevant to $r_1[solution]$ and $r_3[description]$ is relevant to $r_3[description]$, we need not to put $r_3[description, solution]$ in the result set.

Observation 2. *Re-consider Table 1, for attribute description and solvers, some descriptions are similar or same. The solvers from different records have different relationships: (1) superset relation; eg: B, {B,C} and {B,C,D}. (2) overlapped relation; eg: {E,F}, {E,H,G} and {C,D,E}. (3) disjoint relation; eg: A, B, {C,D,E}. Based on these foundations, if B has become our candidate solvers, we do not necessarily consider its superset set such as {B,C} and {B,C,D}. And for overlapped relation: {E,F}, {E,H,G}, we can recommend {E,F} prior to {E,H,G}. So is for the disjoint relation.*

Rule 3. *We choose $r_1[A_M, A_P]$ as the result, if $r_1[A_P]$ is subset of $r_2[A_P]$.*

For example, in Table 1, suppose $r_6[description, solvers]$ is already in the result set. Since $r_3[solvers] \supseteq r_6[solvers]$, we should not put $r_3[description, solution]$ in the result set.

Rule 4. *We chose $r_1[A_M, A_P]$ as the result,*
(i) $r_1[A_M]$ is relevant to $r_2[A_M]$, and
(ii) $r_1[A_P]$ is subset of $r_2[A_P]$.

For example, in Table 1, suppose $r_3[description, solvers]$ is already in the result set. Since $r_6[solvers] \subseteq r_3[solvers]$ and $r_6[description]$ is relevant to $r_3[description]$, we put $r_6[description, solvers]$ in the result set and remove $r_3[description, solvers]$.

Rule 5. *The attribute value $r_1[A_M, A_P]$ ranks higher than $r_2[A_M, A_P]$, if*
(i) $|r_1[A_P]| < |r_2[A_P]|$,
(ii) $r_1[A_M]$ is relevant to $r_2[A_M]$, and
(iii) $r_1[A_P] \cap r_2[A_P] \neq \emptyset$.

For example, suppose $r_1[description, solvers]$ in Table 1 is already in the result set. Since $r_7[solvers]$ and $r_1[solvers]$ are overlapped, and $r_7[description]$ is relevant to $r_1[description]$, we put $r_7[description, solvers]$ before $r_1[description, solvers]$ in the result set.

Rule 6. *The attribute value $r_1[A_M, A_P]$ ranks higher than $r_2[A_M, A_P]$, if*
(i) $|r_1[A_P]| < |r_2[A_P]|$,
(ii) $r_1[A_M]$ is relevant to $r_2[A_M]$, and
(iii) $r_1[A_P] \cap r_2[A_P]$ equals to $\emptyset$.

For example, suppose $r_1[description, solvers]$ in Table 1 is already in the result set. Since $r_2[solvers]$ and $r_1[solvers]$ are disjoint, $r_2[description]$ is relevant to $r_1[description]$ and $|r_2[solvers]| < |r_1[solvers]|$, we put $r_2[description, solvers]$ before $r_1[description, solvers]$ in the result set.

Rule 7. *When a record r satisfies both Rule 4 and Rule 6, Rule 4 takes precedence.*

For example, suppose $r_1[description, solvers]$ and $r_3[description, solvers]$ in Table 1 are already in the result set and $r_6[description]$ is relevant to both $r_1[description]$ and $r_3[description]$. As we known, $r_6[solvers] \subseteq r_3[solvers]$ satisfies Rule 4, whereas $r_6[solvers] \cap r_1[solvers] = \emptyset$ satisfies Rule 6. We use Rule 4 to choose result.

5.3 MI-S Approach

Following the above observations and rules, we improve the MI approach by considering structured data. We call this improved approach MI-S approach. The approach involves in two aspects: (1) *FindTarget* which mainly considers the target attribute A_T and (2) *FindTargetSet* which considers target set attribute A_P.

FindTarget employs Observation 1 to get results. *FindTargetSet* utilizes Observation 2 and the above rules to get results. It is the process of building the map *mappserosn* which finally can return the results to users. Give a query text and each record in A_P. There are some preprocessing work. In order to find the less size of $r_i[A_P]$, we sort the problem records according to the size of $r_i[A_P]$. That is, we put the records whose $r_i[A_P]'s$ size is the smallest in front. At the beginning, we check whether the size of *mappserosn* is null or not. If it is null, we insert the relevant record. If else, we do the following steps. We firstly find that what kinds of relation the checking record has with pairs in *mappserosn* using *Findrelation*. *Findrelation* function finds that all the pairs in have some relation with $r_i[A_M, A_P]$. It returns what kind of relation and *pairset* which is used to store all the pairs existing this relation. Based on rule 6, if there exists several complicated relation, we just choose one. The relevant function is mutual information which we build. Finally, we get $r_i[A_M, A_P]$ pairs and sort the pairs according to the size of $r_i[A_P]$. For the same size of $r_i[A_P]$, we rank the pairs according to the relevance.

These two aspects share the idea which is how to use the additional structured data to improve short text search. But the attributes that they consider are different. The rules vary from the various attributes. And the following experiments are implemented using *FindTargetSet*.

6   Experiments

In order to evaluate the effectiveness of the proposed relevance function and the efficiency of our approach, we have conducted some experiments on HP real data sets. The approach was implemented using MFC and SQL Server 2005. The experiments were run on a PC with an Intel Core 2 Duo CPU E7500 2.93GHz and 4G memory with a 250GB disk, running a Windows XP operating system.

Experiments are conducted to two important aspects: (i) quality of search results and (ii) query performance of search algorithm. We manually constructed twenty queries $(Q_1, Q_2, Q_3, ..., Q_{20})$ for HP data set. These queries are short texts about different problems' description.

6.1   Effectiveness of Our Relevance Function and Our Approach

In order to test the effectiveness of the proposed relevance function and our approach, we use recall, top-k precision and MAP (Mean Average Precision) to do the evaluation with 400 records. The size of 400 records is about 130 KB, including attributes "description" and "solvers". The length of values of "description" in each record ranges from 20 to 10,885. Recall is a ratio of the number of useful results searched over the overall number of results records in the database. Top-k precision is a ratio of the number of returned useful results that are among ranking k results.

We firstly compared the recall between relevance function and Cosin[4] function using queries $(Q_1, Q_2, .., Q_{20})$. The results are shown in Fig.1. From Fig.1, we can see that relevance function is a little better than Cosin function on recall. The recall values of all queries of relevance function can exceed 0.8. It is not surprising that the relevance function can additionally gain some interesting results while Cosin can not. So the recall values of relevance function on some queries can be higher than that of Cosin.

In this part, we test the effectiveness of our approach. We compared top-k precision and MAP between MI-S and MS approach (which does not consider

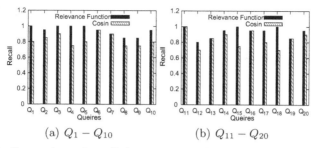

Fig. 1. Comparison of recalls between our relevance function and Cosin

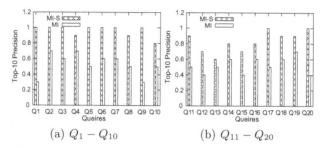

Fig. 2. Comparison top-10 precision between our MI and MI-S approach

the additional structured data and only matches the only value of matching attribute) with the results in Fig.2, Fig.3 and Fig.4, respectively.

Fig.2 and 3 shows that the top-k precision of MI-S approach is more higher than that of MI approach. And as shown in Fig.2(a), using MI-S approach, over 80% queries achieve more than 0.8 top-10 precisions, while using naive, only 20 percent queries can exceed 0.7 about top-10 precisions. The similar results are shown in another three figures. It is because that MI approach is ranking only according to the relevance between the matching attribute and the query. It obviously ignores the structured data. The results which MI approach returned are relevant but not useful results.

Fig.4 describes the comparison on MAP. From Fig.4(a) and 4(b), the MAP values of the queries of MI-S approach are a lot higher than that of MI approach.

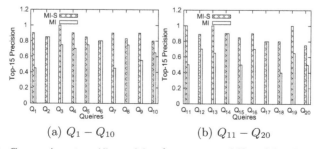

Fig. 3. Comparison top-15 precision between our MI and MI-S approach

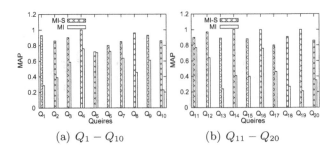

Fig. 4. Comparison MAP between our MI and MI-S approach

It is obvious that the MAP of MI approach is also very low because the ranking is only according to the relevance of the matching attribute. And our MI-S approach synthetically considers the matching attribute and target set attribute to improve the text search. The results are depicted as Fig.4(a) and 4(b). And using MI-S approach, 80% queries achieve more than 0.8 MAP, while using naive, only 30% queries can exceed 0.7 about MAP. We can see that the high recall, top-k precision and MAP can report MI-S approach good ranking results.

6.2   The Effect of Queries on Query Performance

In this section, we implemented our MI-S and MI approach about search time to evaluate the query performance of algorithm with 10,000 records. The size of 10000 records is 2.59 MB, including attributes "description" and "solvers". The length of values of "description" in each record ranges from 429 to 10,000. Fig.5 shows the effect of different queries($Q_1, Q_2, .., Q_{20}$) on the running time. We can see that MI-S approach is more faster than MI approach for all the queries. It is reason that we can use the rules especially the subset relation to quickly filter out a lot of useless results. And the search time of different queries varies from the number of relevant problems. All the query time did not exceed 2 minutes using improved algorithm. That is, the MI-S approach can improve the query performance to some extent.

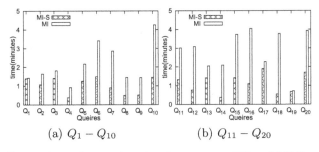

Fig. 5. Comparison of execution time between our MI and MI-S approach

7 Conclusions

In this paper, we studied improving text search on hybrid data. We proposed a novel relevance function which captures the mutual information in certain context and exploited the structured data to make some rules. We combined the relevance function with the rules to propose our approach. We conducted some experiments on our HP data sets and made some comparisons. The experiment results show the relevance function is adequate and proposed approach is effective and efficient.

References

1. Hahn, U., Honeck, M., Schulz, S.: Subword-based text retrieval. System Sciences (2003)
2. Huang, G., Zhang, X., Luoyang: Text Retrieval Based on Semantic Relationship. In: E-Product E-Service and E-Entertainment (ICEEE) (2010)
3. Liu, J., Zhou, H.: Computer Engineering Faculty. Research on the Chinese text retrieval method using context. In: Information Science and Engineering (ICISE), Huaiyin Institute of Technology, Huaian (2010)
4. Lei, J.: A Web Information Retrieval Method Based on Multilayer Vector Space Model. Computer Applications 24, 26–27 (2004)
5. Holt, J.D., Chung, S.M., Li, Y.: Usage of Mined Word Associations for Text Retrieval. In: Tools with Artificial Intelligence. Wright State Univ., Dayton (2007)
6. Strzalkowski, T.: Natural Language Information Retrieval. Text, Speech and Language Technology Book Series, vol. 7 (1999)
7. Rong., F.S., Jun, X.W.: Study on text semantic similarity in information retrieval. Information and Automation (2008)
8. Chen, Y., Wang, W., Liu, Z., Lin, X.: Keyword Search on Structured and Semi-Structured Data. In: SIGMOD (2009)
9. Sahami, M., Heilman, T.D.: A Web-based Kernel Function for Measuring the Similarity of Short Text Snippets. ACM, 1-59593-323-9/06/0005 (2006)
10. Hu, Q., Zhanget, L., et al.: Measuring relevance between discrete and continuous features based on neighborhood mutual information. In: TOC (2011)
11. Li, W.T.: Mutual information functions versus correlation functions. Stat. Phys. 60, 823–837 (1990)
12. Holt, J.D., Chung, S.M., Li, Y.: Usage of Mined Word Associations for Text Retrieval. In: ICTAI (2007)
13. Arslan, A., Yilmazel, O.: A comparison of Relational Databases and information retrieval libraries on Turkish text retrieval. Natural Language Processing and Knowledge Engineering (2008)
14. Chellappa, M., Kambhampaty, S. (Kanishka Syst.): Text retrieval-a trendy cocktail to address the dataworld. In: Computer Software and Applications Conference (1994)
15. Singhal, A.: Modern information retrieval: a brief overview. IEEE Data Engineering Bulletin, Special Issue on Text and Databases 24(4) (December 2001)

Dynamic Table: A Layered and Configurable Storage Structure in the Cloud[*]

Xu Cheng[1,3], Biping Meng[1,3], Yuxin Chen[1,3], Peng Zhao[1,3],
Hongyan Li[2,3], Tengjiao Wang[1,3], and Dongqing Yang[1,3]

[1] Key Laboratory of High Confidence Software Technologies(Peking University),
Ministry of Education, China
[2] Key Laboratory of Machine Perception(Peking University),
Ministry of Education, China
[3] School of Electronics Engineering and Computer Science,
Peking University, Beijing, 100871 China
{chengxu,biping,pengzh,tjwang,dqyang}@pku.edu.cn,
lihy@cis.pku.edu.cn, chenyuxinpku@163.com

Abstract. Big data bring us not only constantly growing data volume, dynamic and elastic storage demands, diversified data structures, but also different data features. Apart from the traditional dense data, more and more "sparse" data emerged and account for the majority of the massive data. How to adapt to the characteristics of the sparse data without losing sight of the traits of the dense data is a challenge. To meet the differentiated storage demands and give a proper way to express the semantic of absent values, we proposed a 3-layered storage structure named "Dynamic Table" to represent the incomplete data. Our approach deliberates on the distributed storage requirements in the cloud and aims to support a hybrid row and column layout, which allows users to mix-and-match the two kinds of physical storage formats on demand. In addition, the original semantic of absent values is divided into two parts with distinct treatments. Specifically a four-valued logic is introduced. Experiments on synthetic and real-world data sets demonstrate that our approach combines the advantages of columnar storage and the merits of row-oriented store. The distinguished semantic of absent values are necessary to describe the missing values in sparse data set.

Keywords: sparse data, absent value, big data, cloud storage.

1   Introduction

As massive data arises in a variety of applications, the constantly emerging huge amount of data brings people not only the data-driven life but also the challenges of the big data age. So, the ability of storing and processing large-scale data has become the

---

[*] This work is supported by National Science and Technology Major Program for Core Electronic Devices, High-end Generic Chips and Basic Software Project of China under Grant No.2010ZX01042-001-003-05, 2010ZX01042-002-002-02, and Natural Science Foundation of China (NSFC) under grant numbers: 60973002, 61170003.

Z. Bao et al. (Eds.): WAIM 2012 Workshops, LNCS 7419, pp. 204–215, 2012.

crux of survival and competition for more and more business applications. With an open architecture, the cloud data store offers almost unlimited storage space and computing power. Compared with other appliances, the cloud storage solution uniquely offers high scalability, dynamic elasticity, in-database processing with MapReduce paradigm and pay-as-you-go service delivery pattern. These all make it possible to cope with the ever-growing data volume and adapt to the real-time variable storage demand.

Due to the fact that the cloud data store is able to supply resources on the fly, it is deployed in a multi-tenant circumstance and served as a shared nothing distributed infrastructure for all kinds of applications. However, different applications usually have datasets with different features. In particular, along with the spring up of the Internet of Things, Social Network and Mobile Internet, lots of data takes on the character of sparsity[1]. This is often the case that people would like to have tuples that range over only a part of the relation schema, and the sheer number of possible attributes is very large. Each object contains a subset of the whole attributes, and distinct object may be possessed of a different fraction of attributes. In contrast with conventional dense data, the characteristics of sparse data can be concluded in three points[2]: (1)Large number of columns. A sparse data set typically has hundreds of or even thousands of attributes[3]. (2)Sparsity. Nulls occupy most of the fields, in other words, most objects have non-null values for only a handful of attributes. (3)Schema evolution. Attributes are allowed to be defined freely[2]. Furthermore, in many cases, the sparse data coexists with the dense data corresponding to distinct subsets of the entity's attributes. It brings about many challenges before us to give a proper way to represent both sparse data and dense data in the cloud.

Existing cloud data stores, either unchanged remain the traditional relational model or do away with it and resort to non-relational data model, are mainly developed for a certain kind of application and are not sufficient to deal with mixed datasets efficiently. Layout methods for dense and sparse data are often seen as two separate problems with its own particular techniques. They didn't supply more than one physical storage layout for choice under a unified logical model.

Intrigued by such an observation, in this paper, we studied how to integrate row and column data-layouts for both dense and sparse datasets in the cloud. Concretely, a 3-layered configurable storage structure named "Dynamic Table" is proposed and implemented in our Massive Unstructured cLoud Data mAnagement System – MULDAS. In this way, a hybrid dataset is represented as a collection of structured (key,value)pairs. Each (key,value)pair corresponds to an attribute value. By sharing the same sub-key values, multiple (key,value) pairs are organized together to describe the same data object. From the perspective of users, they make up of a tabular form. Thus, different kinds of storage policies and data constraints can be defined on this soft schema. Users could organize the columns of the table into different subsets and make choice of row-oriented or column-oriented physical format for them. Moreover, as absent values take majority in the sparse data and most of them are inapplicable to a tuple, continue to use the empty placeholder(nulls) for every missing value may lead to enormous storage space waste and incurred unnecessary I/O cost. So, we distinguished the undefined and inapplicable missing values from the temporary unknown ones. To determine the behavior of absent values participated operators, a 4-valued logic is also introduced.

The rest of this paper is organized as follows. Section 2 presents the related previous work. Section 3 describes our layered and configurable model. In section 4, the differentiated semantic of absent values and a 4-valued logic are introduced. We present an experimental comparison of our model to several open source data stores in section 5. Finally, the Section 6 discusses our conclusions and future work.

2   Related Work

Due to the tremendous increase in the scale of generated sparse data, various methods have been proposed in order to model incomplete data. Based on the adopted logical or physical layout, they can be grouped into two types: (1)The first category tries to capture the sparse characteristic with distinct logic structure. The native idea is to decompose a sparse table into a number of smaller and denser tables. One way is to store a few "dense" attributes that most rows defined in a horizontal table, then relegates the rest of attribute-value pairs to a large text file. It is on the premise that the distribution of the non-null values must conform to this multi-table schema[3]. Another way is to use the 3-ary vertical representation[4]. A single row in a horizontal table is split into as many rows as the number of non-null attributes. The schema evolution is just an addition and deletion of a row. However writing SQL queries on vertical table is much more difficult than on the horizontal table. Decomposed storage model (DSM) [13] and BigTable [5] are also of this kind. (2)The second type try to decouple the logical and physical storage of entities. It remains the logical relational schema of upper layer unchanged and tackles the problem by way of delicate physical storage strategy. For the row-oriented storage[2], we can either use a placeholder to replace each appearance of the absent values or omit the missing values all together. While the former option is a waste of space, the latter one slows down tuple random access[21]. Positional format and interpreted attribute layout[1] belong to this kind. By contrast, in the column-oriented storage, such as C-Store[11], MonetDB[10], the popular way to deal with NULL is to treat it as a special value and resort to different compression methods to compress the values[3]. While improving query efficiency and facilitating schema evolution, this method generally suffers from higher cost of inserts(tuple fragmentation) and record reconstruction[21]. In addition, other works[7][8][9] make choice of an eclectic method to combine the formats of NSM and DSM, also known as hybrid representation. For a given relation, PAX[8] stores the same data on each page as NSM. Within each page, however, it groups all the values of a particular attribute together on a minipage. Similarly, RCFile[7] applies the concept of "first horizontally-partition, then vertically-partition" as well. With the complicated internal structure, it suffers from high overhead of schema evolution and serves as a storage structure for the almost read-only data warehouse system.

3   The Layered and Configurable Storage Structure – Dynamic Table

To address the problem of sparse and dense data representation in the cloud, we resort to a layered and configurable storage structure to describe the dataset. As illustrated in Fig. 1, the storage structure is composed of 3 layers. While the upper two layers give the logical layout of a dataset, the underlying layer defines the physical storage format. When we said the Storage Structure is configurable, we do mean that the tabular soft-schema, top layer of the storage structure, can be customized to the requirements of an application and the underlying physical storage format is optional for different parts of a table. Now, we will discuss them one by one.

3.1   The Structured Key-Value Model

In view of the strong and weak points of the Native Key-Value model, we propose an improved method called Structured Key-Value model, which separates the key space into more levels and add self-defined structures to the value field. To some extent, it can be seen as a multi-dimensional map indexed by distinct subkeys. For the key field, while some subkeys can be thought of as a map of maps and take part in the retrieval of data object, others can serve as a flag to indicate the exclusive access or just defined to save some schema information. As to the value field, we do not assume that all data have the same shape. Other than to provide a fixed layout, some built-in data types and a predefined combination pattern are offered to organize data in a more natural way.

Concretely, for the wide sparse data and dense data, we use a (key,value)pair to represent a basic attribute value. Since an entity may be comprised of a great many of attributes, a data object will be mapped into a collection of (key,value)pairs. Furthermore, for those attributes been frequently accessed together, there ought to be a guarantee that the storage system will assign these pairs to the same node in the distributed environment. As shown by Fig. 2, the key field is composed of four parts. The first segment, known as 'RG', has nothing to do with the data retrieval and is used to lock a (key,value)pair in support of transaction semantic. Furthermore, in order to describe all aspects of information of an entity, the second part 'RK' is reserved to save the object identity. So the (key,value)pairs contained the same identifier all point to the same object. Moreover, we need something that will group some of the attribute values together in a distinctly addressable group. So another subkey 'CG' is required to reference a group of attributes. Finally, because each (key,value)pair corresponds to an attribute of an entity, we store the attribute name in the last segment 'C'. Since the (key,value)pairs are retrieved according to the last three segments of the key field, the combination of subkeys 'RK', 'CG' and 'C' is actually acted as a primary key. In cloud environment, by maintaining several versions of an attribute value, the data store system can effectively avoid read/write conflicts and support high concurrency. Therefore, we would like to provide the ability to keep multiple values for the same attribute. In consideration of the expensive operation cost, we prefer to adopt the flat structure other than the nested structure for the value field. It can be thought of as a binary sequence. Each element consists of a value and a write timestamp and the length of the sequence is not limited. If no data type has been specified, all the values will be treated as an array of bytes.

Fig. 1. The Architecture of Dynamic Table   **Fig. 2.** The Structured Key-Value

3.2   The Tabular Soft-Schema

Despite the structured key-value model inherits fast lookup and high scalability from the native key-value model and provides stratified structure to describe the data structure, there are still some questions have not been solved: ① The simple (key,value)pairs just contain attribute values but not the schema definition itself. ② People do not get comfortable with a set of (key,value)pairs. It lacks a visual representation to organize the dataset. ③It is difficult to define a appropriate query on a group of key-value pairs.

As a consequence, we create a tabular soft schema on top of the schema-less (key,value)pairs, Rather than another newly added storage model, it is more like a visualized view to give an integrated perspective of a collection of related key-value pairs. In contrast with the well-defined relational table, soft schema free people from a fixed structure and serve as an elastic container to hold the wide and sparse data. Its three basic building blocks are illustrated in Fig. 3:

Table. As a container, table is composed of a single column and one or more ColumnGroups. While the individual "RowKey" column stores the keywords of tuples, the ColumnGroup is defined as a subset of columns, and distinct ColumnGroups do not overlap with each other. **ColumnGroup.** In cloud environment, both the data distribution and layout are critical to determine the performance and scalability. Totally random assignment of locations for data segments hinders possible improvements for access locality, the attributes that would be accessed simultaneously are supposed to be located altogether. Like the conception "projections"

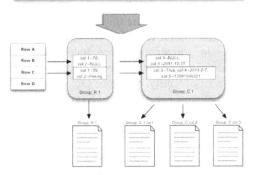

Fig. 3. Dynamic Table

in C-Store[11], We provide two kinds of ColumnGroups, namely ColumnGroup_R and ColumnGroup_C. For the columns belonged to the former one, values are stored in the row-oriented format. As to the ColumnGroup_C contained columns, the storage of data follows the pattern of columnar method. **Column.** Without distinction, column defines the domain of a corresponding attribute. Except for the "RowKey" column, other ones are all belonged to a certain ColumnGroup.

With above definition, a cell in the table is actually a structured (key,value)pair, which contains a chain of sequential multiple values and a timestamp is attached to each version. As to a tuple, it corresponds to a group of structured key-value pairs indexed by the same subkey "RK". For a data object, the attributes values of the same ColumnGroup are promised to be saved at the same node. The tabular soft schema gives users flexible control over the logical layout and physical format of dataset.

3.3   Optional Physical Storage Format

In the face of different data features and variety of query manners, neither the row-oriented method nor the vertical partitioned way can cope with the hybrid datasets. Especially when the table schema is wide and each part of it has its own features, the physical storage need may vary within the same logical layout. Now that no single physical format can take the place of the other one, we try to support both of them in our storage structure.

To organize the physical storage of (key,value)pairs in a table, we exert the horizontal sharding on the table first in order to adaptively adjust to the data updating and then vertically split each segment into several partitions according to the definitions of ColumnGroups. For a certain partition, we have choice of row-stores or column-stores, but the implementation is not directly tangible. As for the row-oriented method, the whole partition will be mapped into a single file. The difficult is that an attribute value of a data object may be missing or has multiple versions. For the traditional way like N-ary in Fig. 4(a), every record is of equal length and cannot save more than one value for an attribute. An alternative solution is to reserve several empty fields for each attribute in advance showed by Fig. 4(b). However, in order to simplify the problem of consistency and achieve the high throughput, the underlying distributed file system in the cloud always follow the way of "write-once and read-many". So we can only append an existing file rather than modifying it. This makes the reserved blanks in record are useless. Another optional method is to link the record by pointers as illustrated in Fig. 4(c). Every newly appended value has a backward link to the previous value. On this occasion, the (key,value)pairs consisted of a record are scattered all over the file. It creates high overhead to retrieve a record. In reality, we resort to the method of piling up the (key,value)pairs continuously in a file. It simply overlooks the absent fields and stores the values in the order of RowKey. For the new inserted and deleted values(write a tomb tag), the system append them in the end of the file, and periodically sweep through it to create a new file. It deletes the obsolete values and merges the related values. As to the column-oriented approach, each column in a partition is saved as a separate file. Due to the dynamic elasticity, the storage location of data is transparent to the applications in the distributed file system. In other words, users cannot assign a file to a specified storage node. This makes it easier to scale-out, scale-back and failover. So how to make sure the associated column files are co-located on the same node is a challenge. We borrow the idea from [14] to solve the problem. By setting the configuration property "dfs.block.replicator.classname" to point to the appropriate class, HDFS allows its placement policy to be changed. This feature has been present since Hadoop 0.21.0 and we implement the logic for our column-oriented storage format. For simplicity, we set the number of storable tuples of a subtable small enough to keep each Column file occupies a single HDFS block. It seems that our column-oriented block placement policy works at the file level and could guarantees that the files corresponding to the different columns of the same partition are always co-located across replicas.

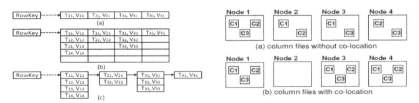

Fig. 4. Row-Oriented Format   **Fig. 5.** Co-Locating Column Files[14]

4   The Differentiated Treatments for Absent Values

Apart from the logical structure and the storage representation of the mixed datasets, a key issue of the sparse data management is to determine how a database behaves in the presence of absent values. In this section, for the purpose of clear description of incomplete data, we classify the meanings of missing values into two kinds to capture the critical distinction between unknown missing and inapplicable missing. Moreover, a simplified 4VL is introduced as an expansion to the original two valued logic.

4.1   The Semantics of Absent Values

Absent values refer to the missing attribute values of data object(tuple). The disappearance of an attribute value comes in many cases. By far, several definitions of absent values have been proposed. A basic problem with missing values is that they have many plausible interpretations[15]. There are many different types of null values, each of which reflects different intuitions about why a particular piece of information is missing[16][17]. In view of the spring up of big data, Our approach is based upon the observation that the distinction between applicable and inapplicable missing data is of great practical value[18]. Most of the above types could be thought of as special cases of 'missing and applicable' and 'missing and inapplicable', and the two explanations seem to capture reasonably well all above manifestations of null[19]. Thus, the various manifestations of nulls can be reduced to two basic interpretations[15],

①*Unknown*: a value exists but it is not known;

②*Nonexistent*: a value does not exist. The attribute is undefined and is not relevant with the data object.

After categorizing all sorts of different situations of the absent values into two meanings, it is time to introduce two formal symbols to represent them respectively. Concretely, a symbol NULL is defined to denote an unknown missing value, and another reserved keyword VACANT is presented to indicate a value is nonexistent. While any occurrence of the value-unknown type of NULL can be replaced in an updating operation, the VACANT field could only be assigned by an explicit insert statement. In addition, neither NULL nor VACANT is a member of any data domain, but all data types are NULL -able and VACANT -able. They are the same for all data types and are not considered two "value", but rather two markers.

①*NULL→Unknown*   ②*VACANT→Nonexistent*

4.2   4-Valued Logic

As the missing values are modeled as *NULL* and *VACANT*, the truth-values of an atomic conditional expression in query clause also need to be expanded. We use symbol *Maybe* to represent the outcome may be either *True* or *False* (The truth-value Maybe is generated whenever Null is compared with any data value, or with another Null.) and add a *Neglect* mark to denote the outcome of the *nonexistent* field participated comparison. So any condition judgment statement may have four possible truth-values: *True*, *False*, *Maybe* and *Neglect*. Therefore, the use of two types of missing values lead to a 4-valued logic(4VL). To echo the previous interpretations, we consider that the 'basic' values of a logical expression are t(true), f(false) and i(inapplicable and undefined), then each truth-value can be assigned to a non-empty subset of the set$\{t,f,i\}$.

$(1)True = \{t\}; \quad (2)False = \{f\}; \quad (3)Maybe = \{t,f\}; \quad (4)Neglect = \{i\}$

Without further discussion, we directly give the truth table of 4VL in Fig. 6.

AND	T	M	F	N	OR	T	M	F	N	NOT	
T	T	M	F	N	T	T	T	T	T	T	F
M	M	M	F	N	M	T	M	M	M	M	M
F	F	F	F	N	F	T	M	F	F	F	T
N	N	N	N	N	N	T	M	F	N	N	N

Fig. 6. Truth Table of The 4VL

Except for the logical operations of the 4VL, to give a complete description of the impact of absent values on the operational semantic, the *NULL* and *VACNAT* involved arithmetical operations as well as the modified relational operators will be further discussed. These beyond the scope of this paper and we will leave the discussions to another one.

5   Experiment Evaluation

In this section, we will evaluate our proposed Dynamic Table for many concerned operations on both wide sparse data and dense data. By making a comparison with the state-of-art systems, we examine the comparative advantages[12] of row-stores and column- stores for Dynamic Table respectively.

5.1   Data Sets and Experiment Settings

Concretely, we used a real-word dataset and a synthetic dataset to evaluate the experiments results. The first one consists of 2 dense tables and is publicly available at (http://an.kaist.ac.kr/traces/WWW2010.html). One of the table *Tweet* contains 4 attributes and 455,606,357 tuples, which occupies 52GB. The other one *TweetRelation* contains 3 attributes and 243,026,950 tuples, which occupies 8GB. The second dataset is a generated hybrid table. Considering the student course registration system, a huge table *Grade* is designed to save all the students course selection information. As the courses are divided into elective courses and required courses, a student is asked to take all the required courses and a certain number of elective courses before graduation. Reference to the data of our university, we defined a wide table contained 32 required courses and 226 elective courses. While the part of required courses attributes is a typical dense table, the rest of the fields are predominantly lack of values. We assume that every student will randomly choose 28 elective courses on average.

As for the baseline systems, we chose 2 traditional relational databases and 2 cloud data stores as follows, ①Open source row-oriented database MySQL-5.0; ②Open source column-oriented database C-Store-0.2[11]; ③HBase-0.90.2[6], an open source non-relational distributed database modeled after Google's BigTable[5]. It is on top of Hadoop and supports random reads and random writes; ④Hive-0.6.0[20], an open source data warehousing solution built on top of Hadoop, the files of which can be selected to be stored as RCFile[7]; ⑤MULDAS.

In consideration of the limited resources, we deploy ③, ④ and ⑤ on the same cluster with 7 nodes connected by a 1Gbit Ethernet switch. Each node has 4 cores (Quad-Core AMD Opteron(tm) Processor 2378MHz), 16GB of main memory, and 1.4TB SATA disks. Besides, we set up a ① cluster consisted of 3 nodes with MySQL Proxy and separately install ② on a single node.

5.2   Results and Discussions

To compare the respective advantages and display the differentiated storage requirements, we would like to validate the performance of Dynamic Table from the following aspects:

(1)Data loading

The first experiment is to verify whether the distinct physical storage formats would have an unequal effect on different kinds of datasets. Except for the two dense tables *Tweet* and *TweetRelation*, we generate two hybrid tables according to the schema of *Grade*: *Grade1*, which contains 271,288 tuples and occupies 267MB; and *Grade2*, which contains 5,826,506 tuples and occupies 5.6GB. For the tow dense tables, we defined only one ColumnGroup_R in a Dynamic Table to save the values. As to the two hybrid tables, a ColumnGroup_R is defined to keep the required courses grades and a ColumnGroup_C for the elective courses grades. So an explicit *NULL* will be inserted into the table when a student' required courses grade is missing. At the same time, the system may simply ignore the absence of an elective courses grade. Since the speed of data loading is most relevant to the number of tuples, in practice, we record the elapsed time required to load a tenth of the dense tables in Fig. 7.

(2)Data retrieval

There are generally two kinds of read operations: while the transactional queries(TQ) are inclined to retrieve the whole tuple, the analytical queries(AQ) usually only interested in a handful of columns. In order to check the difference, we exert TQ on the *Tweet* table and execute AQ on the *Grade2* table respectively. Since Hive[20] is not designed for random access, it is not included in this experiment.

(3)Storage space utilization

To highlight the experimental result, we conduct space utilization experiment on the hybrid table *Grade1* and define the attributes as several distinct types. Moreover, we compare the overhead of storage space in both uncompressed and GZip ("lighter-weight" compression schemes[10]) compressed cases.

(4)Dynamic schema evolution

The problem of schema evolution is not limited to the modification of the schema itself. In fact, it affects the data stored under the given schema and the queries posed on that schema. By inserting a column of data into the table *TweetRelation*, we would like to know its influence on data storage with different structures. Therefore, a row-oriented Dynamic Table and a column-oriented one are used to save the dense table, and we try to examine their schema modification cost.

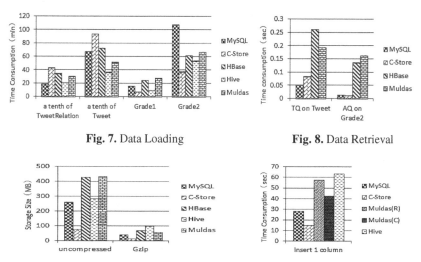

Fig. 7. Data Loading   **Fig. 8.** Data Retrieval

Fig. 9. Storage Space Utilization   **Fig. 10.** Dynamic Schema Evolution

Based on above experiments results, we could draw the conclusion that, in the face of different kinds of datasets and differentiated storage requirements, there is no panacea. While the row-oriented method is fit for the dense data loading, the sparse dataset can be quickly input in the columnar storage format(Fig. 7). To load the dense data by the horizontal way, the attributes of a record(or tuple) are placed contiguously in storage and a single disk write suffices to push all of the fields of a single record out to disk. But for the sparse data, only sporadic values need to be inserted to a table, the vertical method could decrease the network transmission of lots of placeholders. As to the OLTP style queries, which involve most of the attributes, row-stores can avoid the overhead of tuple recombination and have more advantages in retrieving data. But for the analytical queries, when the table schema is wide and a small number of attributes are requested, the vertical partitioned way may skip over irrelevant columns and deliver greater throughput(Fig. 8). With the column-oriented method, the values for each single column (or attribute) are stored contiguously. It supports efficient queries over wide schema, improved bandwidth utilization and cache locality[3]. Only those attributes that are involved in the query need to be fetched from disk. By avoiding bringing into memory irrelevant attributes values, it saves relative expensive caches space and improves the hit rate. A fact that column-stores not only free of the waste of storage space but also improve the compression ratio can be deduced from the comparison of uncompressed space utilization and compressive space consumption. As the column values come from the same attribute domain, compression algorithms work better with vertically partitioned relations(Fig. 9). Although, the hybrid data placement structure, such as RCFile[7], reach a balance between the horizontal and vertical partitioned way and adapted to the dynamic workload patterns, it suffers from frequent schema modifications. By contrast, Dynamic Table has similar performance with column-oriented method(Fig. 10). The addition or subtraction of a column is equal to adding or deleting a separate file without modification of other columns.

Compared with MySQL, HBase[6] and C-Store[11], Dynamic Table provides optional physical storage formats and allows them to be applied separately to a part of the table on demand. As the increase of the data size, the winner has turned from MySQL and C-Store[11] to HBase[6] and Hive[34], and Dynamic Table shows competitive performance benefits of them. Moreover, missing data are a common occurrence and can have a significant effect on the conclusions that can be drawn from the data. By defining an absent value as *NULL* or *VACANT*, users could make the different semantics visible, which help the applications to describe more explicit storage requirements.

6   Conclusions

In this paper, we try to address the problem of setting up a flexible storage structure for datasets with different characteristics in the cloud data store. The goal is to offer a hybrid row and column layout to satisfy the differentiated requirements of wide sparse data and dense data.

The innovative contributions embodied in this paper include:

✓ Proposed a 3-layered storage structure "Dynamic Table". With the middle structured key-value layer, a dataset is mapped into a collection of structured (key,value)pairs. Since the simplicity is inherent to the key-value model, our storage structure is scalable and adapt to the dynamic distributed environment.

✓ With the upper soft schema layer, a group of (key,value)pairs are organized together and presented externally to users as a multi-dimensional table. So, distinct policies could be defined on a subset of logical related (key,value)pairs.

✓ With the underlying physical storage layer, different parts of a table can be configured to be stored in distinct storage formats. It has the advantages of both row-oriented store and columnar storage.

✓ Using *VACANT* and *NULL* as distinct symbols to represent the distinguished two types of semantic of absent values, make it possible to remove the ambiguity of original null-values in sparse data and provide a more explicit description of the incomplete data.

✓ A 4-valued logic is introduced to define the behavior of operations.

References

1. Beckmann, J.L., Halverson, A., Krishnamurthy, R., et al.: Extending RDBMSs to support sparse datasets using an interpreted attribute storage format. In: Proceedings of the 22nd International Conference on Data Engineering, ICDE, p. 58. IEEE Computer Society, Washington (2006)

2. Yang, B., Qian, W., Zhou, A.: Using Wide Table to manage web data: a survey. Frontiers of Computer Science in China 2, 211–223 (2008)
3. Eric, C., Beckmann, J., Naughton, J.: The case for a wide-table approach to manage sparse relational data sets. In: Proceedings of SIGMOD, pp. 821–832. ACM, New York (2007)
4. Agrawal, R., Somani, A., Xu, Y.: Storage and querying of e-commerce data. In: Proceedings of the 27th International Conference on VLDB, pp. 169–180. Morgan Kaufmann Publishers Inc., San Francisco (2001)
5. Chang, F., Dean, J., Ghemawat, J., et al.: Bigtable: A Distributed Storage System for Structured Data. ACM Transactions on Computer Systems 26, 1–26 (2008)
6. Apache HBase, `http://hbase.apache.org/`
7. He, Y., Lee, R.B., Huai, Y., et al.: A Fast and Space-efficient Data Placement Structure in MapReduce-based Warehouse Systems. In: Proceedings of the IEEE International Conference on Data Engineering (ICDE), pp. 1199–1208. IEEE, Hannover (2011)
8. Ailamaki, A., DeWitt, D., Hill, M., et al.: Weaving Relations for Cache Performance. In: Proceedings of the 27th International Conference on VLDB, pp. 149–158. Morgan Kaufmann Publishers Inc., San Francisco (2001)
9. Ramamurthy, R., DeWitt, D.J., Su, Q.: A Case for Fractured Mirrors. The International Journal on Very Large Data Bases 12, 89–101 (2003)
10. Boncz, P., Zukowski, M., Nes, N.: MonetDB/X100: Hyper-pipelining query execution. In: Proceedings of the CIDR 2005, pp. 225–237. VLDB, San Francisco (2005)
11. Stonebraker, M., Abadi, D.J., et al.: C-Store: A Column-oriented DBMS. In: Proceedings of the 31st International Conference on VLDB, pp. 553–564. VLDB Endowment, Trondheim (2005)
12. Abadi, D.J., Madden, S.R., Hachem, N.: ColumnStores vs. RowStores: How Different Are They Really? In: Proceedings of the 2008 ACM SIGMOD International Conference on Management of Data, pp. 967–980. ACM, New York (2008)
13. Copeland, G.P., Khoshafian, S.N.: A decomposition storage model. In: Proceedings of the 1985 ACM SIGMOD International Conference on Management of Data, pp. 268–279. ACM, New York (1985)
14. Floratou, A., Patel, J.M., Shekita, E.J., Tata, S.: Column-Oriented Storage Techniques for MapReduce. Proceedings of the VLDB Endowment 4, 419–429 (2011)
15. Zaniolo, C.: Database Relations with Null Values. Journal of Computer and System Sciences 28, 142–166 (1984)
16. Candan, K.S., Grant, J., Subrahmanian, V.S.: A Unified Treatment of Null Values Using Constraints. Information Sciences 98, 99–156 (1997)
17. Codd, E.F.: Missing Information (Applicable and Inapplicable) in Relational database. In: Margaret, H.E. (ed.) ACM SIGMOD Record, vol. 15, pp. 53–53 (1986)
18. Gessert, G.H.: Four Valued Logic for Relational Database Systems. ACM SIGMOD Record 19, 29–35 (1990)
19. Vassiliou, Y.: NULL values in database management a denotational semantics approach. In: Proceedings of the 1979 ACM SIGMOD International Conference on Management of Data, pp. 162–169. ACM, New York (1979)
20. Thusoo, A., Sarma, J.S., Jain, N.: Hive – A Petabyte Scale Data Warehouse Using Hadoop. In: 2010 IEEE 26th International Conference on ICDE, Long Beach, CA, pp. 996–1005 (2010)
21. Abadi, D.J.: Column Stores For Wide and Sparse Data. In: Proceedings of CIDR, pp. 292–297 (2007)

Fusing Heterogeneous Information for Social Image Retrieval

Xirong Li

The MOE Key Lab of Data Engineering and Knowledge Engineering
Renmin University of China
`xirong@ruc.edu.cn`

Abstract. Given that millions of images are uploaded to the social web in a single day, how to effectively retrieve such increasing amounts of unstructured data becomes crucial. While current social platforms find images on the base of user-contributed tags, the tags are known to be subjective and noisy, and consequently put the effectiveness of social image retrieval into question. Different from existing work which focuses on analyzing individual sources of information such as textual and visual separately, in this paper we propose to fuse the heterogeneous information. We investigate the state of the art weighting methods within a linear fusion framework. Image retrieval experiments on a present day benchmark with 46 visual concept queries show encouraging results. Compared to the best search results obtained by a single source of information, the Uniform method and the Coordinate Ascent method obtain relative performance gain of 3.9% and 5.7% in terms of mean average precision. This work provides some practical guidelines for fusing multiple sources of information for improving social image retrieval.

Keywords: Social image retrieval, tag relevance, information fusion.

1 Introduction

Millions of images are uploaded to social media sharing platforms such as Facebook, Google, and Flickr. Towards making sense of the massive amounts of unstructured visual content, effective social image retrieval is crucial. Present-day search engines retrieve social images by matching textual queries with tags assigned to the images by common users. However, due to varied reasons including batch tagging and the diversity of user knowledge, social tags are known to be subjective and ambiguous [8]. The lack of coincidence between the tags and the visual concepts factually present in the image content puts the effectiveness of image retrieval by tags alone into question.

To improve social image retrieval, a number of algorithms have been proposed, exploiting visual [4,7,10] and textual [13] information separately. For instance, Li et al. proposed a visual neighbor voting algorithm [4], which infers the relevance of a tag with respect to an image by counting the image's visual neighbors labeled with that tag. Liu et al. use a random walk model to propagate tags

Z. Bao et al. (Eds.): WAIM 2012 Workshops, LNCS 7419, pp. 216–225, 2012.

Fig. 1. Examples of social-tagged images. User tags are on the right hand side of each image. Font size reflects the relevance of a tag with respect to the image, automatically estimated by a visual based relevance function [4]. In this example, visual tag relevance yields an incorrect image ranking when searching for "dog", while sorting by the number of tags in ascending order produce a correct image ranking.

between neighbor images [7]. Sun and Bhowmick [10] exploit visual consistency to quantify tag relevance. Zhu et al. [13] consider semantic consistency, measuring a tag's relevance to an image in terms of its semantic similarity to the other tags assigned to the image. By using learned tag relevance values as a new ranking criterion, better image search results are obtained, when compared to search using original tags.

Despite the above progress, we argue that using a single source of information is limited to tackle large variations in both visual content and tags of social images. As illustrated in Fig. 1, when a visual concept is present in a non typical appearance, the visual algorithms may fail to identify the corresponding tag. In this example, simply ranking images by the number of tags is more effective. The semantic algorithm does not work for images labeled with a single tag, because there is no other tags available for computing semantic similarity. Fusing multiple sources of information including meta, semantic, and visual is thus important for social image retrieval.

Information fusion for multimedia retrieval has been studied in the last decade for its importance in boosting the retrieval performance [11]. Learning to rank algorithms which go beyond simple uniform or heuristics weighting strategies have been developed [3,9,12]. In the context of social image retrieval, Li et al. propose multi-feature tag relevance learning which combines tag relevance estimates obtained by different visual features [5]. Later, the authors use the Borda count algorithm to combine image retrieval results obtained by semantic analysis and by multi-feature tag relevance [6]. However, the fundamental question of *how to fuse heterogenous information for social image retrieval* remains open.

The question further leads to a set of sub questions to be answered. Search result scores are often normalized before fusion [11]. *What is the influence of score normalization on fusion for social image retrieval?* Li et al. [5] suggest that for combining visual tag relevance, the uniform weighting scheme is a good choice, compared to supervised alternatives such as RankBoost [3]. *Is uniform weighting also a reasonable choice for combining heterogenous information?* In this paper we aim to answer these questions by proposing a social image retrieval system which fuses meta, semantic, and visual information. To the best of our knowledge, research of this kind has not been done before.

2   The Proposed System

For image retrieval, we have to define a relevance function which measures the relevance of a specific image with respect to a given query. Consequently, we rank the images by the relevance function in descending order. Instances of such a function can be constructed on the base of metadata, semantic analysis, or visual analysis as described in Section 1. Given multiple relevance functions driven by varied sources of information, our goal is to build a social image retrieval system which yields better performance than using the individual functions separately.

To make our discussion more formal, we use x to denote an image. For a given visual concept ω, let $h(x, \omega)$ be a specific relevance function. Shall the system have l functions to combine, we denote them by $\{h_i(x, \omega) | i = 1, \ldots, l\}$. To measure the effectiveness of a specific relevance function, we introduce a performance metric function $E_{metric}(h(x, \omega))$.

2.1   Fusing Multiple Relevance Functions

We choose linear fusion for its widespread use in multimedia retrieval fusion [11]. For each $h_i(x, \omega)$, let λ_i be its weight, and $\Lambda = \{\lambda_1, \ldots, \lambda_l\}$ be the weight vector. Assuming that $h_i(x, \omega)$ is better than random guess and thus adding it will not have negative impact on the retrieval performance, we impose the nonnegative constraint on the weights, i.e, $\lambda_i \geq 0$. Since normalizing weights by dividing by their sum does not affect image ranking, any linear combination with nonnegative weights can be equivalently transformed to a convex combination. Hence, we adopt convex combination, and define the fused function as

$$h_\Lambda(x, \omega) = \sum_{i=1}^{l} \lambda_i \cdot h_i(x, \omega), \tag{1}$$

where $\lambda_i \geq 0$ and $\sum_{i=1}^{l} \lambda_i = 1$. We now formalize our goal as optimizing the weights such as the performance metric is maximized, namely

$$\underset{\Lambda}{\operatorname{argmax}} \, E_{metric}(h_\Lambda(x, \omega)). \tag{2}$$

When no training data is available for solving (2), a sensible choice is to assume that the individual functions are equally effective, and accordingly set $\lambda_i = \frac{1}{l}$. We refer to this weighting method as Uniform.

When well-labeled training data is available, supervised weighting methods can be applied. To that end, we investigate the following three learning to rank algorithms which have been widely used: RankBoost [3], AdaRank [12], and

Coordinate Ascent [9]. RankBoost sequentially combines the output of individual relevance functions with an adaptive weighting scheme to emphasize functions capable of correcting mis-ranking made in previous training rounds. Rather than minimize the pairwise ranking error, AdaRank reinforces relevance functions which answer difficult queries better. The Coordinate Ascent algorithm solves (2) by optimizing one weight in one training round, with the remaining weights fixed. Once the optimal value of the selected weight is obtained, another weight is activated and the same procedure applies. The loop continues until $E_{metric}(h_\Lambda(x,\omega))$ stops increasing.

We have depicted the linear fusion framework. Next, we instantiate the relevance functions which will be fused within the framework.

2.2   Relevance Functions to be Fused

We leverage visual based, semantic based, and metadata based relevance functions which reflect the state of the art of their kinds.

1. Visual based relevance function. We adopt the neighbor voting algorithm [4] which estimates the relevance of a tag with respect to an image by counting the tag's frequency in annotations of the image's k nearest neighbors. The neighbors are found in terms of a specific visual similarity from a large collection of social-tagged images. We express the visual based relevance function as

$$h_{vis}(x,\omega) = \frac{k_\omega}{k} - \frac{N_\omega}{N}, \tag{3}$$

where k_ω is the number of neighbor images labeled with ω, N_ω is the number of images labeled with ω in the entire collection, and N is the size of the collection.

2. Semantic based relevance function. The rationale for this function is grounded on the assumption that the true semantic interpretation of an image is reflected best by the majority of its social tags. Consequently, a tag that is semantically more consistent with the majority is more likely to be relevant to the image. Following [6], we express the semantic based relevance function as

$$h_{sem}(x,\omega) = \frac{\sum_{\omega' \in \mathbf{w}_x} sim(\omega',\omega) \cdot idf(\omega')}{\sum_{\omega' \in \mathbf{w}_x} idf(\omega')}, \tag{4}$$

where $\mathbf{w}_x$ is the set of social tags assigned to x, $sim(\omega',\omega)$ denotes tag wise semantic similarity, and $idf(\omega')$ is the inverse image frequency of ω', reflecting the tag's informativeness.

3. Metadata based relevance functions. We define two metadata based functions: TagNum and Views. An image may be labeled with many irrelevant tags, say to increase the chance of being retrieved. To penalize such over tagged images, we define

$$h_{tagnum}(x,\omega) = \frac{1}{|\mathbf{w}_x|}, \tag{5}$$

where $|\cdot|$ returns the cardinality of a set. Assuming that relevant images are more likely to be viewed when compared to irrelevant images, we define

$$h_{view}(x,\omega) = \frac{views(x)}{MaxViews}, \tag{6}$$

where $views(x)$ is the number of times an image has been viewed, and $MaxViews$ is the maximum view number. Notice that while $h_{vis}(x,\omega)$ and $h_{sem}(x,\omega)$ depend on ω, $h_{tagnum}(x,\omega)$ and $h_{views}(x,\omega)$ are concept independent. So the two metadata based relevance functions act as prior evidence for judging the relevance of an image with respect to (unseen) queries.

2.3 Score Normalization

Since search result scores obtained by different relevance functions may reside at varied scales, score normalization is often conducted before fusion [11]. In order to understand the impact of score normalization on fusion for social image retrieval, we investigate the following two popular methods: MinMax and RankMax. The MinMax normalized $h(x,\omega)$ is defined as

$$\tilde{h}(x,\omega) = \frac{h(x,\omega) - \min(h(x,\omega))}{\max(h(x,\omega)) - \min(h(x,\omega))}, \tag{7}$$

where the min (max) function returns the minimum (maximum) score in the list. The RankMax normalized $h(x,\omega)$ is defined as

$$\hat{h}(x,\omega) = 1 - \frac{rank(x; h(x,\omega))}{MaxRank}, \tag{8}$$

where $rank(x; h(x,\omega))$ returns the rank of an image in a search result list ranked by $h(x,\omega)$, and $MaxRank$ is the list length. Notice that fusing the RankMax normalized functions with the uniform weights is equal to the Borda count method.

3 Experimental Setup

3.1 Data Collections

We select NUS-WIDE [1], a popular benchmark set for social image retrieval. This set has ground truth annotations for 81 visual concepts including a number of objects, scenes, and events. For reasons of efficiency, we took a subset of 20K images randomly sampled from NUW-WIDE for our experiments. Further, we randomly divided the subset into two halves, one for training and the other for testing. By preserving concepts which have more than 100 labeled images in the training set and in the testing set, we construct a query set of 46 concepts, with their visual examples given in Fig. 2.

airplane	airport	animal	beach	bird	boat	bridge	building	car	cat	cloud	cow
dog	fire	fish	flag	flower	food	garden	grass	horse	house	lake	leaf
mountain	ocean	person	plant	reflection	road	rock	sand	sign	sky	snow	sport
street	sun	sunset	temple	tower	toy	train	tree	water	window

Fig. 2. Examples of the 46 visual concepts used in our image retrieval experiments. The concepts correspond to a number of objects, scenes, and events.

3.2 Experiments

Experiment 1. The Impact of Fusing Heterogenous Information. By comparing image search using the individual relevance functions and image search using their combination, we want to understand whether social image retrieval benefits from fusing heterogenous information. For a more comprehensive comparison, we also report a Random baseline which ranks image in a random manner.

Experiment 2. Comparing Different Weighting Methods. By comparing the use of different weighting methods and score normalization methods, we aim to figure out the optimal choice for fusion. With each of the learning to rank algorithms, we learn a single Λ for all concepts on the training data.

3.3 Implementations

Parameters for $h_{vis}(x,\omega)$. We construct four variants of $h_{vis}(x,\omega)$ by using the following four distinct visual features separately: COLOR, CSLBP, GIST, and Dense-SIFT. The four features describe color, texture, spatial layout, and local gradients, respectively. We refer to [6] for the details of the features. We empirically set the number of visual neighbors k in (3) to be 500. For image retrieval runs obtained by the four variants, we name them after their corresponding features as Tagrel-color, Tagrel-cslbp, Tagrel-gist, and Tagrel-dsift.

Parameters for $h_{sem}(x,\omega)$. As an instantiation of $sim(\omega',\omega)$ in (4), we choose the Normalized Google Distance [2], which measures semantic divergence between two tags based on their (co-)occurrence frequency in a collection of 10M social-tagged images. The distance is converted to similarity via a Gaussian function [6].

In total, we have 7 relevance functions for fusion: TagNum, Views, Semantics, Tagrel-color, Tagrel-cslbp, Tagrel-gist, and Tagrel-dsift.

Parameters for the Weighting Methods. We observed that for RankBoost and AdaRank, their performance is sensitive to the number of training iterations. We empirically set the number of iterations to be $2 \times l$. Recall that l is the number of relevance functions.

Evaluation Criteria. We use Average Precision (AP), a popular choice for evaluating visual search engines The overall performance is measured in terms of mean Average Precision (mAP), which is computed by averaging AP scores over all query concepts.

4   Results

4.1   Experiment 1. The Impact of Fusing Heterogenous Information

As shown in Fig. 3, among the individual sources of information, the *Tagrel* which exploits visual information outperforms the *Semantics* (mAP=0.655), the *TagNum* (mAP=0.600), and the *Views* (mAP=0.585). In particular, the *Tagrel-gist* achieves the best performance, with an mAP of 0.707.

When the multiple sources of information are combined, either by uniform weighting or by weights determined by CoordAscent, we observe improvments. At the significance level of 0.001, both *Uniform* and *CoordAscent* are better than *Tagrel-gist*, with relative improvements of 3.9% and 5.7%, respectively. The results allow us to conclude that fusing heterogenous information improves social image retrieval.

For an intuitive understanding of the results, we present in Fig. 4 the top 20 search results of the concept "airplane" by different methods.

4.2   Experiment 2. Comparing Different Weighting Methods

As shown in Table 1, CoordAscent reaches the maximum mAP scores. Given the rank normalized input scores, CoordAscent obtains an mAP of 0.748, while Uniform has an mAP of 0.735. Though the gain seems marginal, the paired t-test reveals that at the significance level of 0.005, CoordAscent is statistically better than Uniform. Also, for 37 out of the 46 concepts, we observe performance increase. These results allow us to conclude that among the weighting methods investigated in this paper, CoordAscent is the best.

What surprised us is the result that neither RankBoost nor AdaRank beats Uniform. We attribute the result to the reason that in contrast to their counterpart in CoordAscent, the optimization goals of RankBoost and AdaRank are loosely correlated with mAP. This suggests the importance of weighting methods which can directly optimize a given performance metric.

Concerning the influence of score normalization, the ineffective TagNum produces higher scores than the Tagrel. Score normalization, either by RankMax or by MinMax, suppress the impact of TagNum. As a consequence, normalization is beneficial for the Uniform fusion, as shown in Table 1. Since CoordAscent can learn proper weights for the individual sources of evidence, normalization is unnecessary for CoordAscent.

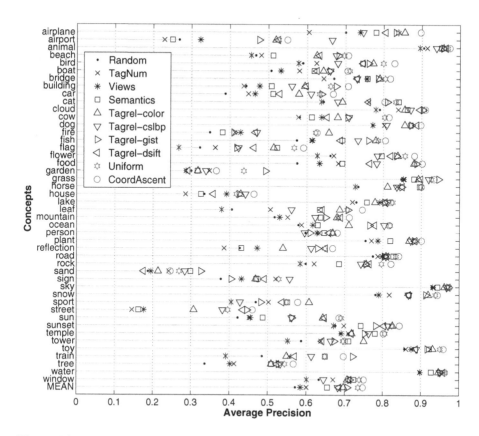

Fig. 3. The Impact of Fusing Heterogenous Information. The input scores for *Uniform* and *CoordAscent* are normalized by RankMax (8). At the significance level of 0.001, both *Uniform* and *CoordAscent* are better than the best single information, i.e., Tagrel-gist, with relative improvements of 3.9% and 5.7%, respectively.

Table 1. Comparing different weighting methods, given different score normalization algorithms. Performance measured in terms of mean average precision. The star symbol * indicates that at the significance level of 0.005, the corresponding weighting method is better than the *Uniform* given the same input for combination.

	Weighting methods			
Input scores	*Uniform*	*RankBoost* [3]	*AdaRank* [12]	*CoordAscent* [9]
Original	0.710	0.660	0.726	0.749*
MinMax normalized (7)	0.736	0.732	0.733	0.745*
RankMax normalized (8)	0.735	0.726	0.733	0.748*

(a) TagNum

(b) Views

(c) Semantics

(d) Tagrel-color

(e) CoordAscent Fusion

Fig. 4. Top 20 search results of concept "airplane" by (a) TagNum, (b) Views, (c) Semantics, (d) Tagrel-color, and (e) CoordAscent fusion. Red borders indicate negative results. Best viewed in color.

5   Conclusions

In this paper we propose to improve social image retrieval by fusing heterogenous information extracted from the metadata, tags, and visual content of social images. Image retrieval experiments on a benchmark set allow us to draw the following conclusions. First, fusing the multiple sources of information improves social image retrieval. Compared to the best single relevance function, the Uniform method and the Coordinate Ascent method obtain relative performance gain of 3.9% and 5.7% in terms of mean average precision. Second, among the three supervise weighting methods in consideration, Coordinate Ascent is most effective. The paired t-test reveals that the improvement of Coordinate Ascent over Uniform is statistically significant. The gain, of course, comes with the cost of acquiring training data for optimization. Finally, we find that score normalization is necessary for Uniform and unnecessary for Coordinate Ascent. In sum, for social image retrieval fusion, when no training data is given, we recommend the Uniform weighting method after score normalization. When well-labeled training data is accessible, we suggest Coordinate Ascent, without the need of score normalization.

Acknowledgements. This work was partly supported by the HGJ program (2010ZX01042-002-002-03).

References

[1] Chua, T.-S., Tang, J., Hong, R., Li, H., Luo, Z., Zheng, Y.-T.: NUS-WIDE: A Real-World Web Image Database from National University of Singapore. In: ACM International Conference on Image and Video Retrieval (2009)

[2] Cilibrasi, R., Vitanyi, P.: The Google Similarity Distance. IEEE Trans. on Knowl. and Data Eng. 19(3), 370–383 (2007)

[3] Freund, Y., Iyer, R., Schapire, R., Singer, Y.: An efficient boosting algorithm for combining preferences. J. Mach. Learn. Res. 4, 933–969 (2003)

[4] Li, X., Snoek, C., Worring, M.: Learning social tag relevance by neighbor voting. IEEE Trans. Multimedia 11(7), 1310–1322 (2009)

[5] Li, X., Snoek, C., Worring, M.: Unsupervised multi-feature tag relevance learning for social image retrieval. In: ACM International Conference on Image and Video Retrieval (2010)

[6] Li, X., Snoek, C., Worring, M., Smeulders, A.: Harvesting social images for bi-concept search. IEEE Trans. Multimedia 14(4), 1091–1104 (2012)

[7] Liu, D., Hua, X.-S., Yang, L., Wang, M., Zhang, H.-J.: Tag ranking. In: International Conference on World Wide Web (2009)

[8] Matusiak, K.: Towards user-centered indexing in digital image collections. OCLC Sys. and Services 22(4), 283–298 (2006)

[9] Metzler, D., Croft, B.: Linear feature-based models for information retrieval. Inf. Retr. 10(3), 257–274 (2007)

[10] Sun, A., Bhowmick, S.: Quantifying tag representativeness of visual content of social Images. In: ACM International Conference on Multimedia (2010)

[11] Wilkins, P.: An investigation into weighted data fusion for content-based multimedia information retrieval. PhD Thesis, Dublin City University (2009)

[12] Xu, J., Li, H.: AdaRank: a boosting algorithm for information retrieval. In: ACM SIGIR (2007)

[13] Zhu, S., Wang, G., Ngo, C.-W., Jiang, Y.-G.: On the sampling of web images for learning visual concept classifiers. In: ACM International Conference on Image and Video Retrieval (2010)

Mining Rules to Predict Anomalies in the Field of Insurance Industry from Unstructured Data Based on Data Mining

Shengfei Shi[*], Yue Wu, and Hao Zhang

Harbin Institute of Technology, Harbin, China
shengfei@hit.edu.cn, wuyuehit@gmail.com, zhanghaohit@126.com

Abstract. In order to help insurance industry to predict anomalies in new transactions through mining rules from a great deal of raw unstructured data, a complete, effective rule mining system is needed. In this paper, unstructured data is processed into feature vectors, which are then clustered. Clusters are used to construct a tree classifier, which contributes to reprocessing the feature vector and extracting anomaly rules. Besides, considering the weakness of a single process system, we adopt an iteration idea, in other words, we iterate the above steps for several times, thus guaranteeing the quality of the rules mined. Differently, we only focus on the information we need of much unstructured data, which avoid dealing with the whole unstructured data. Besides, combined with data mining, algorithm to extract unstructured data can achieve better effect.

Keywords: unstructured data, cluster, classifier, rule mining, insurance.

1 Introduction

At present, in the case of processing of unstructured data (documents), most of the research is mainly based on the extraction of the frequency of certain feature words, topic words, etc. [1-4] Sequence of feature words extracted by this method can represent the documents' subject to a certain extent, which has a very good effect for the management of unstructured data.

The procedure of current unstructured data processing method is as follows:

- Document pre-processing, including word segmentation, filtering stop words, word tagging, etc;
- Word frequency statistics;
- Feature words selected. It mainly bases on word frequency, the inverse text frequency, topic information;
- Feature words extracted from the document are then processed by the follow-up application, instead of the document.

This unstructured data processing method has the following deficiencies:

- The quality of feature words selected largely depends on the adequacy of the document preprocessing, in other words, if the original document is not fully and

---

[*] Corresponding author.

Z. Bao et al. (Eds.): WAIM 2012 Workshops, LNCS 7419, pp. 226–239, 2012.

properly preprocessed, thus leading to some bad words ignored and left, it will have great negative effect on the word frequency and feature words selected.

- Although sequence of feature words extracted from the unstructured document can represent the original document to some extent, it ignores the semantics and the integrity of the full document.
- Word with higher frequency does not mean higher possibility that it reflects the topic of the document, namely, that word is not necessarily a good feature word.
- Different application fields may have different needs for the processing of the unstructured data. So the method depending on the idea of "word list representing document" is not applicable in many application scenarios.

In the special research field of anomalies' rules mining in insurance industry, most of the materials about the anomaly cases are unstructured, which makes it more important to focus on the processing of unstructured data. It is not possible to discover rules about anomalies from unstructured data only relying on the above method. On the one hand, a rule cannot be described thought a sequence of feature words. On the other hand, not all the content of unstructured materials in insurance is relevant, in other words, much of the data in a document is irrelevant and can be ignored, which will overlook the important part and waste time if we process the unstructured data in the original way. So, considering the special application need in rule mining in insurance, we do some research and come up with a innovative method, which not only overcome the shortcoming of the current research but also apply to the field of rule mining in insurance. The general look of this paper is as follows.

Combined with experience from domain expert, the rule mining system can discover rules to predict anomalies in insurance automatically from large scale of unstructured data. In consideration of the special application, the attributes we need can be defined by domain experts in advance, thus restricting the set of useful attributes we want to extract from the original unstructured documents. The obvious benefit is that we do not need to care about the irrelevant information in the document and the quality of feature extracting does not depend on the preprocessing sufficiency. [5-8]

The general procedure is as follows:

- Preprocess the original document, including word segmentation, filtering, etc.
- Construct the feature vector according the defined set of attributes (or features) and extract all the attributes or features in the feature vector from the document preprocessed.
- Cluster according to the weighted feature vectors and generate k clusters, tagged with Ci respectively. Then each vector is added a class attribute with the value of corresponding Ci.
- Construct a tree classifier using the decision tree algorithm according the k clusters, thus generating the characteristics of each cluster.
- Reprocess the feature vector, deleting the bad attributes and adding some good attributes according to the tree classifier and some dictionaries.

- Repeat the step 2,3,4,5 until the number of iterations meets the need or the difference of two contiguous classifiers is small.
- Extract k rules from the final tree classifier, supposing that there are k clusters. The format of rule can be defined according to the following application.

The sketch map of the overall procedure described above can be expressed as follows:

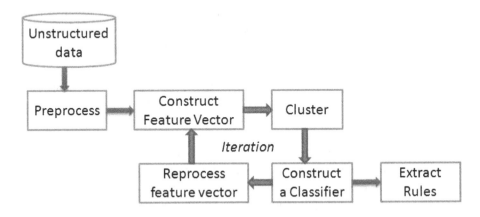

Fig. 1. The overall procedure of the rule mining system

2   Extracting Feature Vector from Document

2.1   Feature Vector

According to the experience in insurance industry, we can define the set of attributes or features needed. The original set of attributes includes two parts, the feature vector attributes and extension attributes.

Definition 1

The feature vector attributes are the attributes in the feature vector, which is the contemporary vector in the current iteration. We assign the set VA to feature vector attributes.

Definition 2

Extension attributes are the attributes for feature extension in the latter procedure. In fact, these attributes are stored in a dictionary and will be used in the "Reprocess the feature vector" section. We assign the set EA to extension attributes. EA mainly contains two kinds of attributes. One is decomposition attributes called by DA for short, and the other is induction attributes called by IA. So EA = $DA \cup IA$.

Definition 3

Decomposition attributes (DA) refers to pairs of attributes with a relation between a so-called source attribute and some more detailed attributes, where the source attribute can decompose more detailed attributes. For example, the attribute "time (2011.11.12 9:00)" can be decomposed into "date (2011.11.12)" and "moment (9:00)".

Definition 4

Induction attributes (IA) refers to a relation between a so-called source attribute and some other attributes, where the source attribute can induce some other attributes. For example, the attribute "time" can induce "season".

Just as a slight clarification, the feature vector attributes are in-use attributes during extracting features from document and clustering. Extension attributes are only used in the "Reprocess the feature vector" section.

2.2   Document Feature Extraction

As we focuses on the anomalies in insurance industry and the set of attributes we need has already be defined in 2.1 part, we are going to take advantage of the simple and direct way –retrieval and matching in the feature extraction step. Just as the name implies, "retrieval and matching" is attempting to extract feature or pattern we need directly from the original document, similar to the information retrieval.

The advantages of the "retrieval and matching" method:[9-12]

- Avoid processing the full document by ignoring the unnecessary content;
- Time complexity is proportional to the dimension, less influenced by the length of a document;
- Improve the speed of extracting features from document, with complex preprocessing and feature clustering;
- Avoid clustering words blindly, which wastes a lot of time.
- The disadvantages of this method:
- If the dimension of the feature vector is too high, it would be very complex to construct a matching pattern for each attributes or feature in the vector.
- The time complexity is quite high to algorithms of extracting same features frequently.
- In order to eliminate the above two disadvantages, we take the following two modifications:
- The conception of extension features restricts the dimension of the feature vector to a great extent;
- Simple preprocessing can improve the speed of iteration and not increase the complexity of document processing.

Document feature extraction contains 3 steps: preprocessing, matching pattern constructing, feature extracting.

Preprocessing. Preprocessing has two parts:

- Word segmentation.
  Segment the original unstructured document into words or phrases.
- Stop words filtering
  This part needs a dictionary of stop words. The process of filtering is, in fact, searching the dictionary and matching the word. If the word or phrase is matched, it will be filtered and deleted.

Definition 5

Stop words refer to the words that have no meaning and should be filtered, such as "is", "but", etc. [13]

Construction of Matching Pattern. Matching pattern is used for extracting feature from unstructured document. The matching pattern is not limited to regular expression, and each feature has its own matching pattern. It should be noticed that only features in the set of "feature vector attributes" need to be constructed patterns for.

For example, the feature or attribute "location" should have its own matching rule used during extracting the values of "location" from the unstructured document. We can define a presupposed list of places such as {Peking, London, Washington ...}. Thus, the process of matching "location" is a process of searching the place list. For another example, the matching pattern for "time" can be defined as a series of regular expressions, such as "/d/d/d/d","/d-/d-/d-/d", etc, thus making the values of feature "time" can be identified from the unstructured document.

Feature Extracted. By using the matching patterns constructed in section "Construction of matching pattern", we extract features and fill the feature vector from the document preprocessed in section "Preprocessing", which is actually a pattern matching process.

- Extract feature
  In this step, we only extract features or attributes in the set of "feature vector features". Considering the fact that the number of words in the document is much more than the dimension of feature vector, we adopt a parallel method to matching feature in the unstructured documents. In detail, every matching pattern tries to match the word, which begins from the first word in the document preprocessed. If it matches, then comes to the next word. Otherwise, every pattern will have a try until the last pattern.
- Fill the feature vector
  A single feature in the feature vector may have more than one value. For example, the feature "location" may have values such as "Peking, London, Washington", If so, to the feature vector, the value should be multiple-valued.

3   Clustering and Classifier Construction

3.1   Clustering

The algorithm of clustering we use is based on distance. In consideration of the rule mining research field, no limit is set on the number of clusters, which means the final number of clusters will automatically generated without depending on the manual setting. Besides, considering the specialty of the feature vector we define, the distance function and the representative of a cluster should be also chose particularly.[14-15]

Distance Function. The fact that every feature in the feature vector may be multi-valued should be taken into consideration in determining the distance functions. The distance functions are as follows:

$$D = (\sum_{i=1}^{m} d_i * w_i) / m$$

(1)

$$d_i = dn * wn + ds * ws + dp * wp$$

(2)

$$dn = | n_1 - n_2 | /(n_1 + n_2)$$

(3)

$$ds = \sum_{j=0}^{j=min(sum(fea1),sum(fea2))} similar(fea1_j, fea2_j)$$

(4)

$$similar(fea1_j, fea2_j) = \begin{cases} 1, fea1_j = fea2_j \\ 0, fea1_j \neq fea2_j \end{cases}$$

(5)

In this, D represents the distance of two feature vectors; d_i represents the distance between the respective first i attributes from two feature vectors; m is the dimension of feature vector, w_i is the adjusting factor of the first i attribute. dn, ds and dp are the number distance, similarity distance and particular distance respectively. dn and ds have their uniform functions, but dp should be set according to particular attribute. Besides, wn, ws and wp are the adjusting factors of dn, ds and dp. n_1 and n_2 are the numbers of the two feature vectors to calculate. $fea1_j$ and $fea2_j$ are the values of the respective first i attributes of the two feature vectors to calculate.

Just as a slight clarification, w, wn, ws and wp are set manually according to some experience and experiment records. What's more, the adjusting factors are not limited to single value, namely, they can be multi-valued, which means that in different generation of iteration, the adjusting factors can have corresponding values according to the predefined configuration, aiming at focusing on different object in different generation.

Clustering. The algorithm adopted in this paper is based on distance without limit on the number of clusters. The distance function has been defined in (1)-(5). The particularity of research in this paper results in the particularity of the feature vector. The traditional clustering algorithms, such as K-MEANS, K-MEDOIDS, which select a representative from a cluster, are not applicable. [16-19] So, we come up with a new concept – average distance to resolve the problem. The algorithm is as follows:

- Given a feature vector set, $V = \{V(1), V(2),...,V(n)\}$, $\forall v \in V$, v is initially regarded as a cluster. Let cluster set $C = \{C1, C2,.., Cn\}$, in this $C1 = \{V(1)\}$, $C2=\{V(2)\}$, ..., $Cn=\{V(n)\}$.
- $\forall ci, cj \in C$, calculate the average distance $\overline{D}(ci, cj)$.Choose the two clusters Ci, Cj with the max "average distance" into one cluster Cij, thus the cluster C becoming $\{C1, C2, ..., Ci-1, Cij, Ci+1, ..., Cj-1, Cj+1, ..., Cn\}$.
- Repeat the second step until |C| < C_min (the preset minimum cluster threshold) or the max "average distance" < D_min (the preset minimum average distance threshold).

Definition 6

Average distance is the average of all the distances between two arbitrary feature vectors in different clusters. Only the vectors in different clusters need to be calculated their distance. The function expression is as follow:

$$\overline{D} = \sum_{\substack{Cii \in Ci \\ Cjj \in Cj}} D(Cii, Cjj) / count(Ci) * count(Cj) \tag{6}$$

In this, D() is the distance function (1), and count() represents the number of feature vectors in a cluster.

3.2   Classifier Construction

Supposing that we get k clusters after clustering, we regard the k clusters as k classes, each of which is tagged with Ci. Thus each feature vector is added with a new feature, the class feature with its corresponding value Ci. So the set of feature vectors can be seen as a training set, used to construct a tree classifier in a decision tree way. So we can get the characteristics of cluster Ci, in other words, what kind of documents can fall into cluster Ci. Detailed describe of the algorithm will be introduced below.[20-21]

In this, we adopt the existing algorithm – ID3 to construct the tree classifier: [22]

- Initiate the clusters C. For each feature vector V in a cluster Ci ($Ci \in C$), it is added a class attribute with its corresponding Ci.
- Feature selection.
  Choose a feature A from m+1 features(m original features in the vector and 1 class feature added) in a feature vector to make the gain ratio maximum Gain(S, A). In this, S is the set of feature vectors. Then the feature A is used as a node, which classify the feature vectors into different parts as different subtrees.

- Repeat the second step until the feature vectors in a leaf node have the same class feature (or allow some few differences). In other words, for each feature vector v in the set of vectors in a leaf node, their class label (attribute) is the same.

In this, the emphasis of the algorithm is the functions of entropy and gain ration, which measure the effect of a feature to the classifier.
   The entropy function is as follow:

$$Entropy(S) = -\sum_{i=1}^{m} P(u_i) \log P(u_i) \qquad (7)$$

In this, S is the set of feature vectors, P(ui) is the probability that a feature vector belongs to Class Ci :

$$P(u_i) = \frac{|u_i|}{|S|} \qquad (8)$$

The function of gain ratio is as follow:

$$Gain(S, A) = Entropy(S) - \sum_{v \in Value(A)} \frac{|S_v|}{|S|} Entropy(S_v) \qquad (9)$$

In this, A represents a feature or an attribute. Value(A) is the set of possible values of feature A. v is one of the values of A and S_v is part of S which has A feature with value v. $|S_v|$ is the number of feature vectors in S_v.

4   Reprocessing the Feature Vector

Reprocessing the feature vector is to redefine the set of vector features based on the classifier, including four key points as follows:

- The attributes in the set of feature vector attributes, which, however, are not used in classifiers, should be deleted;
- Merge the paths, which have no branches in the decision tree, into one node. That is to say, merge several attributes into one.
- The Top-level nodes in the decision tree represent more important attributes, which should be refined or decomposed by searching the set of extension features.
- The Top-level nodes in the in decision tree represent more important attributes, which can induce other good features by searching the set of extension features.

A new set of feature vector attributes can be redefined by the steps above-mentioned and will be used in the next iteration cycle.

4.1   Iteration and Rule Extraction

After the reprocessing of the feature vector, we should extract features from the documents to be processed and reconstruct feature vectors for each document. The attributes which have been extracted in the last generation can be used directly, while the newly added attributes in step 4 (reprocessing the feature vector) should be extracted again. Then cluster, construct classifiers, reprocess feature vectors, etc. Go on the iteration until it reaches the iteration times set in advance or there are few differences between the new classifier and that in the last generation. The method about the classifier difference will be introduced in 5.1 section.

4.2   Classifier Difference Analysis

Once a new classifier is constructed, it needs to be compared with the last tree classifier generated in the last iteration or generation. This process operates in a recursive manner. The following describes the algorithm used to compare two tree classifiers, TC1 and TC2:

- Let the queue Q1 contain only the top node n1 of TC1 and let the queue Q2 contain only the top node of n2 of TC2;
- Initiate the difference degree to be 0;
- If the queues Q1 and Q2 are both empty, then terminate;
- If either of the two queues is empty while the other is not, the difference degree is added by a constant C1;
- Take the element n1 from queue Q1 and the element n2 from Q2;
- If one of n1, n2 is a leaf node, while the other is not, the difference degree is added by a constant C2, and go to Step 3;
- Compare gain ration for n1 and n2, and the difference degree is added by this difference value;
- If n1 and n2 are both non-leaf nodes, insert the children of n1 at the end of Q1 and the children of n2 at the end of Q2;
- Go to Step 3;

After the above steps, a difference degree will be calculated. If the difference degree is large than the given threshold, the iteration will go on. Otherwise, it will be terminated and proceeds to "Rule Extraction" step.

4.3   Rule Extraction

After the tree classifier is finally constructed, we need to extract the rules related with insurance anomaly from the classifier. Rules can be extracted based on the cluster tag or the leaf node of the tree classifier. If it is based on the cluster tag, one rule can be generated from a cluster. Otherwise, if it is based on the leaf node, several rules can be extracted from the same clusters. The expression of these rules should be determined by the specific follow-up applications and will not be introduced in this paper.

5   Experiment

In this part, we have tested the results of the system from unstructured data to feature vectors. It shows that the effectiveness of the algorithm used in the paper is quite good.

In the experiment, all we need are as follows:

- The original unstructured data, namely the insurance fraud cases(can be get from internet);
- The tool of segmenting words (can be get from internet);
- Filtering system;
- Matching patterns for each feature to extract;
- The extracting program;

In order to improve the matching accuracy and the actual feasibility, we narrow the insurance field to auto insurance in the experiment. To other subfields of insurance, what we need to do is just adding matching patterns in these fields.

The important part is the construction of matching patterns for each feature. Aiming just to test the feasibility and effectiveness, we initially construct matching patterns for only ten important features, which are time, place, driven age, time claimed before, used age, police recognition protocol, relation between the beneficiary and the driver, repair shop, owned age, insurance company.

Take a sample unstructured document about an auto insurance fraud case for example. The case is a report about an auto insurance fraud case in Jiangsu Province, China.

Through segmenting, filtering, searching features and filling the feature vector, the feature vector can be expressed as the following figure:

Table 1. Feature Vector

Feature	Value1	Value2	Value3	Value4
Time	20050920	1:50	20050919	1:48
Place	Yudong School Haimen City	County road		
Drive_age	2.5			
Owned_age	0.1			
Used_age	5			
Before_claim	3			
Police_protocol	0			
Relation	friends			
Repair	4S			
Insurance_company	PICC Property and Casualty Company Limited			

Table.1 shows that there are 10 features or attributes in the feature vector and the values of each feature may be a list instead of a single value. The feature vectors extracted from unstructured documents will be stored in a file, which will be used in the subsequent clustering step.

So as to the time cost of the feature extraction algorithm, the following three figures, Fig2, Fig3 and Fig4, describe the time consumed in three aspects, namely, number of features in the feature vector(VA) , lines and words of a document, respectively. Fig 5 describes the time cost graph of the traditional algorithm –TFIDF feature vector construction. Compared with traditional algorithm of processing unstructured documents, the algorithm in this paper is much better in time cost.

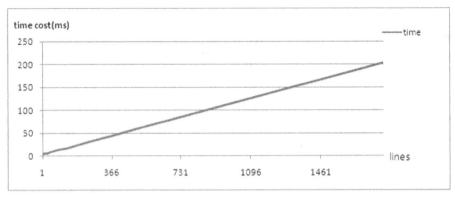

Fig. 2. Line chart describing the relation between time cost and lines of a document (10 features of a vector)

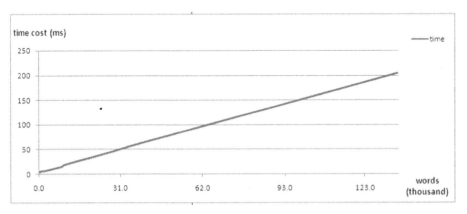

Fig. 3. Line chart describing the relation between time cost and words of a document (10 features of a vector)

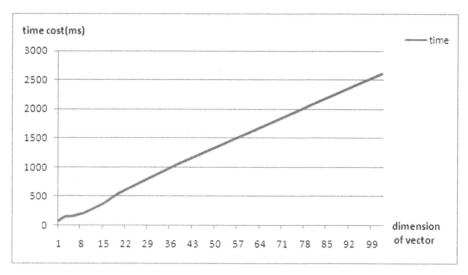

Fig. 4. Line chart describing the relation between time cost and dimension of a feature vector (document with 1792 lines, 135.3 thousand words – much more than normal)

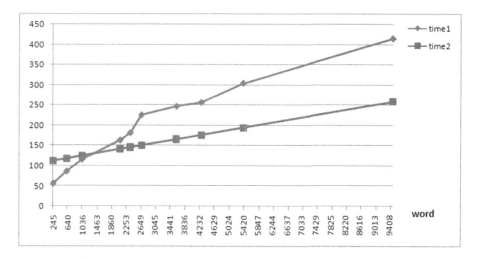

Fig. 5. Line chart describing the relation between time cost and words of a document for traditional algorithm (time1) and algorithm present in this paper (time2)

Just as the above four charts shows, the time cost grows with the increase of the scale of the unstructured document, such as the increase of words or lines of the document. But, just as figure 3 and figure 4 show, it takes only about 200 milliseconds to extract features from a document with 1792 lines and 135.3 thousand words, whose scale is much larger than normal document's. What's more, we also pay attention to the effect that the dimension of feature vector plays on the time cost. Figure 5 shows that the time cost is proportional to the dimension of feature vector. Even so, we do not worry about the bad effect. On the one hand, the dimension of feature vector is controlled as a result of the particular application field restricted and the subsequent process methods. On the other hand, as figure 5 shows, it takes no more than 3 seconds to extract a feature vector with more than 100 dimensions from a document with 135.3 thousand words in a normal PC computer with low allocation, which means that it may only takes less than 1 second to process a normal document. Fig 6 describes a comparison in time cost between traditional algorithm of processing unstructured data using TFIDF and the algorithm used in this paper. Obviously, our algorithm of extracting features from unstructured data is superior to the traditional algorithm in the aspect of time cost. The time cost (time2 line in the chart)of our algorithm present a linear growth with a low growing coefficient, while time cost (time1 line in the chart) of the traditional algorithm presents a partial exponential growth and partial linear growth with a relatively high coefficient.

6   Conclusion and Future Work

The rule mining system for predicting anomalies in insurance industry based on unstructured data also need some future work to improve its effectiveness and practical applicability. If we want to apply effectively, we also need some experts in insurance, who can help us define the features needed and assist us construct the matching patterns. As to the clustering, construction of classifier and iteration, some more experiments are also needed, which depend on the quality of feature vectors.

References

[1]   Jing, T.: The research on the structural model and algorithm of web text mining system based on inherent mechanism of knowledge discovery system. Beijing University of Science and Technology (2003)

[2]   Ronen, F., James, S.: The Text Mining Handbook: Advanced Approaches in Analyzing Unstructured Data, vol. 1. Cambridge University Press (2006)

[3]   Weiss, S., Indurkhya, N., Zhang, T., Damerau, F.: Text Mining: Predictive Methods for Analyzing Unstructured Information. Springer (2004)

[4]   Kao, A., Poteet, S.: Text Mining and Natural Language Processing-Introduction for the Special Issue. SIGKDD Explorations 7(1), 1–3 (2004)

[5]   Changkai, D.: Spatial data mining and knowledge discovery, pp. 20–23. Wuhan University Publishing House, Wuhan (2001)

[6]   Yang, B.: A driving force of knowledge discovery in database main stream - double bases cooperating mechanism. In: IC-AI 2002, Las Vegas, USA (2002)

[7]   Jiawei, H., Micheline, K.: Data Mining: Concepts and Techniques, vol. 2. Morgan Kaufmann Publisher (2006)

[8]   Gharehchopogh, F.S.: Approach and Developing Data Mining Method for Spatial Applications. In: Proceedings of International Conference on Intelligent Systems & Data Processing (ICISD), India, pp. 342–345 (2011)

[9]   Gupta, V.: A Survey of Text Mining Techniques and Applications. Journal of Emerging Technologies in Web Intelligence 1(1), 60–76 (2009)

[10]   Berry Michael, W.: Automatic Discovery of Similar Words, pp. 24–43. Springer, USA (2004)

[11]   Baeza-Yates, R.: Challenges in the Interaction of Information Retrieval and Natural Language Processing. In: Gelbukh, A. (ed.) CICLing 2004. LNCS, vol. 2945, pp. 445–456. Springer, Heidelberg (2004)

[12]   Mugunthadevi, K., Punitha, S.C., Punithavalli, M.: Survey on Feature Selection in Document Clustering. International Journal on Computer Science and Engineering 3(3), 1240–1241 (2011)

[13]   Popowich, F.: Using Text Mining and Natural Language Processing for Health Care Claims Processing. SIGKDD Explorations 7(1), 59–66

[14]   Cutting, D.R., Karger, D.R., Pedersen, J.O., Tukey, J.W.: Scatter/Gather: A Cluster-based Approach to Browsing Large Document Collections. In: Fifteenth Annual International ACM SIGIR Conference, pp. 318–329 (June 1992)

[15]   Jain, A.K., Dubes, R.C.: Algorithms for Clustering Data. Prentice Hall, Englewood Cliffs (1988)

[16]   Liu, X., He, P.-L.: A Study on Text Clustering Algorithms Based on Frequent Term Sets. In: Li, X., Wang, S., Dong, Z.Y. (eds.) ADMA 2005. LNCS (LNAI), vol. 3584, pp. 347–354. Springer, Heidelberg (2005)

[17]   Fung, B.C.M., Wang, K., Ester, M.: Hierarchical Document Clustering. In: Encyclopedia of Data Warehousing and Mining, pp. 555–559 (2005),
doi: 10.4018/978-1-59140-557-3.ch105

[18]   Hung, C., Xiaotie, D.: Efficient Phrase-Based Document Similarity for Clustering. IEEE Transaction on Knowledge and Data Engineering 20, 1217–1229 (2008)

[19]   Jing, L.P.: Survey of Text Clustering, pp. 3–4. The University of Hong Kong, Hong Kong (2005)

[20]   Shiqun, Y., Yuhui, Q., Jike, G., Fang, W.: A Chinese Text Classification Approach Based on Semantic Web. In: Fourth International Conference on Semantics Knowledge and Grid, pp. 497–498 (2008)

[21]   Shiqun, Y., Gang, W., Yuhui, Q., Weiqun, Z.: Research and Implement of Classification Algorithm on Web Text Mining. In: 2007 IEEE Third International Conference on Semantics Knowledge and Grid, pp. 446–449 (2007)

[22]   Decision Tree Algorithm,
http://wenku.baidu.com/view/3a0ad06c1eb91a37f1115c4c.html

A Classification Framework for Similar Music Search[*]

Jing Zeng[1], Zhenying He[1], Wei Wang[1], and Hai Huang[2]

[1] School of Computer Science, Fudan University, Shanghai, China, 201203
{11210240038,zhenying,weiwang1}@fudan.edu.cn
[2] Cultural Institutions Security Branch of Shanghai Public Security Burea,
Shanghai, China, 200000
huanghai@163.com

Abstract. This paper has concentrated on how to retrieve a list of songs from music database similar to the specific one. Content-based retrieval of music is one of the most popular research subjects, which mostly focuses on querying the exactly one from database by humming a tune or submitting a recording of music. However, getting some songs similar to, but not exactly the given one could be also interested by people. In this paper, we propose a classification framework to solve this problem using string-based methods. Introducing string-based similarity measure, our framework has lower computational complexity and better effect. We also developed a new distributed clustering algorithm under MapReduce framework, which performed well for massive audio data. Experiments are performed and analyzed to show the efficiency and the effectiveness of our proposed framework.

1 Introduction

As the speed and capacity of computers and networks increase, more and more songs and other music data are stored in databases. Since rapid access of such data is also strongly desired for large music databases, traditional information retrieval techniques can be applied to these data as well. However, the applications of music data will be limited if the database lacks appropriate ways to manage the data. Therefore, many research works were done on the content-based retrieval of music data in recent years, such as querying by rhythm, querying by humming (QBH), etc.

All the work mentioned above aims at searching a desired piece of music by a record of song or humming, which is very useful when you want to find a song from music library but forget its title or artist. It could be the most important purpose for people to use music databases and thus a traditional requirement as well. But in some cases, one may just want to know all the similar songs sharing the given piece of tune, not just the exact one containing the given record. First of all, one would be interested in different versions of the same song retrieved by a given query. Secondly, it is

---

[*] National Natural Science Foundation of China under Grant No.61170007; National Major Research Plan of Chinese Infrastructure Software under Grant No.2010ZX01042-002-003-004.

Z. Bao et al. (Eds.): WAIM 2012 Workshops, LNCS 7419, pp. 240–251, 2012.

useful to get a result set involving different songs, since all the results from one query may be similar to each other and perhaps share the same music style. Researchers then could go into some deeper work with these query results. Thirdly, through this, it would be an easy job to judge whether the author of a given song have plagiarized others to make his work.

It is natural to consider this new problem as classifying songs in databases relevant to the given query or not. Simply, we can achieve the goal by directly compare the whole content of the given query and that of all the songs in the database. Although useful, it is not realistic to do such comparison for each query since its high computational complexity would call for expensive response time. Thus, we turn our thoughts to compare their features to decide which class the given query belongs to. In our framework, features of the songs in databases can be extracted offline. Then an offline clustering process divides these features into numbers of classes, followed by an indexing step, which is also offline. Given a query with a record of one song or a piece of humming, we can extract its feature in a short time. The real work then is to do, which will compare the query feature with prior features in databases to classify the given query and return the songs in the same class as a result set. It is obvious that using our framework, time cost is dramatically reduced since most of work is done offline.

Although many meaningful feelings can be comprehended by human beings when listening to a song or a piece of music, however, it is very difficult to automatically extract these feelings from an audio object. In fact, these meaningful feelings come from rhythm, pitch interval, pitch contour and some other content features, except for some descriptive attributes such as name, file format, or sampling rate. Considering most of these music content features, our feature extraction method is to transform a song into a multi-dimensional vector representing different features in each dimension. In addition to completeness, time-saving should be the other important characteristic of our feature extraction method, since this job is not only involved in offline work, but also an essential step when a query comes, which expects the response time to be as short as possible.

The following clustering algorithm is another challenge in our framework. Since music data maintains so much melody information, each feature vector extracted at former step has dozens of dimensions. Our clustering algorithm should be able to deal with high-dimensional calculation first. Furthermore, this clustering algorithm would be incremental for starting over the clustering process is not efficient when new songs are imported into databases. In this paper, we propose a modified K-Means algorithm for clustering high-dimensional vector on this problem.

The remainder of the paper is organized as follows. Related work is summarized in Section 2. Then we formulate our problem and introduce our framework in Section 3. In Section 4 we introduce some feature extraction methods we used. In Section 5 we study the clustering process, and the classifying and indexing process are introduced in Section 6. In Section 7 we introduce our online retrieval procedure. Experimental results are presented in Section 8. Finally, concluding remarks are made in Section 9.

2   Related Work

There are several kinds of previous work on music query. In some works, songs are searched by some descriptions attached to them, such as file names or keywords excerpted from lyrics, the words of a song. But nowadays content-based method has become one of the most popular research subjects. Some early works on content-based audio or music retrieval are totally based on signal processing and using acoustic similarity, such as in [1,2]. In [3], music query problem is transformed into the substring matching problem by extracting thematic feature strings from original music objects. Another popular content-based retrieval method is query by humming (QBH). Most of recent works on QBH is focused on melody representations, similarity measures and matching processing. In some works, such as [4], only pitch contour is used to represent melody. When rhythm and pitch interval is considered, more complex similarity measure and matching algorithm should be used. [5] uses two-dimensional augmented suffix tree to search the desired song. In [2], a new melody representation and new hierarchical matching method are proposed.

Retrieving a specified singer's music recordings from an unlabeled database by submitting a record of music as a query to the system is similar to our task. But previous work only used the characteristics of the singer's voice from a music recording. In this paper we almost make full use of music features to retrieve similar result set of songs based on their content, which we consider as a novel study.

Music clustering has been studied for a long time. The main purpose is to solve the automatic genre classification problem. Some work, as [6], studies the problem of identifying similar artists by classifying songs in databases. Contrary to this, [7] processes clustering of popular music recordings based on singer voice characteristics. [8] uses additional information to help clustering, one making use of the advice and recommendations posted by humans, while the other considering the background knowledge and the user access patterns. To automatically organize a collection of music files according to their musical genre and sound characteristics, [9] propose a two-stage clustering procedure grouping music records according to their similarity, followed by a clustering of compositions according to the record similarities.

Our work is quite different from these works. First, the clustering procedure is just a part of our framework, though indispensable. It provides intermediate results, which is to help classify songs sharing the same feature into the same cluster, instead of return final results directly. Second, our clustering method is applied in high-dimensional features, and other constraints besides music content itself are not involved.

3   The Classification Framework

There could be a set of songs sharing the same feature in a music database. People submitting a record or humming query might want to retrieve all these songs instead of getting the exact one maintain the given feature. To satisfy this kind of information

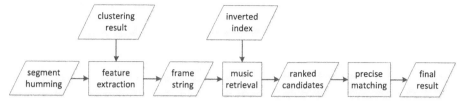

Fig. 1. Classification Framework

need, it is necessary to find out songs related to the given query based on their features in a time rapidly and accurately. We summarize the problem as follows.

Formally, given a record of song or that of humming as a query, we wish to automatically retrieve the relevant features similar to the feature extracted from the given query. Songs maintaining one or more relevant features could be a desired result set to users. To approach this problem, we model it in a classification framework. The framework is shown in Figure 1.

The framework works as follows. For clustering, features of all the songs in the database are extracted, represented as multi-dimensional vectors maintaining most of the melody information of each frame of songs, such as rhythm, pitch, tune, etc. The following clustering process is a modified K-Means algorithm, which automatically classify the extracted features into different classes after several iterators, with some original points as input. The clustering results are something like document data, each tuple containing a song ID and class numbers of all the frames in the song. Thus inverted indexes are generated from these clustering results, considering it as a document retrieval problem. At this point, offline part of work is done in our framework. Given a query, feature extraction is applied on it first. The extracted feature then is classified into the class which it most likely belongs to, based on prior classes resulted from prior clustering process. Using features most related to the given one within the same class, the result set of songs sharing the same feature and relevant to each other is returned, along with their ranking information according their feature similarity.

4   Feature Extractions

Many feature extraction methods have been studied in audio data analysis. Noise reduction is the process of removing noise from a signal. All recording devices, either analogue or digital, have traits which make them susceptible to noise. To achieve noise reduction, what features and how to extract them from music have been discussed a lot in previous work. In addition to noise reduction, features maintaining melody information such as tune or rhythm are also essential to extract for music type classification. Using all these feature extraction methods together to transform a song into a feature vector is very helpful for classifying music data within our framework. In this section, we briefly introduce the features we have used in our framework, and the procedure of transforming a song into a feature vector is shown in Figure 2.

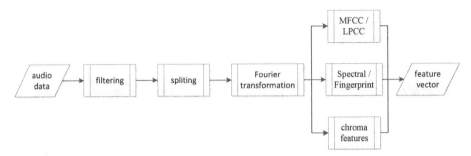

Fig. 2. Feature Extractions

- Spectral Characteristics[10]
  Voice spectrums of human and instruments are generally steady. In other words, each musical note has fundamental frequency. Effective audio data thus could be extracted by using Fourier transformation in noise elimination.
- Mel-Frequency Cepstral Coefficients (MFCC)[11]
  Mel-frequency cepstral coefficients (MFCCs) are one kind of stored templates used as voiceprints in speaker recognition. MFCCs are increasingly used in music information retrieval such as genre classification, audio similarity measures, etc.
- Linear Predictive Cepstral Coefficient (LPCC)[12]
  Since a specific speech sample at the current time can be approximated as a linear combination of past speech samples, linear predictive (LP) coefficients are useful in estimating the basic parameters of speech but typical show high variance. LP cepstral coefficients (LPCCs) are much more robust.
- Chroma Features[13]
  Knowing the distribution of chroma, which represents the 12 distinct semitones of the musical octave, can give useful musical information about the audio. Chroma features are almost entirely independent of the spectral features.
- Fingerprint Characteristics[14]
  In content-based audio retrieval, an audio fingerprint is a compact content-based signature that summarizes an audio recording. Fingerprint characteristics allow the identification of audio to be independent of its format, without the need of meta-data or watermark embedding.

5   Clustering

A feature vector of each frame of every song extracted at former step is represented as $v(v_1, v_2, ..., v_N)$, where N is the number of features we decide to extract from a frame at the very start. Thus, after the feature extraction process, a vector set V of features is generated, where $|V|$ can be the total number of frames in the music database.

To process our clustering analysis on these feature vectors efficiently, we have to consider several factors.

- **Size of a feature vector could be large.** Because of our combining use of several feature extraction methods on one frame, each feature vector would has dozens of elements. Our clustering algorithm should be able to deal with such data first.
- **Number of elements in the vector set is amazing.** In a music database, it is normal to store tens of thousands of songs in it. Each song could have thousands of frames, and tens of features could be extracted from each frame. Clustering analysis on so much data can be challenging.
- **Isolated points might exist.** It is unavoidable that there are some isolated feature vectors in the set since it is such a staggering set. To achieve higher accuracy, the clustering algorithm should not be sensitive to the isolated points.

As massive data sets are usually stored on distributed systems, our clustering algorithm should be designed under distributed framework. Hadoop, a software framework that implements a computational paradigm named MapReduce, enables applications developed using it to work with thousands of computational independent computers and petabytes of data. Based on this framework, researchers have done much work in the past few years and many applications have been developed. Mahout is one of them. It is an Apache project to produce free implementations of distributed or otherwise scalable machine learning algorithms on the Hadoop platform. And K-means, which is an efficient method of cluster analysis aiming to partition n observations into k clusters in which each observation belongs to the cluster with the nearest mean, is implemented in Mahout. However, K-means algorithm is not suitable for our framework due to following reasons.

- The implement of K-means in Mahout does not fully support high-dimensional data computing, which leads to bad performance in our framework with the fact that feature vectors are high-dimensional data.
- K-means algorithm uses the average value of objects as cluster centers which are abstract points, not the concrete objects. In addition, the K-means algorithm is sensitive to the isolated point.

Algorithm 1. map (*key*, *value*)

Input: Global variable centers, the offset key, the sample value
Output: <*key'*, *value'*> pair, where the *key'* is the index of the closest center point and *value'* is a string comprise of sample information
1 Construct the sample instance from value;
2 *minDis* = Double.MAX;
3 *index* = -1;
4 *for* i = 0 *to* centers.length *do*
5   dis = ComputeDist(*instance*, *centers[i]*);
6   *if* dis < *minDis* *then*
7   *minDis* = dis;
8   *index* = i;
9 *key'* = *index*;
10 Construct *value'* as a string comprise of the values of different dimensions;
11 output < *key'*, *value'*> pair;

Compared to K-means, K-medoids algorithm uses the actual objects as cluster centers, as well as solute the isolated point question of K-means algorithm and enormously enhances the precision of clustering. Based on above observations, we are to implement a modified K-medoids algorithm using the Hadoop framework to process our clustering procedure.

There are two main steps in our K-medoids algorithm. First, initialization process is done to select some feature vectors as the medoids. Then it comes to the iterative clustering work. Repeatedly, each feature vector is associated to the closest medoid, each time forming new clusters and new medoids as well, until there is no change in the medoids.

To deal with massive data in clustering, we implement the K-medoids algorithm on the Hadoop platform. First, in the initialization process, K original center points are randomly chosen from the feature vectors. Then we come to the real iterative clustering work. In iteration, two main functions are included. The **map** function performs the procedure of assigning each feature vector to the closest center while the **reduce** function performs the procedure of updating the new centers.

On the Hadoop platform, the input dataset is stored on HDFS as a sequence file of *<key, value>* pairs, each of which represents a record in the dataset. In our framework, the key could be the index of a frame the feature vector representing, and the value is the content of this vector, in the string. The map function takes two inputs. One is a *<key, value>* pair as a sample point. The other is the original center points, also in a <key, value> form, with the index of a cluster as a key and the center information as a value. Given these, the function computes the closest center point for the sample, then output a new <key', value'> pair, where key' is the index of a cluster, and value' is the sample information. The pseudo-code of map function is shown in Algorithm 1.

The input of the reduce function is the data obtained from map functions, which is also <key, value> pairs. The key here is the index of a cluster, while the value is the list of samples belongs to this cluster. Formally, we represent the input as <key, V> pairs. With such input, the reduce function would find the median of the samples in V, then output a new <key', value'> pair whose key' is the index of the cluster and value' is the median of the samples. The output then is treated as new center points and is used for next iteration. Algorithm 2 shows the pseudo-code of reduce function.

In Algorithm 2, FindMedian(A, M) is used to return the number in A which is the M-th in its size. Its expected running time is linear because each recursive call takes on the average linear time, and each recursive call reduces the size of the problem, by a constant factor.

Algorithm 2. reduce (*key, V*)

    Input: *key* is the index of the cluster, *V* is the list of the partial sums from different host

  Output: < *key', value'*> pair, where the *key'* is the index of the cluster, *value'* is a string representing the new center

1 *center* = FindMedian(*V, | V | / 2*);

2 *key' = center*;

3 Construct *value'* as a string comprise of the center's coordinates;

4 output < *key', value'*> pair;

The MapReduce job iterates until the new center points are nearly the former center points in last iteration. Then we successfully divide the feature vectors into K clusters, knowing the center point of each cluster.

6   Indexing

With the fact that a song is made up of many frames, we can also use a vector to represent a song. Based on clustering results, each frame of a song can be a number, indicating which cluster it belongs to. In this way, a song can be represented as a list of numbers, which can be linked to form a string.

As songs are represented by strings through above method, inverted index could be efficient for our music retrieval, which is now something like document retrieval. The inverted index in our framework now is not full inverted index, which means that index does not contain any position information. It is reasonable because size of index with position would be too large, and position information has been proved to be not so important win our framework. So that the inverted index data structure is made up of two parts. The key is a cluster ID which a frame belongs to, while the value is a list of IDs of songs containing this frame.

Although inverted index have much smaller size than original music data, efficient storage and rapid access of index are still an important problem in our framework. To achieve this goal in a distributed system, we introduce two speed-up technologies: Cassandra and Memcached.

Cassandra is a highly scalable, eventually consistent, distributed, structured key-value store. It brings together the distributed systems technologies from Dynamo and the data model from Google's BigTable. Inverted index in our framework is first stored in Cassandra. Using this technology, the whole index can consistently exist in distributed systems. Meanwhile, the access of index is proved to be of high-speed and high-accuracy.

Memcached is a general-purpose distributed memory caching system. It is often used to cache data and objects in RAM to reduce the number of times an external data source (such as a database or API) must be read. In order to speed up index access for common queries, the latest visited index in Cassandra will be inserted into Memcached. This step is taken at the same time of the query process. And it leads to a significant speed-up for repeated queries.

7   Online Retrieval

With all the offline jobs above completed, online retrieval is naturally designed as Figure 3 shows.

It takes roughly two steps to accomplish the online retrieval. First, based on the clustering results, the record or the humming of a song as a query is transformed to a string after classifying each frame of the query. Taken this string and the inverted index as input, ranked candidate songs are retrieved.

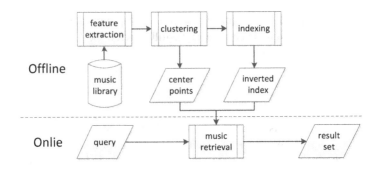

Fig. 3. Online Retrieval Procedure

Given ranked candidates, precise matching is processed on them. Most of the string searching algorithm can be used in this process under our framework. And edit distance, which is a common string metric for measuring the amount of difference between two sequences, is proved to be also efficient in this process. After this process, the final result set of songs sharing the same feature of the query is returned.

8   Experiments

We next present an experiment study of our classification framework. We will first introduce our experimental setting, and then discuss the experimental results.

8.1   Experimental Setting

Based on the framework, we implemented a prototype system. In this system, the feature vector of each song has 15 dimensions, implying that 5 feature extraction methods are applied to one song. The modified K-medoids algorithm and classifying indexing method are implemented. The inverted index is stored in Cassandra, while Memcached is also used. At online retrieval step, Levenshtein distance is our measure of the similarity between two strings. The result set of this system is a top-10 list of songs sharing of the feature of the query.

Our database consists of about 100,000 MP3 songs downloaded from the Internet, in which most of them are pop songs and the rest are classical and folk music. 20,000 songs of them were chosen to be a test set, while 57 pop songs a test query set. The recordings of these songs then were used as queries. We had 2 kinds of recordings, some were 10-second long, and others were 30-second long as a comparison.

The experiments were run on a cluster of 14 computers, each of which has two Pentium(R) Dual-Core (2.70GHz) CPU E5400 and 4GB of memory, using Linux. Hadoop version 0.20.3 and JAVA 1.6.0_15 are used as the MapReduce system for all experiments.

8.2   Experimental Results

Exp-1: Accuracy. In the first set of experiments, we evaluated the accuracy of our music retrieval framework using real data. We also investigated the impact of length of queries on our framework, by comparing the accuracy of queries of two different lengths.

As mentioned, queries we used are records with length of 10-second and 30-second respectively, each of which including 57 queries. For each query, we considered only top-10 results returned by our music retrieval prototype system. Figure 4(a) measures the accuracy by the percentage of queries returning the accurate result at top-1 among the whole test set. It indicates how many queries can retrieve the correct song at a given ranking level. From Figure 4(a), it is shown that, for about 93% 10-second queries, the correct song can be listed among the first 2 matches; and among the first 5 matches for 96% queries. It can be also seen that, for nearly 90% 10-second queries and all 30-second queries, their corresponding correct song can be retrieved as the first match!

In the second experiment, 10 queries with length of 30-second each were chosen. We evaluated the top-10 result sets returned by our music retrieval prototype system using Normalized DCG (NDCG) measure. In Figure 4(b), nDCG values at each rank position are shown, with the maximum one, minimum one and the median among 10 queries respectively. From Figure 4(b) we can tell that the perfect result can be always retrieved at position 1, just as Figure 3(a) indicates. And nDCG values of top-10 results stay greater than 0.70 and the median value stay no less than 0.85. But we can see that the range of nDCG values becomes larger as the rank position increases and drops a little as well.

Exp-2: Time Efficiency. We evaluated the time-efficient music retrieval step using real data.

Fixing the size of each query to 5KB, we varied the number of queries from 5 to 45. The result of execution time is reported in Figure 5(a), in which y-axis indicates various execution time with their values ranged from 0 to 120s. The results in Figure 5(a) show the running time with respect to the number of queries when the size of each query remains the same. We can clearly see a linear increase in the curve which indicates that query time correlate with the number of queries.

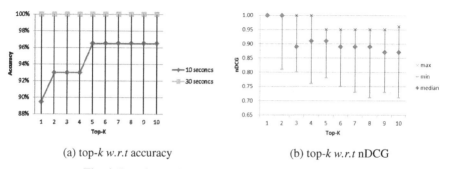

(a) top-k w.r.t accuracy   (b) top-k w.r.t nDCG

Fig. 4. Experimental Results on Music Retrieval Accuracy

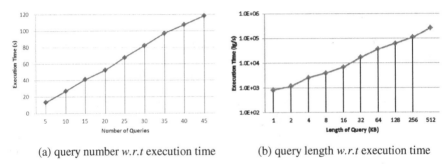

(a) query number *w.r.t* execution time   (b) query length *w.r.t* execution time

Fig. 5. Experimental Results on Time Efficiency

On the other hand, varying the length of query from 1KB to 512 KB, Figure 5(b) indicates the query time with respect to the size of query when the number of query is the same. Only one query was submitted each time in this experiment. We use the logarithmic scale on the execution time (*y*-axis). Figure 5(b) provides a clear look that the query time has a linear growth with the length of query increasing.

Thus this set of experiments help to conclude that the expansion ratio of our music retrieval prototype system is linear.

Exp-3: Memcached *vs.* Cassandra. In the third set of experiments, we used real data to evaluate the benefits of using Memcached and Cassandra. Figure 6 compares the performance of querying with Memcached and Cassandra respectively on the same dataset. Varying the size of each query, querying with Memcached performed much better than that with Cassandra in all cases.

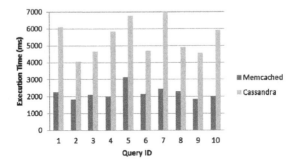

Fig. 6. Experimental Results on Memcached *vs.* Cassandra

9   Conclusion

In this paper, a classification framework for similar music search is proposed and proved to be feasible. Compared with traditional music retrieval systems, our framework not only build inverted index in offline job, but also cluster and classify the songs in the database as well. It is more scalable and practical to complete more work offline. Meanwhile, popular distributed technologies are used in our framework, which makes the system compatible and extensible for massive data.

References

1. Foote, J.T.: Content-Based Retrieval of Music and Audio. In: Kuo, C.-C.J., et al. (eds.) Proc. of SPIE Multimedia Storage and Archiving System II, vol. 3229, pp. 138–147 (1997)
2. Liu, M., Wan, C.: A study of content-based retrieval of mp3 music objects. In: Proc. of the Int'l. Conf. on Information and Knowledge Management, Atlanta, Georgia, pp. 506–511. ACM (2001)
3. Liu, C.C., Hsu, J.L., Chen, A.L.P.: 1D-List: An Approximate String Matching Algorithm for Content-Based Music Data Retrieval (submitted for publication)
4. Ghias, A., Logan, J., Chamberlin, D., Smith, B.C.: Query by humming—musical information retrieval in an audio database. In: ACM Multimedia 1995, San Francisco, USA (1995)
5. Chen, A.L.P., Chang, M., Chen, J.: Query by Music Segments: An Efficient Approach for Song Retrieval. In: Proc. of IEEE International Conference on Multimedia and Expo. (2000)
6. Li, T., Ogihara, M., Peng, W., Shao, B., Zhu, S.: Music clustering with features from different information sources. IEEE Transactions on Multimedia 11(3), 477–485 (2009)
7. Tsai, W.-H., Wang, H.-M., Rodgers, D., Cheng, S.-S., Yu, H.-M.: Blind Clustering of Popular Music Recordings Based on Singer Voice Characteristics. In: Proc. of ISMIR 2003, pp. 167–173 (2003)
8. Cilibrasi, R., de Wolf, R., Vitanyi, P.: Algorithmic Clustering of Music Based on String Compression. Computer Music J. 28(4), 49–67 (2004)
9. Frühwirth, M., Rauber, A.: Self-Organizing Maps for Content-Based Music Clustering. In: Proceedings of the Italian Workshop on Neural Nets (2001)
10. Jiang, D.-N., Lu, L., Zhang, H.-J., Tao, J.-H., Cai, L.-H.: Music type classification by spectral contrast feature. In: Proceedings of the IEEE International Conference on Multimedia and Expo (ICME), Lausanne, Switzerland (August 2002)
11. Davis, S.B., Mermelstein, P.: Comparison of parametnc representahons for monosyllabic word recognition in continuously spoken sentences. IEEE Trans. Acoust., Speech, Signal Processing ASSP-28(4), 357–366 (1980)
12. Yuan, Y., Zhao, P., Zhou, Q.: Research of speaker recognition based on combination of LPCC and MFCC. In: IEEE International Conference on Intelligent Computing and Intelligent Systems (ICIS), October 29-31, vol. 3, pp. 765–767 (2010)
13. Ellis, D.: Classifying music audio with timbral and chroma features. In: International Symposium on Music Information Retrieval (ISMIR 2007), pp. 339–340 (2007)
14. Cano, P., Batle, E., Kalker, T., Haitsma, J.: A review of algorithms for audio fingerprinting. In: Processing 2002 IEEE Workshop on Multimedia Signal Processing, December 9-11, pp. 169–173 (2002)

Managing and Collaboratively Processing Medical Image via the Web

Hualei Shen, Dianfu Ma*, Yongwang Zhao, Chunyao Yang,
Sujun Sun, and Bo Lang

State Key Laboratory of Software Development Environment,
Beihang University, 100191, Beijing, China
{shenhl,zhaoyw,yangch,sunsujun}@act.buaa.edu.cn,
{dfma,langbo}@buaa.edu.cn

Abstract. This article presents a strategy for medical image management. Based on the database named AUDR (Advanced Unstructured Data Repository) developed by Beihang University, a web-based system for medical image management and collaborative processing is built. Different from other applications, the system is accessed via web browser, and allows both text-based and content-based retrieval of medical images in DICOM format. Besides, the system allows geographically separated radiologists to process and interpret the same medical image simultaneously. To import medical image into AUDR, a strategy for DICOM image import is proposed. To implement the system, a collaboration framework and a method for image transmission is proposed. System evaluation in terms of functionality and performance has been performed in order to prove the availability of the system.

Keywords: Unstructured Data, Tetrahedral Data Model, AUDR, Medical Image, Collaboration.

1 Introduction

The amount of medical unstructured data produced by hospitals is amazing, which is mainly medical images. Just take China as an example, there are more than 2,000 cities across the country, each city has 2 to 4 hospitals. With the economy development of China in recent years, the radiology departments in hospitals have been equipped with modern medical imaging instruments, such as magnetic resonance imaging(MRI) scanners, computerized tomography (CT) scanners and positron emission tomography(PET) injectors, etc. Everyday, these hospitals produce large amounts of medical images.

Medical image is stored in DICOM (digital imaging and communications in medicine) format[1][13]. Typically, hospitals rely on electronic picture archiving and communication systems (PACS) to manage DICOM image[3][6][12]. Via PACS system, users can store, retrieve, transport and electronically display medical image. Usually, method for PACS to retrieve medical image is naive: only

---

* Corresponding author.

Z. Bao et al. (Eds.): WAIM 2012 Workshops, LNCS 7419, pp. 252–263, 2012.

text-based retrieval is supported. Users of PACS, such as radiologists or physicians, can only achieve medical images by providing keywords to PACS.

To overcome these limitations, researchers try to introduce content-based image retrieval (CBIR) techniques into PACS [9]. Several online medical image repositories are built [9]. In these online environment, medical images are usually stored in common formats (such as JPEG, PNG, etc) other than DICOM format. There is no textual description attached to medical images in common formats, so information contained in the header of DICOM image is lost.

Teleradiology via the web has been researched in recent years[2][10][11]. The web provides a natural infrastructure for interpreting and annotating medical images by users at different locations. Via the web, intelligence of distributed experts can be brought together. Current teleradiology system depends on a particular client application, which can not be accessed directly via web browser.

In this paper, we propose a solution to manage medical images based on a new unstructured database named Advanced Unstructured Data Repository (AUDR), which is developed by Beihang University[8]. Depending on the solution, we develop a new teleradiology system which enables users to manage and collaboratively process medical image via the web. Compared with existing solutions, the system has the following features: (1)Supporting both text-based and content-based retrieval of medical images in DICOM format; (2)Providing a platform for teleradiology which could be accessed via web browser running on different terminals; (3)Integrating functionalities of medical image processing into teleradiology, which enables users to process medical images via web browser.

The rest of the paper is organized as follows. Section 2 introduces the solution we proposed for AUDR-based medical image management. Section 3 presents the system for medical image management and collaborative processing. Section 4 discusses evaluations of the system in terms of functionality evaluation and performance evaluation. Section 5 concludes the paper.

2 AUDR-Based Medical Image Management

2.1 Tetrahedral Data Model

Unstructured data cannot be understood and processed directly by computers. Following the paradigm of describing the data and then using the descriptive information to implement data operations, Li Wei et al.[7] argue that unstructured data is composed of four aspects: basic attributes, semantic features, low level features and raw data. They propose the tetrahedral data model to characterize unstructured data based on these four aspects. For detailed description of tetrahedral data model, please refer to [7].

2.2 AUDR

Based on the tetrahedral data model described above, an unstructured database named AUDR is developed[8]. AUDR supports distributed unstructured data storage and computing under multiple operation system platforms. AUDR also implements unstructured data automatic association, classification, clustering and summarization[8][5][4]. Please refer to [8] for detailed description of AUDR.

2.3   Importing DICOM Image into AUDR

Medical images in DICOM format can not be directly managed by AUDR. As shown in Figure 1, we propose the following four steps to import DICOM images into AUDR:

- Step 1: Getting basic attributes, semantic features and raw data of DICOM images stored in PACS.
  Basic attributes refers to attributes of DICOM image, including creation date, modality, institution name, operator's name, patient's name, patient's age and patient's sex. *Semantic features* includes radiologist's diagnoses, annotation, etc. *Raw data* refers to volume dataset (header and data block) contained in DICOM image.
- Step 2: Converting DICOM image to JPEG format, and extracting low level features of the JPEG image.
- Step 3: Organizing the information obtained above based on tetrahedral data model.
- Step 4: Storing the tetrahedral data model based information into AUDR.

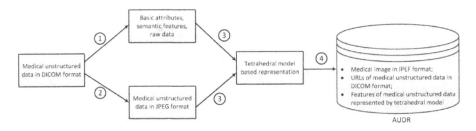

Fig. 1. The proposed scheme for importing medical unstructured data into AUDR

3   System

3.1   System Architecture

Figure 2 details the architecture of the system we built for medical image management and collaborative processing. From bottom to top, the system is divided into the following three layers:

- AUDR: Which includes three sub-layers: distributed storage, tetrahedral data model and medical image retrieval.

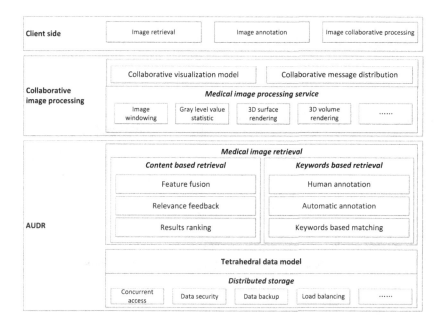

Fig. 2. System architecture

- Distributed storage: The distributed storage layer provides functionalities for conveniently and safely store and retrieve medical images. Techniques such as concurrent access, data security, data backup and load balancing are provided.
- Tetrahedral data model: The tetrahedral layer is used to abstract and present information about medical images stored in the distributed storage layer.
- Medical image retrieval: In the medical image retrieval layer, content-based and keywords-based retrieval are provided. Techniques including feature fusion, relevance feedback and results ranking are used to enhance the performance of content based retrieval. Keywords-based retrieval utilizes the technique of keywords based matching. Besides tags contained in the header of DICOM image, we use the methods of human annotation and automatic annotation to add tags to medical image.
- Collaborative image processing: The collaborative image processing layer provides medical image processing services which can be accessed via web browser.
- Client side: The client side provides functionalities which can be accessed directly via web browser. These functionalities includes image retrieval, image annotation and image collaborative processing.

3.2   Collaboration Framework

In this section, we introduce the framework for collaborative medical image processing. We propose a collaboration framework for collaborative medical image processing as shown in figure 3.

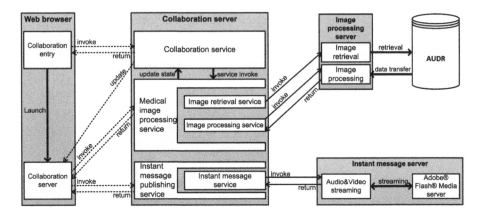

Fig. 3. Framework for collaborative medical image processing

In Figure 3, dashed lines are HTTP/HTTPS connections for HTML, XML and image transmission, solid lines are TCP connections. At web client side, the application runs as Ajax application on web browser. The application consists of a collaboration entry and a collaboration client. The collaboration entry is the entrance for user, which offers entry for medical image annotation. Collaboration client is the interface between user and collaboration server. The client consists of a user interface and an Ajax engine. The engine includes a group of Flex functions which can construct a SOAP message sent to collaboration server and parse the responding SOAP message.

The collaboration server is the bridge between clients and back-end servers (rendering server and instant messaging server). The collaboration server publishes three kinds of web services: *collaboration service*, *medical image processing service* and *instant messaging publishing service*. The collaboration service implements collaboration schemes among clients. The medical image processing service encapsulates image retrieval service, image processing service into web service. And the instant messaging publishing service publishes instant messaging service.

Image processing server is used to retrieve and process medical images stored in AUDR. It has TCP socket as network interface. Web browser can display DICOM image stored in image repository by invoking interface provided by image processing server to render DICOM image into JPEG format. Instant messaging server runs Adobe® Flash® Media Server to provide instant messaging service for clients.

3.3   Data Transmission

We adopt the strategy that the image processing server renders DICOM image into JPEG format, which can be displayed directly in web browser, and streams them to clients. The strategy enables the system running in low bandwidth network environment. As shown in Figure 4, the procedure of data transmission between server and collaborative web clients includes the following steps:

- Step 1: Operator within the collaboration group interacts with web client to get new image from the system. Web client encapsulates the request into SOAP message, and sends the message to collaboration server;
- Step 2: Collaboration server decodes the received SOAP message to get image processing command. The command is sent to image processing server via socket;
- Step 3: According to the received image processing command, image processing server performs image processing on the retrieved image, and renders DICOM image into JPEG format. URLs of JPEG image is sent to collaboration server via socket;
- Step 4: Collaboration server encapsulates received image URLs into SOAP message, and sends the message to all users within the collaboration group;
- Step 5: After receiving image URLs, each client sends image download request to image processing server via HTTP protocol;
- Step 6: The image processing server sends image to each client;
- Step 7: Each client displays the received images in web browser.

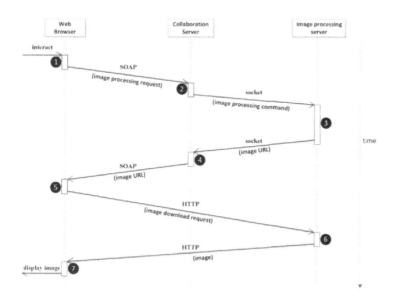

Fig. 4. Procedure of data transmission

3.4   User Interface

In this section, we briefly introduce user interface of the system, which includes the interface of medical image retrieval and collaborative medical image processing.

Medical Image Retrieval. The system provides three kinds of retrieval methods: semantic retrieval, instance retrieval and joint retrieval. Figure 5(a) shows the client user interface of medical image retrieval. Figure 5(b) shows the results of image retrieval. Where retrieved images are returned to users.

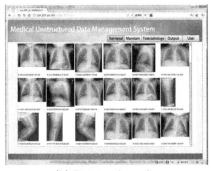

(a) Interface of medical image retrieval   (b) Retrieval results

Fig. 5. User interface of medical image retrieval

Collaborative Medical Image Processing. Collaborative medical image processing includes 2D medical image processing and 3D medical image processing. Figure 6 shows the client user interface of collaborative medical image processing. Where the interface can be divided into 5 sub-regions. Region 1 is used for image display. Region 2 includes interface for medical image processing. Region 3 provides interface for collaboration. Region 4 provides navigation buttons, which can navigate to 2D and 3D medical image processing. Region 5 includes auxiliary functionalities for collaboration, which includes buttons for text chat, audio chat and video chat.

4   System Evaluation

We evaluate the system in terms of functionality evaluation and performance evaluation. Functionality evaluation aims to evaluate the results of image retrieval, collaborative medical image processing and the auxiliary functionality of the system. Performance evaluation aims to evaluate the response delay of collaborative medical image processing.

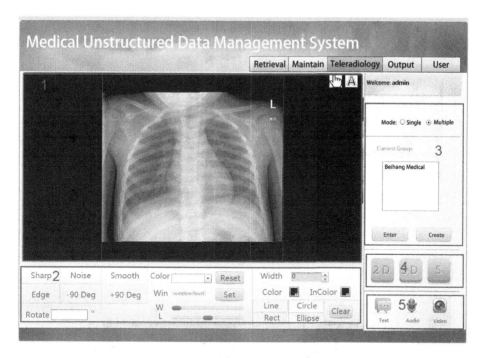

Fig. 6. User interface of collaborative medical image processing

4.1   Functionality Evaluation

Image Retrieval. Figure 7 demonstrates the results of semantic retrieval and instance retrieval. For semantic retrieval, users can feed keywords concerning *basic attributes* and *semantic features* about DICOM image (detailed in Section 2.3) to the system. For instance retrieval, users offer one sample image to the system, and the visual similar images stored in the system will be returned. In Figure 7(a), medical images imported into AUDR with date "2012-05-08" are retrieved. In Figure 7(b), all images having similar visual content with a chest image are retrieved.

2D and 3D Medical Image Collaborative Processing. Figure 8 shows the results of 2D and 3D medical image processing. Where Figure 8(a) shows the original image, Figure 8(b) shows results of original images after lookup table transformation (LUT). Figure 8(c) shows 3D surface rendering of a multi-slice CT image. Figure 8(d) shows 3D volume rendering of the CT image.

Auxiliary Functionality. The system provides auxiliary functionalitics including text chat, audio chat and video chat to facilitate different clients to collaboratively process medical image. Figure 9 shows the snapshots of collaborative medical image processing. It can be seen that the auxiliary functionalities provides better user experience for different clients.

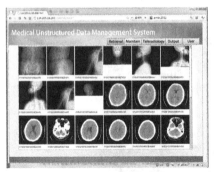 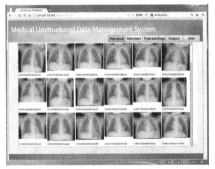

(a) Demonstration of semantic retrieval (b) Demonstration of instance retrieval

Fig. 7. Demonstration of medical image retrieval

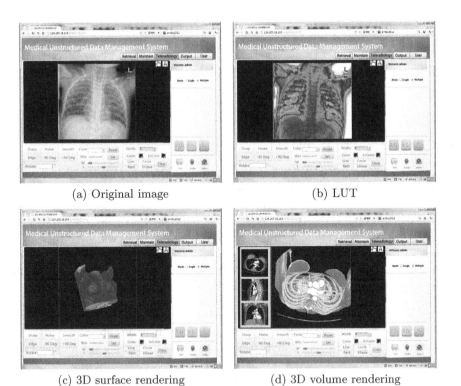

(a) Original image

(b) LUT

(c) 3D surface rendering

(d) 3D volume rendering

Fig. 8. 2D&3D medical image processing results

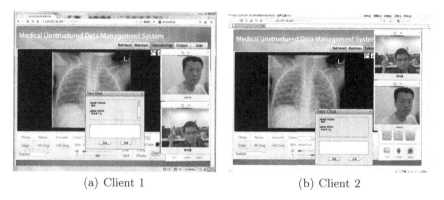

(a) Client 1   (b) Client 2

Fig. 9. Auxiliary functionality for collaborative medical image processing

4.2   Performance Evaluation

Evaluation Setup. We deploy the system on servers located at State Key Laboratory of Software Development Environment, Beihang University. We test the system by providing services to the Internet. The test venues are:

- *Server:* We use a HP ProLiant ML 570 server (CPU: 2 Intel E5320; RAM: 16 GB; OS: Windows Server 2008) as AUDR server and instant messaging server. We use a HP Z800 workstation (CPU: 2 Intel X5690; RAM: 24 GB; OS: Windows Server 2008; hard disk: 6 TB) as medical image processing server. A Levovo WanQuan T350 server (CPU: 2 Intel E5320; RAM: 24 GB; OS: Windows Server 2008) is used as the collaboration server. All servers are connected to the Internet via a gigabit router.
- *Client:* Three peers located at Beihang University running Windows 7, Web browser: Google Chrome 18. All clients are connected to the Internet via a gigabit router.

We evaluate system performance in terms of 2D and 3D medical image processing. We use HP LoadRunner to simulate users' concurrent requests, and a 568 MB DICOM image to perform 2D and 3D medical image processing. All concurrent requests are equally sent from three clients located at Beihang University as detailed above.

Average Response Delay for a Single Group. We firstly consider the simplest condition that there is only one collaboration group using the system at a time. To evaluate the system performance under this condition, we consider user numbers from 10 to 200. Figure 10(a) depicts the average response delay for different number of users. It shows that for single group, the number of users within the group affects system performance. Where there are more users within the group, there are higher image processing delay. When the user number is 50, the time delay for 2D processing is about 0.7 seconds, and the time delay for 3D processing is about 0.8 seconds.

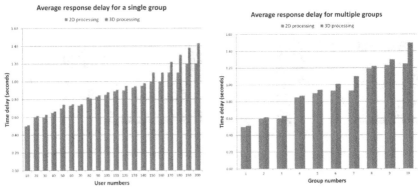

(a) Average response delay for a single group

(b) Average response delay for multiple groups

Fig. 10. Response delay in terms of 2D and 3D collaborative medical image processing

Average Response Delay for Multiple Groups. We then consider the second condition that there are many groups using the system concurrently, and each group process different medical images. Following the test setup mentioned above, we test the system performance with group numbers from 1 to 10, each group has 10 members. Figure 10(b) depicts the results. It shows that the response delay is proportional to the number of concurrent groups. When the group number is 5, the response delay for 2D image processing is about 0.85 seconds, while the response delay for 3D image processing is about 0.9 seconds.

5   Conclusion

In this paper we propose a solution for managing and processing medical images via the web. Based on the AUDR developed by Beihang University, we build a web based system for medical image management and processing. Compared with other similar solutions, our implementation has some significant features. To enable efficient retrieval of medical images, information contained in DICOM image is extracted and organized according to tetrahedral data model, which is managed by a database named AUDR. To enable collaborative medical image processing via web browser, the system uses a collaboration framework and a strategy for data transmission. Evaluation results show the availability of the system.

Acknowledgments. This work was supported by the National Major Research Plan of Infrastructure Software under Grant No.2010ZX01042-002-001 and National Natural Science Foundation of China (NSFC) under Grant No.61003017.

References

1. Bidgood Jr., W.D., Horii, S.C., Prior, F.W., Van Syckle, D.E.: Understanding and using DICOM, the data interchange standard for biomedical imaging. Journal of the American Medical Informatics Association 4(3), 199–212 (1997)
2. Bui, A.A.T., Morioka, C., Dionisio, J.D.N., Johnson, D.B., Sinha, U., Ardekani, S., Taira, R.K., Aberle, D.R., El-Saden, S., Kangarloo, H.: OpenSourcePACS: An extensible infrastructure for medical image management. IEEE Transactions on Information Technology in Biomedicine 11(1), 94–109 (2007)
3. Choplin, R.H., Boehme, J.M., Maynard, C.D.: Picture Archiving and Communication Systems: an overview. Radiographics 12(1), 127–129 (1992)
4. Guha, S., Rastogi, R., Shim, K.: CURE: an efficient clustering algorithm for large databases. Information Systems, 35–38 (2001)
5. Hartigan, J.A.: A k-means clustering algorithm. Applied Statistics, 100–108 (1979)
6. Hernandez, J.A., Acuna, C.J., de Castro, M.V., Marcos, E., López, M., Malpica, N.: Web-PACS for multicenter clinical trials. IEEE Transactions on Information Technology in Biomedicine 11(1), 87–93 (2007)
7. Li, W., Lang, B.: A tetrahedral data model for unstructured data management. SCIENCE CHINA Information Sciences 53, 1497–1510 (2010)
8. Liu, X.L., Lang, B., Yu, W., Luo, J.W., Huang, L.: AUDR: An advanced unstructured data repository. In: 6th International Conference on Pervasive Computing and Applications (ICPCA), pp. 462–469. IEEE press, New York (2011)
9. Müller, H., Michoux, N., Bandon, D., Geissbuhler, A.: A review of content-based image retrieval systems in medical applications–clinical benefits and future directions. International Journal of Medical Informatics 73(1), 1–23 (2004)
10. Neri, E., Thiran, J.P., Caramella, D., Petri, C., Bartolozzi, C., Piscaglia, B., Macq, B., Duprez, T., Cosnard, G., Maldague, B.: Interactive DICOM image transmission and telediagnosis over the European ATM network. IEEE Transactions on Information Technology in Biomedicine 2, 35–38 (1998)
11. Noro, R., Hubaux, J.P., Meuli, R., Laurini, R.N., Patthey, R.: Real-time telediagnosis of radiological images through an asynchronous transfer mode network: The ARTeMeD project. Journal of Digital Imaging 10, 116–121 (1997)
12. Pilling, J.R.: Picture archiving and communication systems: the users' view. British Journal of Radiology 76(908), 519 (2003)
13. Pianykh, O.S.: Digital Imaging and Communications in Medicine (DICOM): A Practical Introduction and Survival Guide. Springer, Heidelberg (2011)

Improving Folksonomy Tag Quality of Social Image Hosting Website

Jiyi Li, Qiang Ma, Yasuhito Asano, and Masatoshi Yoshikawa

Department of Social Informatics, Graduate School of Informatics, Kyoto University,
Yoshida-Honmachi, Sakyo-ku, Kyoto 606-8501, Japan
jyli@db.soc.i.kyoto-u.ac.jp,
{qiang,asano,yoshikawa}@i.kyoto-u.ac.jp

Abstract. Social image hosting websites such as Flickr provide services to users for sharing their images. Users can upload and tag their images or search for images by using keywords which describe image semantics. However various low quality tags in the user generated folksonomy tags have negative influence on the image search results and user experience. To improve tag quality, we propose three approaches with one framework to automatically generate new tags, and rank the new tags as well as the existing raw tags, for both untagged and tagged images. The approaches utilize and integrate both textual and visual information, and analyze intra- and inter- probabilistic relationships among images and tags based on a graph model. The experiments based on the dataset constructed from Flickr illustrate the effectiveness and efficiency of our approaches.

Keywords: Tag Quality, Image Search, Graph Model.

1 Introduction

On social image hosting websites, e.g. Flickr[1], users can upload and tag their images, to share them with other people. As which has been investigated in [1], in all social tags generated by users, tags which are used to describe image content and semantics occupy the largest proportion; in all queries that users use for search images, queries which are related to image content have the largest proportion. It is to say that many social tags can be used to index image content from semantic aspect; and users can search images by using this kind of social tags as keywords. Although user generated tags are useful for social image management and sharing, there is a problem that they are folksonomy tags [1]. Compared with taxonomy tags, they have an open vocabulary and very free on type, form and content. It results in various low quality tags, i.e., missing tags, imprecise tags, meaningless tags, unranked tags and so on.

We use Fig. 1 as an example to illustrate various low quality tags. On one hand, because choosing proper tags manually for large amount of images is so time consuming, many users may miss some important tags which describe image content when they assign tags to images, e.g. the missing tag "grass" in

---

[1] http://www.flickr.com

Z. Bao et al. (Eds.): WAIM 2012 Workshops, LNCS 7419, pp. 264–275, 2012.
© Springer-Verlag Berlin Heidelberg 2012

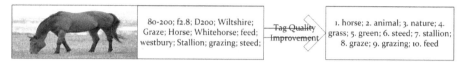

Fig. 1. Social Image, Raw Tags, Tag Quality Improvement

Fig. 1. The special case of missing tags is no tags which means that users do not assign any tag to images. Missing tags causes that this image cannot appear in the corresponding search results. On the other hand, for tagged images, in many cases, the assigned tags are neither precise nor meaningful enough for reflecting image semantics, e.g., the imprecise tag "whitehorse" and image-semantics meaningless tag "D200" in Fig.1. It results in that this image will appear in the improper keyword search results or it will not appear in the proper keyword search results. Furthermore, Fig. 1 also shows that there is no information to describe the importance of tags and the raw tags are unranked. It causes that this image may have an improper rank value in the corresponding search results. As a result, low quality tags will decrease the search results quality; and users who want to share or search images will fail to reach their purposes.

To improve search results quality as well as user experience, one of the solutions is to improve tag quality automatically. We generate precise and meaningful tags which can reflect the objective content of images or how most of users understand the image content from a statistical viewpoint. We propose a solution with various functions to generate new tags, and rank the new tags as well as the existing raw tags automatically, for both untagged images and tagged images.

Some work can be applied to provide some of these functions and solve the problems, such as tag recommendation, image annotation, tag ranking and so on. The tag recommendation approaches using textual information only, e.g. [2], cannot work automatically and depend on the initial tag set assigned by users too much. The image annotation approaches using visual information only [3] have been originally proposed for classifying images into a small number of concepts. They are proper to taxonomy keywords, but not proper enough to folksonomy tags which has a large concept space. They have a separate training stage and need to construct a training set manually which is time consuming. Furthermore, [4] proposed by Liu et al. can ranks the raw tags of images. It is not related to generate new tags and ranking them with raw tags.

The contributions of our work are as follows. We improve image search results on social image hosting websites by improving tag quality. We propose three different approaches with one framework to solve the problems. These approaches aggregate above-mentioned functions to meet various cases, with mixed usage of textual and visual information, based on a image-tag graph model. Our approaches do not have the disadvantage of too depending on the initial tag set. We improve tag quality based on user generated folksonomy tags directly, which have a large concept space, without constructing and using manual training set.

In these three approaches, the first one utilizes textual co-occurrence information among all candidate tags. The second one has a mutual reinforcement process and mixes both textual and visual information. The third one is based on random walk method and use visual similarity and textual co-occurrence for constructing the transition matrix. The experimental results show that all of them can improve tag quality prominently, while compared with other two approaches, the second one has better performance on time cost. On the other hand, our approaches can also be used for the task of ranking raw tags only without adding new tags. For this task, all our approaches can improve tag quality. It also shows that for improving tag quality, compared with ranking raw tags only, adding new tags has a better performance on the MAP metric.

The remainder of this paper is organized as follows. In section 2, we introduce the related work. Section 3 presents our approaches for improving tag quality. In section 4, we report the experimental results. Section 5 presents the conclusion.

2   Related Work

We concentrate on the related work on automatic tag recommendation for images since they are most related to our work. The approaches proposed in this area can be divided into several categories based on the information they use, i.e. text-based approaches and visual-based approaches.

The text-only-based approaches only use textual co-occurrence relationships for tag analysis. [2] is a typical text-only-based approach. It uses two tag co-occurrence measures, and aggregate three types of tags with two strategies. This approach which only uses textual information has the deficits that it cannot work automatically and depends on the initial tag set. It assumes that a user assigns a few candidate tags to the input image manually. It cannot recommend tags to an untagged image automatically. It also does not have a good characteristics on the initial tag fault tolerance. A fault initial tag set will be certain to generate fault result. Furthermore for different images with same initial tag sets, it will generate same results. It cannot generate diverse enough results for different images. Our approaches, because of the usage of visual information, do not so depend on the initial tag set and can handle untagged images as well as tagged images. They can generate better results even if the initial tag set is wrong, and generate different results for different images even the initial tag sets are same.

The visual-only-based approaches only use image content for tag analysis. [5] uses an image annotation approach. It defines and learns 62 concepts, and annotates new images with the top-n concepts. It has a learning process and the size of concepts space is fixed and small. The approaches only using visual information have been originally proposed for classifying images into a small number of concepts which can be regarded as tags. They need to create the training set manually which costs lots of time and labor. Adding a new concept for the classification takes computational time for reconstructing the classifiers. Such approaches are not suitable for user generated folksonomy tags. Furthermore the performance of content-based approaches for information retrieval nowadays is still not better than text-based approaches. In our work we propose approaches which utilize user generated tags directly, without training set construction

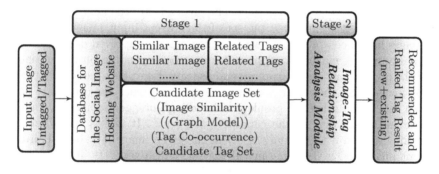

Fig. 2. Overview of Framework

manually and training process. They make handling large concept space possible and easier. We also integrate textual information with visual information.

Several approaches using both textual and visual information have also been proposed. [6] generates a ranking feature for tags with textual and visual modality respectively, and then use Rankboost algorithm to learn an optimal combination. It just combines the results of two kinds of information. The approaches we propose mix both textual and visual information with each other.

3   Tag Quality Improvement

In this section we first propose the framework of three approaches for improving the tag quality. After that we introduce the graph model constructed for relationship analysis among images and tags. At last we propose the three approaches in details respectively.

3.1   Approach Framework

The relevance relationships between images and tags can be presented in probability. In our work, for an input image q, a tag t has a probability of $p(t|q)$ to represent its semantics. This probability depends on the tag co-occurrence relationship $c(t)$ (textual information) and image similarity $s(q)$ (visual information): $p(t|q) = f(c(t), s(q))$. Fig. 2 shows the framework of our approaches. It has two main stages.

Stage 1 (Candidate Set Construction): Given an input image q, find the top-k (our current work, $k = 100$) similar tagged images in the database P and construct a candidate image set A with them. Construct a candidate tag set T with user generated tags of these images. For image a_i, its own tag set is T^{a_i}. The similarity between q and a_i is s_i. We denote t_j as a tag.

Stage 2 (Relevance Relationship Analysis): Analyze the image-tag relationship on candidate set A and T. After that, we can get the relevance probability of each tag to q and tag rank list for q. We do not update the original database after this stage immediately. The processing of the next input image is not influenced by the improved tag set of the current input image. After all of the images have been processed, we get a tag quality improved database.

This framework makes our approaches not too depending on the initial tag set of the given input image from several aspects. First, because it uses the top similar images and their user generated tags to construct the candidate image and tag set, it can solve the problem of handling untagged images. Second, we consider all of the tags in the candidate tag set T. It makes the approaches still available even if the initial tag set of the given image is wrong. Third, for different images with the same initial tag set, it can generate different results because the top similar images of these different images are different.

In this paper, we propose three different approaches with this framework. All of them are same in stage 1, while they are different in stage 2. These approaches are based on the following graph model in the image-tag relationship analysis.

3.2   Image-Tag Relationship Analysis Model

Fig. 3 shows the graph model we proposed. It includes both intra- and inter-relationships among images and tags. It is composed of several parts, image similarity graph, tag complete graph and image-tag bipartite graph. First, q denotes current input image which can be untagged or tagged images, a_i denotes one of other images. The links among images denote image similarity. For q, there is also a link point to itself, the similarity it denotes is equivalent to 1. These images and links construct the image similarity graph. Second, t_x denotes a tag. The links among tags denote tag co-occurrence. These tags and links construct the tag complete graph. Third, the links between images and tags denote image-tag annotation relationships and construct an image-tag bipartite graph.

The visual parameters s and textual parameters c in the model is computed as follows. We use image similarity s_i ranging from 0.0 to 1.0 as visual parameters.

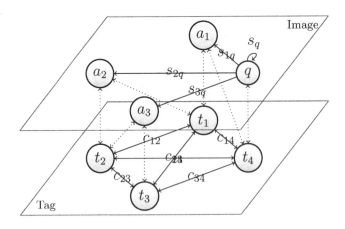

Fig. 3. Image-Tag Relationship Analysis Model

For each tag in the database, we compute two textual parameters. For each pair of tag t_u and t_v, $c(u, v)$ is computed to evaluate their co-occurrence degree. It is a symmetric-asymmetric pairwise co-occurrence parameter. $c(u, v) = |t_u \cap t_v|/|t_u| + |t_u \cap t_v|/|t_v|$, $|t_u|$ means the number of images that contain t_u in the database, $|t_u \cap t_v|$ means the number of the images that contain both of t_u and t_v. For each t_u, we compute a single co-occurrence parameter $c_u = \sum_v c(u, v)$. Both of these two textual parameters for all tags are computed in advance.

3.3   MCKA: Modified Collective Knowledge Approach

First we propose an approach that only uses textual information in stage 2. It is only based on the tag complete graph in the model in Fig. 3. For each tag t_u in the candidate tag set T, we find the top-k most co-occurring tags t_v in T to construct the top co-occurrence tag list T_{uk}. We compute the score for each t_v and aggregate them. To promote the better tags, we also uses three *promotion* functions, *rank*, *descriptive* and *stability*, which are based on the tag frequency and tag position in the top co-occurring tag list, to generate the final result. These *promotion* functions and their value of the parameters follow the definition in [2]. The algorithm is list in Algorithm 1. Because we consider all tags in the candidate tag set T to find their top k most co-occurring tags, this approach is still available even if the initial tag set of the given image is wrong. A brief description of the formulas is as follows.

$$score(t_v) = \sum_{t_u \in T} vote(t_u, t_v) * promotion(t_u, t_v), vote(t_u, t_v) = \begin{cases} 1 \ if t_v \in T_{uk} \\ 0 \ otherwise \end{cases}$$

$$promotion(t_u, t_v) = rank(t_u, t_v) * descriptive(t_v) * stability(t_u)$$

Algorithm 1. Modified Collective Knowledge Approach

1: Input: Candidate Image Set A, Candidate Tag Set T
2: Output: Rank List *score* for T
3: Initial: zeros(*score*)
4: **for all** $t_u \in T$ **do**
5:     $stability(t_u) = k_s/(k_s + abs(k_s - log|t_u|))$
6:     $C(u, T) = \{c(u, v)|t_v \in T\}$
7:     $T_{uk} = \{t_v|c(u, v) \in Topk(C(u, T))\}$
8:     **for** $t_v \in T_{uk}$ **do**
9:         $rank(t_u, t_v) = k_r/(k_r + (r - 1)), r = the\ position\ of\ t_v\ in\ T_{uk}$
10:         $score(t_v) + = stability(t_u) * rank(t_u, t_v)$
11:     **end for**
12: **end for**
13: **for all** $t_v \in T$ **do**
14:     $score(t_v) * = descriptive(t_v), descriptive(t_v) = k_d/(k_d + abs(k_d - log|t_v|))$
15: **end for**

3.4   MMRA: Mixed Mutual Reinforcement Approach

MCKA only uses image similarity in stage 1. It does not use visual information in stage 2. We propose an approach that also integrate visual information with textual information in the image-tag relationship analysis.

The iteration process of this approach is mainly based on the image-tag bipartite graph, which denotes the inter- relationships of the graph model in Fig. 3. The candidate image set and tag set construct two disjoint sets of the bipartite graph. This approach also considers the information on tag complete graph and image similarity graph which denote the intra- relationships. MMRA is based on a basic *assumption*: a high quality tag for q is a tag that point to many high quality images; a high quality image is an image that is pointed to by many high quality tags. A high quality tag means this tag has a high relevance probability to be a tag of the input image. A high quality image can be regarded as an image that has a high semantic similarity with the input image. We initiate the score $Q(t)$ of the tags with the textual single co-occurrence parameters and the score $Q(a)$ of the images with the image similarities. $Q_0(a_j) = s_j$, $Q_0(t_i) = c_i$.

After the initialization, we have an iterative computation to compute $Q(t)$ and $Q(a)$. When we design the iteration formulas, we need to make it following the requirements of the above basic assumption. A normalization is also necessary to solve the convergence problem of this iterative computation. Based on these conditions of designing the approach, we come to the following iterative formulas.

$$Q_k(t_i) = \alpha + (1 - \alpha) \sum_{\forall j: i \rightarrow j} Q_{k-1}(a_j), \ 0 \leq \alpha \leq 1$$
$$Q_k(a_j) = \beta + (1 - \beta) \sum_{\forall i: i \rightarrow j} Q_k(t_i), \ 0 \leq \beta \leq 1$$
$$Q_k(t_i) = Q_k(t_i)/\sqrt{\sum_r Q_k(t_r)^2}, \ Q_k(a_j) = Q_k(a_j)/\sqrt{\sum_r Q_k(a_r)^2}$$

This approach has similar ideas with some approaches in the link analysis area, e.g. HITS method [7]. There is another common idea in this area. It is the random walk method.

3.5   RWBA: Random Walk Based Approach

The iteration process of this approach is mainly based on the tag complete graph in the model in Fig. 3. It also uses the information of image similarity relationship and image-tag annotation relationship. It utilizes the random walk method in the iteration process. We use image similarity information as well as image-tag annotation relationship to construct the transition matrix, and integrate tag co-occurrence information in the iteration process.

This approach is based on an assumption. For an input image, a high quality tag has a high co-occurrence with other high quality tags. Suppose that when we want to generate tags for an image, if we have confirmed a tag t and then we want to generate more tags, the tags which is most associated with t will be the good candidates and we will assign them with high relevance probability.

We initial the score of tags with the tag single co-occurrence parameters. The iterative formulas for computing the score of the tag are as follows.

$$P_k(t_u) = \gamma P'(t_u) + (1 - \gamma) \sum_v (P_{k-1}(t_v) P(t_v \rightarrow t_u))$$

$$P'(t_u) = \frac{c_u}{\sum_v c_v}, \quad P(t_v \to t_u) = \frac{r(v,u)}{\sum_w r(v,w)}, r(v,u) = e^{-\frac{\sum_{t_v \in T^{a_i}} \sum_{t_u \in T^{a_j}} \|s_i - s_j\|}{|t_v|_c * |t_u|_c}}$$

$P'(t_u)$ denotes the probability that we may not turn to recommend other tags and keep the state at current tag. The state transition matrix $R = \{P(t_v \to t_u)\}_{m*m}$ is decided by visual similarity information and image-tag annotation relationship. T^{a_i} is the corresponding tag set of image a_i, and a_i is an image in the candidate image set A. $|t_v|_c$ is the number of images in A that are pointed to by tag t_v.

This approach is similar but different from tag ranking approach [4]. Both RWBA and tag ranking use the random walk method. But they have the following difference. First, the tag ranking approach is originally for ranking the existing raw tags of the images. After the database has been modified into a database with the tags ranked. It applies this updated database for tag recommendation as an application of their work. The tag recommendation no longer uses random walk method and just uses the tag co-occurrence. Second, the detailed computation of these two approaches such as initialization and transition matrix are different.

4   Experimental Results

4.1   Dataset Settings

We set several rules to filter the original tag set before computing textual parameters. We delete the space in tags and convert them into lower case format so that these tags can be regards as a single tag. We eliminate the tags that are not existing in WordNet, because we consider that most of them are misspelled or irregular tags and have little contribution to the keyword search. We also eliminate the numeric tags. Furthermore, we eliminate the tags with too low frequency, which are misspelled tags, too special tags and so on. This processing can reduce the noises in the tag set and the time cost in the computation of image-tag relationship analysis. We do not eliminate the tags with too high frequency. Although lots of previous work think that this kind of tags is too common to represent the semantic of content. We think that this kind of tags can reflect important semantic and user tagging behavior with image categories.

The dataset we use for experiment is MIR Flickr 25000 [11] which is constructed by images and tags downloaded from Flickr. It has 25000 images and provides the raw tags of these images on Flickr as well as a manual reference tag set. Based on our tag filter rules, we set the low frequency threshold as 20 and the setting of WordNet search as NOUN. There are 808 unique tags left after filtering. In our work, the version of the reference tag set we use is v080[2]. There are 24 unique reference tags in total. Since four of them never appear in the raw tag set, which means that it will also never appear in our results. We use the left 20 unique reference tags for evaluation. Because we improve tag quality utilizing

---

[2] There are two sub versions for some tags in this version, e.g., flower and flower_r1. We use the "r1" version for them.

the user generated raw tags directly, not the reference tags, we do not need to divide this dataset into training set and testing set, and do not need to train on the training set. It is different from the image annotation task on this dataset.

We use two kinds of visual feature for searching the top similar images. One is a 1024-Dimension color histogram feature on the HSV color space. The other one is the SIFT feature [9] with the visual words model. The codebook for the visual words of SIFT is from the dataset in the ImageNet Large Scale Visual Recognition Challenge 2010 [10]. It's a visual vocabulary of 1000 visual words based on a k-means clustering on a random subset of 10 million SIFT descriptors.

4.2 Metrics and Approach Parameters

We set several metrics to evaluate the statistical performance of our approaches. Each metric can be regarded as one aspect of the tag quality. If a metric is improved, the tag quality on the aspect of this metric is improved. The metrics are as follows:

MAP: The Mean Average Precision is a good metric for the performance evaluation of the rank result in the information retrieval area. In this experiment, specially, for the raw tag set, because it originally doesn't have the rank information. Without loss of generality, we simply use the position of these tags in the raw tag list as their rank value when we compute MAP. We evaluate MAP on the whole dataset.

MPFRR and MPLRR: Mean Precision at the First Relevant Rank (PFRR) is the mean of precision at the position of the first relevant tag in the rank list. Mean Precision at the Last Relevant Rank (PLRR) is the mean of precision at the position of the last relevant tag in the rank list.

NDCG: Normalized Discounted Cumulative Gain (NDCG) [12]. The ground truths on 20 unique reference tags are limited to reflect the performance of tag quality improvement on the large amount of images and their tags in the dataset. We therefore construct another ground truth for evaluation on NDCG, which evaluates the rank results of all tags in the candidate tag set. We randomly select 100 images, a subset of the dataset, and label the relevance degree of all tags in the candidate tag sets of these images on the HSV feature[3] by human beings. The range of relevance degree is from 0(irrelevant) to 4(relevant). We evaluate NDCG on this subset of the dataset.

We set the parameters of the approaches as follows. For MCKA, $k = 25$, and the parameters in the promotion functions follow the value in [2], $ks = 9$, $kd = 11$, $kr = 4$. For MMRA, to decide the iteration parameters of α and β, we choose their candidate values by an interval of 0.1 and get 121 groups of candidate values. We run the MMRA with these candidate parameter groups on 1000 images in the dataset, based on the HSV feature, and observe the performance on the metric of MAP. According to Fig. 4, we choose (α, β) as (0.3, 0.9) in

---

[3] The ground truth of NDCG for different feature need to be labled seperately because the candidat image and tag sets are different

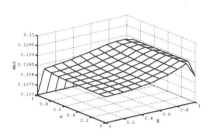

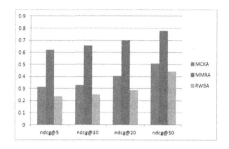

Fig. 4. Parameters for MMRA   **Fig. 5.** NDCG (by HSV)

our following experiment to evaluate the performance of this approach on the whole dataset and all metrics. Furthermore from Fig. 4 we can find that if not choosing the boundary value, the performance difference of difference parameters is small. So actually choosing (α, β) randomly without such a decision process is also acceptable. For RWBA, the value of the damping factor parameter γ follows the general choice of the damping factor in PageRank [8], which could also be regarded as a random walk based method, $\gamma = 0.15$.

4.3   Approach Evaluation

Table 1 shows the experimental results on the metrics of MAP, MPFRR and MPLRR. Considering the columns of "Raw" and "*:All" first, all of the approaches have improved the tag quality on MAP prominently. All of the approaches can also improve tag quality on MPFRR and MPLRR. Furthermore, Fig. 5 shows that MMRA performs best in our experiment on NDCG metric for the subset of dataset among these three approaches.

Although our approaches are originally designed to recommend new tags, and rank the new tags as well as the existing raw tags. They are very easy to be applied to the task of ranking raw tags only, by extracting and sorting the raw tags according to their rank sequence in the whole candidate tag set directly. Note that it's not a pure "ranking raw tag only" results. When we rank these raw tags, we also utilize the information of other new tags. Table 1 also provides

Table 1. Statistical Results on MAP, MPFRR, MPLRR

Metric	Raw	MCKA:All	MMRA:All	RWBA:All
MAP	0.0304	0.144(0.154)	0.110(0.134)	0.153(0.157)
MPFRR	0.210	0.321(0.345)	0.224(0.273)	0.363(0.377)
MPLRR	0.028	0.039(0.036)	0.036(0.032)	0.032(0.030)
Metric	Raw	MCKA:Raw	MMRA:Raw	RWBA:Raw
MAP	0.0304	0.067(0.068)	0.063(0.067)	0.068(0.068)
MPFRR	0.210	0.449(0.451)	0.441(0.449)	0.452(0.452)
MPLRR	0.028	0.071(0.071)	0.071(0.070)	0.071(0.070)

Cell Format: by HSV (by SIFT)

the result on this issue. Comparing the columns of "Raw" and "*:Raw", we can find that only ranking the raw tags can also improve tag quality from on MAP, MPFRR, MPLRR. The "*:Raw" results for all three approaches are similar.

We also investigate another issue that if we only rank the existing raw tags without including new tags, how the tag quality could be improved and what's the difference between including new tags and not. Comparing the columns of "*:Raw" and "*:All", we can find that recommending new tags and ranking them as well as the raw tags always has better tag quality on MAP metric, but worse on MPFRR and MPLRR metrics because more irrelevant tags are included.

4.4   Time Complexity

Table 2 provides an overview of the time complexity comparison. Here we just consider the time complexity of image-tag relationship analysis of stage 2 in Fig. 2, because for these approaches, the time cost of other stages are the same. In this table, n is the size of the candidate tag set, and m is the size of the candidate image set. MMRA and RWBA need an iteration process. For MMRA, which is based on the bipartite graph, for each tag, the iteration computation uses the information of the annotation relationships among images and tags. For RWBA, which is based on the tag complete graph, for each tag, the computation uses the information that refers to all of the other tags. Actually the time complexity for the iteration process of MMRA and RWBA are both $O(eI)$, where e is the number of links need to be analyzed in the graphs, and I is the iteration times. In MMRA the e_{MMRA} is equal to μmn, μ is a positive number but much smaller than 1. In RWBA, e_{RWBA} is equal to $(n-1)^2$ (for each pair of tags, two edges on two directions) because of the tag complete graph. For example, for the sample image which is used for computing and illustrating the running time in table 2 (Our experiment environment is on Ubuntu with CPU Intel Core i7 920.). It has $m = 100$ and $n = 599$, and its $e_{MMRA} = 963$ and $e_{RWBA} = 357604$. Both of their iteration times are less than ten times. We can get that $O(\mu mnI) \ll O(n^2 I)$ and the MMRA is much faster than the RWBA.

MMRA doesn't need additional computation in the initialization step. RWBA needs to compute the transition matrix which depends on the candidate tag and image set. Even RWBA uses a fixed transition matrix so that there is no additional computation in the initialization, it still costs much more time than MMRA. MCKA does not have an iteration process. It needs to select top k most co-occurring tags for each tag in the candidate set. In total, the MMRA has a better time complexity comparing to other two approaches.

Table 2. Time Complexity Comparison

Approach	Time Complexity		Time
	Initialization	Iteration	Cost(s)
MCKA	$O(kn^2)$		0.62
MMRA	0	$O(\mu mnI)$	0.09
RWBA	$O(n^2)$	$O(n^2 I)$	7.22

In conclusion, for our experiments, all of the three approaches we proposed can provide a prominent improvement on tag quality, for ranking raw tags only or for adding new tags and ranking new tags as well as raw tags. Compared with other two approaches, MMRA has better performance on time cost. Furthermore, for the task of ranking raw tags only, the three approaches have similar performance. Adding new tags has a better performance than ranking raw tags only, on the MAP metrics, but worse on the metrics of MPFRR and MPLRR.

5 Conclusion

In this paper, we improve user generated folksonomy tag quality on social image hosting websites, to improve social image search result as well as user experience. For this topic, we propose three different approaches with one framework. The computational experiments reveal that our approaches are effective and efficient.

References

1. Bischoff, K., Firan, C.S., Nejdl, W., Paiu, R.: Can all tags be used for search? In: CIKM 2008, pp. 193–202 (2008)
2. Sigurbjörnsson, B., van Zwol, R.: Flickr tag recommendation based on collective knowledge. In: WWW 2008, pp. 327–336 (2008)
3. Datta, R., Joshi, D., Li, J., Wang, J.: Image retrieval: Ideas, influences, and trends of the new age. ACM Comput. Surv. 40(2), 1–60 (2008)
4. Liu, D., Hua, X.-S., Yang, L., Wang, M., Zhang, H.-J.: Tag ranking. In: WWW 2009, pp. 351–360 (2009)
5. Chen, H.M., Chang, M.H., Chang, P.C., Tien, M.C., Hsu, W.H., Wu, J.L.: Sheep-Dog: group and tag recommendation for flickr photos by automatic search-based learning. In: MM 2008, pp. 737–740 (2008)
6. Wu, L., Yang, L.J., Yu, N.H., Hua, X.S.: Learning to tag. In: WWW 2009, pp. 361–370 (2009)
7. Kleinberg, J.M.: Authoritative sources in a hyperlinked environment. J. ACM 46(5), 604–632 (1999)
8. Brin, S., Page, L.: The Anatomy of a Large-Scale Hypertextual Web Search Engine. Computer Networks and ISDN Systems 30(1-7), 107–117 (1998)
9. Lowe, D.G.: Distinctive Image Features from Scale-Invariant Keypoints. IJCV 60(2), 91–110 (2004)
10. Deng, J., Dong, W., Socher, R., Li, L.J., Li, K., Li, F.F.: ImageNet: A large-scale hierarchical image database. In: IEEE Computer Society Conference on CVPR 2009, pp. 248–255 (2009)
11. Huiskes, M.J., Lew, M.S.: The MIR Flickr Retrieval Evaluation. In: MIR 2008 (2008)
12. Jarvelin, K., Kekalainen, J.: Cumulated gain-based evaluation of IR techniques. ACM Transactions on Information Systems (TOIS) 20(4) (2001)

An Effective Top-*k* Keyword Search Algorithm Based on Classified Steiner Tree

Yan Yang, Mingzhu Tang, Yingli Zhong, Zhaogong Zhang, and Longjiang Guo

School of Computer Science and Technology, Heilongjiang University,
150080 Harbin, China
Key Laboratory of Database and Parallel Computing, Heilongjiang University,
150080 Harbin, China
yangyan@hlju.edu.cn

Abstract. Keyword search has become one of hot topics in the field of information retrieval. It can provide users a simple and friendly interface. But the efficiency of some existing keyword search algorithms is low and there are some draws in sorting results. Most algorithms are suited for either unstructured data or structured data. This paper proposes a new kind of top-*k* keyword search algorithm. No matter the data is unstructured, semi-structured or structured, the algorithm is always effective. It introduces the concept of neighbor sets of nodes and uses set join algorithm to narrow the search space. We also propose the definition of classified Steiner tree, which can reduce the draw phenomenon in results. In addition, the algorithms can output the results of the classified Steiner tree at the same time of computing them. So it can reduce the waiting time of the users and improve the efficiency of keywords search.

Keywords: keyword search, top-*k,* classified Steiner tree, neighbor sets of nodes.

1   Introduction

In recent years, keyword search in the Internet has become more and more mature. Users only need to input keywords through search engine interface and they can get the search results through links. This technology has been quite perfect and convenient for users to get what they want. Furthermore, the benefits of keyword search have been recognized and introduced into relational database, XML databases, and graph databases. According to the needs of users, keyword search problem over unstructured data, semi-structured data and structured data arises at the historic moment. But with the development and application of information retrieval technology, it also brings us a series of problems and contradictions. For example, in the databases, structured query language like SQL is not convenient for users without professional computer knowledge. We can see that the SQL language cannot satisfy the needs of ordinary users for accurately and efficiently retrieving structured data in the databases. So it is necessary to do research on keyword search to support

Z. Bao et al. (Eds.): WAIM 2012 Workshops, LNCS 7419, pp. 276–288, 2012.

unstructured data, semi-structured data and structured data, by means of which users can get the keyword search results through the user interface.

This paper studies the related problems about keyword search, gives the search model a formal definition and proposes a new top-*k* keyword search algorithm based on set join. Compared with existing algorithms, this algorithm can be suitable for unstructured data, semi-structured data and structured data. It uses the neighbor sets of nodes to narrow query space and orders the results at the same time of calculating them, then returns top-*k* results to users, by which it reduces the waiting time of users. This paper puts forward the concept of classification of Steiner trees for the first time, which can break the draw phenomenon among results and the experimental results verify the performance of the algorithm.

2   Related Work

For structured data, the target is not only looking for tuples containing the given keywords, but also finding the semantic relations between tuples containing the keywords [1]. There are many research work [2, 3, 4, 5, 6] use database graph method for keyword search. The main works that use schema graph method are [7, 8, 9]. Reference [3] proposes a kind of dynamic programming method and it can approximately give the top-*k* tuple connection trees. Discover [7] first puts forward a keyword search algorithm which uses the database schema for the expansion of the database. For existing keyword search algorithms, they are not very efficient for top-*k* keyword search. Only after all the results are calculated according to a score function, they are returned to the user in turn as a batch, which greatly increases the waiting time of users. Discover considers the number of tuples in the trees as an evaluation standard. If the result tree contains fewer tuples, the score of the tree is higher. But they do not consider how to break the draw phenomenon of results.

XRANK [10] and XSEarch [11] are systems keyword search systems for XML data. XRANK proposes a ranking function for the XML result trees, which combines the scores of the individual nodes of the result tree. XSEarch pays attention to semantics. It uses an all-pairs interconnection index to check the connectivity between nodes. XKeyword [12] is a system that offers keyword proximity search over XML documents that conform to an XML schema.

Keyword search has been well studied for unstructured data. For example [13] presents the Google search engine. Reference [14] offers an overview of current Web search engine design. It also introduces generic search engine architecture and covers crawling and indexing issues.

Up to now, reference [15] proposes a keyword search method for document data, XML data and relational databases. But it may produce duplicate results. Very few existing methods could be universally applied to unstructured data, semi-structured data and structured data. This paper proposes a method which can be used for the three types of data effectively. In this paper, an algorithm based on classification of Steiner tree is designed. The algorithm outputs one classification just when it is calculated, which reduces the waiting time of users greatly. The concept of classified Steiner tree is also proposed.

3   Problem Definition

Next we take structured data for example to describe our problem. The first step of processing a keyword search is to consider how to model the data. This section first describes the database graph used in this article and then gives the definition of query result, which is Steiner tree. Finally the formal definition of top-k keyword search is given.

3.1   Data Model

The modeling methods for relational databases can be divided into database graph method and schema graph method. In this paper we adopt database graph method.

Definition 1 (Database Graph) [1]. Suppose the database contains n relations R_1, $R_2, \ldots, R_n$. Model the database as an undirected graph G(V, E). V is the set of tuples in the database, and E is the set of edges between nodes based on the foreign key references between the corresponding tuples in the database. Set a non-negative weight to each edge, denoted as $w_G(e)$. The weight of database graph G is $w(G) = \sum_{e \in E} w_G(e)$, which is the sum of the edge-weight in the graph. In this paper, all the edge-weight are set to 1 for convenient. A simple DBLP database is shown in Figure 1. Figure 1(a)-(d) show the four relations in the database and 1(e) illustrates database graph for the simple DBLP database.

For unstructured data and semi-structured data, we also model them as graph, with nodes being documents or elements respectively, and edges being hyper-links or parent-child relationship respectively.

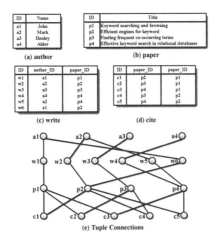

Fig. 1. DBLP Database

3.2   Query Model

When using database graph to model the database, Steiner tree is usually used to represent the keyword search results. The relationship between tuples containing keywords can be shown in Steiner trees with the least amount of nodes and edges. In the following, the related definitions of Steiner tree, minimum Steiner tree and top-*k* Steiner tree are given.

Definition 2 (Steiner Tree) [1]. Given a graph G(V, E) and a set of nodes $V' \subseteq$ V. If T is a connected sub-tree in G and T contains all the nodes in V', T is called a Steiner tree of V' in G. We use ST for short.

Definition 3 (Minimum Steiner Tree) [1]. Suppose T is a ST of V' in G. $c(T) = \sum_{e \in E(T)} w_G(e)$ represents the weight of T, where E(T) is the set of edges in G, $w_G(e)$ is the weight of e. T is the minimum Steiner tree if $c(T)$ is the minimum one among all the ST of V' in G. We use MST for short.

For a given set of nodes $V' \subseteq$ V, how to find the MST of V' is a problem. If the first *k* MST are required, the top-*k* ST is defined as follows.

Definition 4 (Top-*k* Steiner Tree) [1]. Given a set of nodes $V' \subseteq$ V, the top-*k* ST is the top-*k* MST $T_1, T_2 \dots, T_k$ of V', which are sorted by their weights and $c(T_1) \leq c(T_2) \leq \cdots \leq c(T_k)$.

Because top-*k* ST is strictly sorted according to their weights, it can correspond to top-*k* search results. In addition, ST can express the joins between nodes very well, therefore, we choose top-*k* ST to represent keyword search results.

3.3   Problem Definition

According to the former introduction in 3.2, it is clear about how to represent the keyword search results. Below is the formal definition of top-*k* keyword search, which is just the problem we will solve.

Definition 5 (Top-*k* Keyword Search) [1]. Given a keyword set $K = \{k_1, \dots, k_n\}$, $V' \subseteq$ V is the set of nodes containing at least one keyword in K. Finding top-*k* ST $T_1, T_2 \dots, T_k$ of V' in database graph G(V, E), where $c(T_i) = \sum_{e \in E(T_i)} w_G(e)$ ($i = 1, 2, \dots \dots, k$) is the score of each ST, satisfying $c(T_1) \leq c(T_2) \leq \cdots \leq c(T_k)$.

The problem to be solved in this paper comes down to finding correlative nodes that meet the conditions of ST and return top-*k* ST to users.

4   Keyword Search Algorithm Based on Classified Steiner Tree

In this section, we also take structured data for example to introduce our method. In this article, database graph method is adopted for modeling the database. Assuming that database graph G (V, E) has existed, then we use database graph G for top-*k* keyword search. In this algorithm, parameter *Tmax* is set as the maximum join number in output results, which is also the largest number of edges in ST. In the following, the edge-weight in G is set to 1.

4.1   Classification of Steiner Tree and Sort of Results

In many works, the order of search results is determined by the number of joins. In keyword search, there are many results with the same number of joins, which makes the sorting method meaningless. So, it is necessary to search for new ways to break the draw. The results are sorted according to the classification of ST.

4.1.1   Classification of Steiner Tree

For a given keyword set $K = \{k_1, \dots, k_n\}$, the nodes in database graph are divided into $2^n - 1$ tuple sets according to the keywords it contains. The tuple sets are denoted as $g(K_i)$ ($K_i \subseteq K, i = 1, 2, \dots, 2^n - 1$), where the serial number of the sets are denoted as O_i, that is, $O_i = i$. Suppose the set of keywords is $K = \{A, B, C\}$, the nodes in database graph are divided into $2^3 - 1 = 7$ groups, $g(K_1), \dots, g(K_7)$. The serial number O_i of a group in binary and it shows which keywords it contains. The binary value of O_1 is 001, which means the tuples in $g(K_1)$ only containing the keyword C. According to the above methods, tuples can be classified as follows:

$g(K_1) = \{C\}, O_1 = 1; g(K_2) = \{B\}, O_2 = 2; g(K_3) = \{B, C\}, O_3 = 3$
$g(K_4) = \{A\}, O_4 = 4; g(K_5) = \{A, C\}, O_5 = 5; g(K_6) = \{A, B\}, O_6 = 6;$
$g(K_7) = \{A, B, C\}, O_7 = 7$

In this paper, set $\{A, B, C\}$ is used to represent all the tuples containing keywords A, B and C. Similarly, the set $\{A, B\}$ means all the tuples containing and only containing keywords A and B.

So, $g(K_7)$ can be output first. We define the results containing all the keywords as ST of class 1. Such results are ST with only one tuple, but the tuple contains all the keywords. The weight of the tree equals to the minimum weight, 0. So ST of class 1 is output first.

ST of class 2 is the MST formed by two tuples from two tuple sets respectively. If the union of two tuple sets contains all the keywords, join the two sets as ST of class 2. According to the binary coding rules, if the result of OR of two binary codes is 111, join the corresponding two tuple sets. In the above example, join the tuple sets $g(K_5)$ and $g(K_6)$, $g(K_3)$ and $g(K_6)$, $g(K_1)$ and $g(K_6)$, etc. What needed to emphasize is that, join the sets means joining the nodes corresponding to the sets.

ST of class 3 is the MST formed by three tuples from three tuple sets respectively. If the union set of three tuple sets contains all the keywords, then join the three sets as ST of class 3. According to the binary coding rules, if the result of OR of three binary codes is 111, join them. In the example above, join the corresponding nodes in $g(K_1)$, $g(K_2)$ and $g(K_4)$ to form the Steiner tree of class 3.

According to the rules of classification of ST, the ST of class i is defined as follows.

Definition 6 (Steiner Tree of Class i). If there are i tuple sets whose union set can contain all the keywords the users given, the MST formed by i tuples from i tuple sets respectively is called ST of class i, denoted as Steiner (i).

4.1.2 Sort of Results

In this paper, we sort the results according to the classification of ST. STs are output to the users in the order they are calculated. This section first presents the definition of prior relationship of the results, and then gives the method of how to classify and sort the results.

Definition 7 (Prior Relationship). Let $\leq$p represents the prior relationship. Prior relationship of the results is defined as follows:

(1) Steiner(i) has priority over Steiner(i+1), that is Steiner(i+1) $\leq$p Steiner(i).
(2) In the same classification of ST, the higher the frequencies of keywords appear, the more preferred the result is. When the frequency of keywords is the same, consider the number of joins. The less number of joins, the more preferred the result is.

The definition of priority relationship is based on the following two points. First, in the same classification of ST, if the intersection of some sets is not null, it means there will be keywords that appear equal to or greater than two times. Keywords appear more frequently, the results will be closer to the need of users. At the same time, the more the elements in the intersection sets, the higher the frequency of keywords, the results will have higher priorities. Second, if the number of keywords in a certain set is more than one, the tuples in the set must contain more than one keyword, which means it doesn't need to join tuples containing the keywords, because they are in the same tuple. So the relationship between the keywords is closer. Therefore, the results with above two features are better.

Still use input keywords $K = \{A, B, C\}$ as an example. Steiner tree of class 1 is the nodes containing keywords A, B and C. Steiner tree of class 2 should include the following joins: $g(K_1)$ and $g(K_6)$; $g(K_2)$ and $g(K_5)$; $g(K_3)$ and $g(K_4)$; $g(K_3)$ and $g(K_5)$; $g(K_3)$ and $g(K_6)$; $g(K_5)$ and $g(K_6)$

Steiner(3) should be the join results of $g(K_1)$, $g(K_2)$ and $g(K_4)$. According to the definition 7, we can judge that Steiner(3) $\leq$p Steiner(2).

If a user input i keywords, the algorithm will output the search results with the order of class 1, class 2... class i. Within each class, the results are first sorted according to the frequency of keywords appears in it. If draw appears, the number of joins is considered. In this way, it reduces the problem of the same number of the joins in the results, which causes draw in sorting results.

4.2   Search Algorithm

It has been clear about the representation of search results and the output order about the classification of ST. Next we discuss how to calculate the ST of a certain classification. In the existing algorithms, most algorithms adopts Dijkstra algorithm to compute the shortest path between two nodes. Because of the large scale of the database graph, Dijkstra algorithm always obtains low efficiency. What's more, Dijkstra algorithm only returns the shortest path rather than produce top-k results. In this paper, the set of neighbor nodes is introduced to reduce search space. In the beginning of the algorithm, indices are built for each character attribute. According to the keywords user input, we get all the tuples containing keywords and classify them according to the number of keywords they contain as section 4.1. Steiner(1) can be directly output. When calculating the Steiner(2), joins of two sets are calculated. Next, we will introduce how to find the sets for joining. We use Order (i) to represent the sets for joining when calculating the Steiner(i). First define the set Order (i).

Definition 8 (Order (i)). Order(i) is the set for joining when calculate the Steiner(i). Steiner(i) corresponds to the set Order(i). An element in Order(i) is a combination of O_{k_j}, ($j = 1,2,...,i$), where O_{k_j} is the serial number of $g\left(K_{k_j}\right)$. Given the sets of $g(K_1), g(K_2), ..., g(K_i), ..., g(K_{ni})$, Order(i) $= \{(\quad O_{k_1}, O_{k_2}, ..., O_{k_i}),$ $(O_{k_{i+1}}, O_{k_{i+2}}, ..., O_{k_{2i}}), ..., (O_{k_{i+n}}, O_{k_{i+n+1}}, ..., O_{k_{ni}})\}$, satisfying the OR of the i serial numbers in one combination is $2^n - 1$ and the OR of any i-1 serial numbers in it is not $2^n - 1$.

Still take the input keywords $K = \{A, B, C\}$ for example. Steiner(1) is just one set. Each element in the set is a tuple, which contains all the keywords. So Order(1) $= \{(7)\}$ and output the set $g(K_7)$. The OR of O_i and O_j is $2^n - 1$ means the join of $g(K_i)$ and $g\left(K_j\right)$ contains all the keywords. For the Steiner(2), it needs the join of two sets. Order(2) $= \{ (1,6), (2,5), (3,4), (3,5), (3,6), (5,6)\}$, in which (1,6) represents $1|6 = 2^3 - 1$, means the join of $g(K_1)$ and $g(K_6)$ containing all the keywords. For the Steiner tree of class 3, it needs the join of three sets. To calculate Order(3) $= \{ (1,2,4)\}$, the join of sets $g(K_1)$, $g(K_2)$ and $g(K_4)$ is needed.

Next is the definition of neighbor sets of nodes.

Definition 9 (Neighbor Sets of Nodes). For the two nodes k and u in the database graph, the shortest path between them is denoted as dis(k, u). If dis(k, u) $\leq Tmax$, we say u is a neighbor of k. All the nodes u which satisfy the above condition forms the neighbor set of node k, denoted as neighbor(k).

We take $i=3$ as an example to discuss the generation process of Steiner(3). Steiner(3) is the join of three sets, suppose they are {A}, {B} and {C}. According to the introduction above, we first find the sets of $g(K_4)$, $g(K_2)$ and $g(K_1)$ respectively, then calculate the join of them. We are faced with an important task, that is, how to construct all the STs.

When calculate the join, we first select one element respectively from {A}, {B} and {C}. In this assumption, suppose:

$$g(K_4) = \{A\} = \{a_1, a_2, a_3\}, g(K_2) = \{B\} = \{b_1, b_2, b_3, b_4\}, g(K_1) = \{C\} = \{c_1, c_2, c_3\}$$

Assume a_1, b_1, c_1 are selected for joining respectively. Next, one node among the three nodes is chosen as the node in ST, for example, a_1. Then the neighbor set of a_1 is examined for if it contains b_1 and c_1. If contains, put b_1 into ST. Then consider the neighbor set of b_1, judge whether the set contains c_1. If so, put c_1 into ST. Now it forms a ST. Consider all the elements in {A}, {B} and {C} and it can generate all the STs related to {A}, {B} and {C}. All the STs must ensure that the number of joins is less than or equal to $Tmax$. Below, we outline our algorithm for top-k keyword search.

Algorithm 1.   Top-*k* Search Algorithm of   Classification of Steiner tree

---

Input: database graph G(V, E), keyword set $K = \{k_1, k_2, \ldots \ldots, k_n\}\}$, *Tmax*, *k* in
      top-*k*.

Output: top-*k* Steiner tree.

```
1 : Use index to find the set of nodes containing
keywords, g(K_i) (i = 1,2,...,2^n - 1);

2 : Result= Ø, Count=0;

3 : for i=1 to n do //find the ST of class i

4 :     find the set for joining Order(i)

5 :     While Order(i) != Ø do

6 :          if i==1 then

7 :                Result←(Order(1).currentElement);

8 :          else

9 :                Result←SetJoin(Order(i).currentElement, k);

10 :        Order(i)=Order(i)-{Order(i).currentElement};

11 :        Output Result;

12 :        Count = |Result|;//the number of results

13 :        if k==Count then

14 :             return;

15 :        else

16 :             k=k-Count;

17 : end.
```

In the beginning, according to keywords the user input, divide the nodes containing keywords into $2^n - 1$ groups. Next, the third line comes into a loop and calculates the Steiner(*i*). Line 4 finds the sets for joining Order (*i*). Line 6-7 finds the Steiner(1). Next computes the Steiner(*i*) and judges whether the number of results has come to *k*. If so, output the first *k* results. Return the results to users according to the sorting method in section 4.1.2.

Algorithm 2.   SetJoin(Order(i).currentElement, k)

Input: the sets $S_1 \ldots \ldots S_n$ in Order(i).currentElement, k in top-k.
Output: join results of n sets Result.

```
1: Result = ∅;  i=0;

2: traversal Sets S₁ ...... Sₙ  , get a node from each set
to add to v[1...n];

3:    add v[1] into Steiner tree STᵢ;

4:    for all vᵢ∈v[2...n]

5:     for all(tⱼ∈STᵢ)do

6:        if (tⱼ ∈v[1...n] &&vᵢ ∈neighbor(tⱼ))

7:         add vᵢ and nodes in the path from vᵢ to tⱼ to STᵢ;

8:    Result = Result ∪ STᵢ;

9:    i++;

10:  if i==k return Result;

11:return Result.
```

Algorithm 2 is a set join algorithm which is used to calculate a certain classification of ST. In line 2 of the algorithm, we choose a random element from each candidate join set and add them to v[1...n]. Let one element to be the node of ST. Line 3 to line 10 is a loop. If a node from the rest elements in v[1...n] can be joined with the ST, add the node and the nodes in the path into the ST. If a ST with all the keywords and the number of edges is no larger than *Tmax* is constructed until all the nodes in v[1..n] are considered, the ST is added to Result. So repeatedly, until calculate k search results which meet the conditions. For unstructured data and semi-structured data, our algorithm is also suitable.

5   Experiments

5.1   Experimental Parameters and Settings

All of the experiments are conducted on a 2.80GHz CPU and 4GB memory PC running Windows XP. Oracle9i is used in our experiments. All the algorithms are written in Java, and Oracle's interface is JDBC.

The algorithms are tested with detailed performance evaluation on DBLP database. The size of the database depends on the different experimental requirements. For DBLP, there are 4 relations as shown in Figure 2: Author (Aid, Name), Paper (Pid, Title, Other), Write (Aid, Pid, Remark), Cite (Pid1, Pid2). We create full text indices on Author attribute and Paper attribute. All character attributes are created indices. We have do experiments on unstructured data, semi-structured data and structured data respectively. Due to the space, we only show experiments on structured data. For the other two types of data, our algorithm is also effective.

We adopt [7] (denoted as Discover) and [16] (denoted as PDk) as a comparison with the results in our paper (denoted as SJ). Discover and PDk are keyword search algorithms over relational databases consistent with our problem. Therefore, in order to verify the classification of Steiner tree in this paper, we select Discover and PDk as a contrast. In order to test the performance of our algorithms, we implement six experimental groups. In particular, Discover uses the schema graph method. PDk uses database graph method, takes multi-core sub-graphs as keyword search results and adopts the shortest path algorithm of Dijkstra.

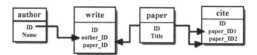

Fig. 2. DBLP schema graph

5.2   Experimental Results and Analysis

The Effect of Classification Sorting on Experimental Results. We evaluate the percentage of draw in the final results with different size of datasets in this experiment. Figure 3 shows that, with the dataset increases, the percentage of draws in results of three algorithms does not change much. But Discover and PDk have larger percentage of draw results. In this paper, the classification of Steiner tree is introduced. Definition 7 ensures that there is no draw phenomenon between different classifications of Steiner tree. Only in the same classification of Steiner tree, it may have a sort draw phenomenon with the same number of joins. The experiment results show that the algorithm in this paper greatly reduces draw phenomenon which caused by the same number of joins in sorting results.

The Time of the First Result Returning to the User. We evaluate the time of the first result returning to the user with 100M, 200M, 300M, 400M and 500M datasets in this experiment. Figure 4 shows that, the time of the first result returning to the user of three algorithms is very different. The algorithm in this paper outputs the results when they are calculated. However, Discover and PDk uniformly output results after calculating all the results. It is shown that the time of the first result returning to the user in our algorithm is much shorter than Discover and PDk.

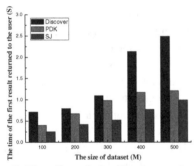
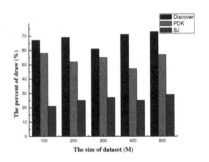

Fig. 3. The effect of classification sorting on experimental result

Fig. 4. The time of the first result returning to the user

The Search Time for Different Keywords. We evaluate the search time for different keywords with the size of 100M dataset in this experiment. The keywords and their frequency of occurrence values KWF are shown in Table 1. Obviously, different keywords cause different effects on the running time. If the frequency of certain keywords in database is high, the final sets needed to join are more, so it costs more time. In addition, the number of keywords also impacts on the running time. Figure 5 describes different keywords impact on the running time of algorithms. Figure 5 shows that, using the same keywords to query, Discover spends more time, PDk and our algorithm spend less time than Discover.

Table 1. Keywords List

KWF	Keywords
.0003	Scalable, protocols, distance, discovery
.0006	Space, graph, routing, scheme
.0009	Environment, database, support, development, optimization, fuzzy
.0012	Dynamic, application, modeling, logic
.0015	Web, parallel, control, algorithms

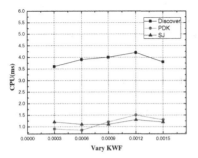
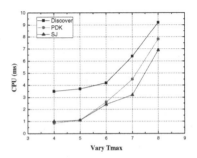

Fig. 5. The search time for different keywords

Fig. 6. The search time for different *Tmax*

The Search Time for Different *Tmax*. We evaluate the search time for different *Tmax* values in this experiment. As described above, the value of *Tmax* means the maximum number of joins in the final result, i.e. the maximum number of edges in the STs. In the experiment, we set *Tmax* to be 4, 5, 6, 7 and 8 to examine its impact on running time of the algorithms. Figure 6 shows that, with *Tmax* values increasing, the running time of three algorithms is also increasing. The running time increases rapidly when *Tmax* is larger than 5. The difference of the running time among the three algorithms is not significant. But they all have an increasing trend.

6   Conclusion

In this paper, we focus on the keyword search related problems over unstructured data, semi-structured data and structured data. A top-*k* keyword search algorithm based on classified Steiner tree and set join is proposed in this paper. The algorithm outputs the search results when they are calculated. It can greatly reduce the waiting time of users when the size of database is large. This paper presents the concept of classification of Steiner tree for the first time, which effectively reduces draw phenomenon caused by the same number of joins in the search results. Our algorithm is effective for unstructured data, semi-structured data and structured data. We have conducted a thorough experimental study of the algorithm. The results proved the efficiency of our algorithm is high.

Acknowledgments. This work is supported by the National Natural Science Foundation of China under Grant No.60973081, by the Natural Science Foundation of Heilongjiang Province under Grant No. F201011, by Scientific Research Foundation of Heilongjiang Provincial Education Department under Grant No.11551352 and No. 12511401, and by Heilongjiang University Foundation for Young Scholar under Grant No. QL201029.

References

1. Lin, Z., Yang, D., Wang, T.: Keyword Search over Relational Databases. Journal of Software 21(10), 2465–2466 (2010)
2. Kimelfeld, B., Sagiv, Y.: Finding and approximating top-k answers in keyword proximity search. In: Vansummeren, S. (ed.) Proc. of the 25th ACM SIGACT-SIGMOD-SIGART Symp. on Principles of Database Systems (PODS 2006), pp. 173–182. ACM Press, Chicago (2006)
3. Liu, F., Yu, C.T., Meng, W.Y., Chowdhury, A.: Effective keyword search in relational databases. In: Chaudhuri, S., Hristidis, V., Polyzotis, N. (eds.) Proc. of the 2006 ACM SIGMOD Int'l Conf. on Management of Data (SIGMOD 2006), pp. 563–574. ACM Press, Chicago (2006)
4. Ding, B., Yu, J.X., Wang, S., Qin, L., Zhang, X., Lin, X.M.: Finding top-k min-cost connected trees in databases. In: Proc. of the 23rd Int'l Conf. on Data Engineering (ICDE 2007), pp. 836–845. IEEE ComputerSociety Press, Istanbul (2007)

5. Tao, Y.F., Yu, J.X.: Finding frequent co-occurring terms in relational keyword search. In: Kersten, M.L., Novikov, B., Teubner, J., Polutin, V., Manegold, S. (eds.) Proc. of the 12th Int'l Conf. on Extending Database Technology (EDBT 2009), pp. 839–850. ACM Press, Saint Petersburg (2009)

6. Kmelfeld, B., Sagiv, Y.: Efficiently enumerating results of keyword search over data graphs. Information System 33(4-5), 335–359 (2008)

7. Hristidis, V., Papakonstantinou, Y.: DISCOVER: Keyword search in relational databases. In: Proc. of the 28th Int'l Conf. on Very Large Data Bases (VLDB 2002), pp.670–681. Morgan Kaufmann Publishers, Hong Kong (2002)

8. Luo, Y., Lin, X.M., Wang, W., Zhou, X.F.: Spark: Top-k keyword query in relational databases. In: Chan, C.Y., Ooi, B.C., Zhou, A.Y. (eds.) Proc. of the 2007 ACM SIGMOD Conf. on Management of Data (SIGMOD 2007), pp. 115–126. ACM Press, Beijing (2007)

9. Markowetz, A., Yang, Y., Papadias, D.: Keyword search on relational data streams. In: Chan, C.Y., OoiBC, Z.A. (eds.) Proc. of the 2007 ACM SIGMOD Conf. on Management of Data (SIGMOD 2007), pp. 605–616. ACM Press, Beijing (2007)

10. Guo, L., Shao, F., Botev, C., Shanmugasundaram, J.: Xrank: Ranked keyword search over XML documents. In: SIGMOD, pp. 16–27 (2003)

11. Cohen, S., Mamou, J., Kanza, Y., Sagiv, Y.: Xsearch: Asemantic search engine for XML. In: VLDB (2003)

12. Kacholia, V., Pandit, S., Chakrabarti, S., Sudarshan, S., Desai, R., Karambelkar, H.: Bidirectional expansion forkeyword search on graph databases. In: VLDB, pp. 505–516 (2005)

13. Brin, S., Page, L.: The Anatomy of a Large-Scale Hypertextual Web Search Engine. In: WWW Conference (1998)

14. Arasu, A., Cho, J., Garcia-Molina, H., Paepcke, A., Raghavan, S.: Searching the web. Transactions on Internet Technology (2001)

15. Li, G.L., Ooi, B.C., Feng, J.H., Wang, J.Y., Zhou, L.Z.: EASE: An effective 3-in-1 keyword search method for unstructured, semi-structuredand structured data. In: Tsong, J., Wang, L. (eds.) Proc. of the 2008 ACM SIGMOD Conf. on Management of Data (SIGMOD 2008), pp. 903–914. ACM, Vancouver (2008)

16. Qin, L., Yu, J.X., Chang, L., Tao, Y.: Querying communities in relational databases. In: Proc. of ICDE 2009 (2009)

Effective Keyword Search with Synonym Rules over XML Document

Linlin Zhang, Qing Liu, and Jiaheng Lu

School of Information, Renmin University of China, Beijing, China
{linlinzhang,qliu,jiahenglu}@ruc.edu.cn

Abstract. Keyword search is a friendly way for user to find the information they are interested in from XML documents without having to learn a complex query language or needing prior knowledge of the structure of the underlying data. However, the existing methods are usually limited to the input keywords. In this paper, we introduced the notion of synonyms, acronym, abbreviations and so on to capture user query intentions. We propose a SLCA based keyword search with synonym rules over xml documents which are orthogonal to various of xml keyword search techniques. In addition, we also use this to give a effective and efficient slca based keyword search.

1 Introduction

It is becoming increasingly popular to publish data on the Web in the form of XML documents. Keyword search is well-suited to XML trees. It allows users to find the information they are interested in without having to learn a complex query language or needing prior knowledge of the structure of the underlying data [3, 6, 7, 10, 18, 19]. XML keyword search enforces a conjunctive search semantics (i.e. all the keywords should be covered in each query result), such as LCA (Lowest Common Ancestor) [7] and its variants [6, 18, 19]. Among those proposals, SLCA(Smallest LCA) is widely adopted [19], where each SLCA result contains all query keywords but has no descendant whose subtree also contains all keywords.

Unfortunately, all of them assume each keyword in the query is intended as part of it. However, a user query may often be an imperfect description of their real information need. Even when the information need is well described, a search engine may not be able to return the results matching user's query intention as illustrated by the following example.

Example 1. Consider a query $\mathcal{Q}$="Jennie paper" issued on a bibliographic document in Figure 1 which is modeled using the conventional labeled tree model. The query most likely intend to find all papers written by Jennie. According to SLCA semantics defined in [18], it will return the most specific relevant answers - the subtrees rooted at nodes 0.1.2 and 0.3.1. However, terms *inproceedings* and *article* are synonyms of *paper*. Therefore, we should also take nodes 0.1.1, 0.1.3 and 0.3.2 into consideration to predict user's intention.

Z. Bao et al. (Eds.): WAIM 2012 Workshops, LNCS 7419, pp. 289–298, 2012.
© Springer-Verlag Berlin Heidelberg 2012

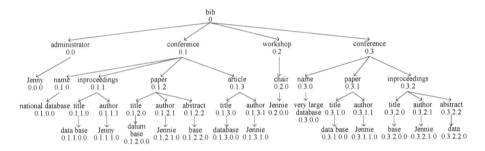

Fig. 1. bib.xml

According to the phenomenon illustrated in Example 1, we should take advantage of the equivalence between strings to improve the quality of keyword search over xml documents. There are many cases where strings that are syntactically far apart can still represent the same real world object. This happens in a variety of conditions such as synonyms, acronyms, abbreviations and so on. For instance, "bike" is a synonym of "bicycle", "ICDE" is an acronym of "International Conference on Data Engineering" while "DB" is an abbreviation of "database". We just use "synonym" as a placeholder for any kinds of equivalence expressions.

In this paper, we assume there exists a collection of predefined synonym rules. Such rules can be obtained from users' existing dictionaries, explicit inputs, by data and text mining [15] or query log analysis [12]. For ease of presentation, we will describe the rules without instantiating with any particular types in this paper. The following example shows how to use these rules to rewrite input queries thereby capturing user's potential purpose.

Example 2. Consider the query in Example 1, the answer should be [0.1.2, 0.3.1] according to the semantic of SLCA. Given a set of synonyms which are of the form *paper* → *inproceedings* and *paper* → *article*. Informally, the first rule means that an occurrence of *paper* can be replaced by *inproceedings*. Therefore, the query can be expanded to three equivalent strings - { "Jennie paper", "Jennie inproceedings", "Jennie article"}. Thus, the answers should be [0.1.2, 0.1.3, 0.3.1, 0.3.2].

In this paper, we study how to design a effective framework with synonym rules to not only support XML keyword search but also improve the quality of existing methods. we take SLCA as the underline keyword search semantic but our framework is orthogonal to other xml keyword search methods. While the synonym rules is expressive and powerful, it also introduces many new challenges.

Our major contributions towards xml keyword search with synonym rules are summarized as follows:

- We introduce a effective SLCA based keyword search over xml documents with synonym rules to capture user's potential query intentions.
- We give a formal definition of expansion rules. Meanwhile, we also explain how to rewrite a query based on given synonym rule sets and deeply analyze all the possible transformations.

- we theoretically discuss how to use this framework to support SLCA based xml keyword search and give some optimization techniques. we also design a Transformation Matching based IL Algorithm(TM-IL algorithm) which could efficiently and effectively return SLCA of input keywords.

The rest of the paper is organized as follows. In Section 2, we formally define synonym rules and show how to use these rules to rewrite user's queries. Section 3 and Section 4 present how to enhance the quality of SLCA based XML keyword search which show that our expansion based framework is orthogonal to the choice of the underlying XML keyword search technique which are not limited to SLCA or it's variations but also related ranking strategies. In section 5, we give a briefly conclusion and show the future of our work.

2   Query Rewriting

This section presents how we get and use rules to expand our queries. As we know, user input query is a string which can be modeled as a ordered sequence of *grams* where each grams is a (smaller) string. We can easily convert a string into a ordered gram sequence by splitting it based on delimiters(white spaces) or regular expression. i.e. the string "International Conf on Data Eng" can be converted to ordered sequence of grams <International,Conf,on,Data,Eng>.

2.1   Synonym Rules

A synonym rule consists of a pair of strings which is of the form $lhs \rightarrow rhs$ denoted by r which means they are equivalent with bias. Each of lhs and rhs is a string and could not be empty.

Example 3. Some example of synonym rules

- $William \rightarrow Bill$
- $bike \rightarrow bicycle$
- $VLDB \rightarrow VeryLargeDataBases$

Here we use "$\rightarrow$" to refer to the relevance between lhs and rhs which means lhs could be replaced by rhs. however, the inverse is not true due to abbreviations and acronyms might lose information compared with original strings.

There are multi-ways to obtain synonym sets. First, we can easily extract synonyms from existing data including various domains such as biology[1], address postal service(e.g., USPS[2]), academic publications(e.g., computer science[3] and medicine[4]) and so on. Second, synonyms can be obtained by applying data mining methods on datasets to get *equivalence classes* which represent the same

---

[1] http://www.expasy.ch/sprot
[2] United States Postal Service, http://www.usps.com
[3] http://academic.research.microsoft.com/CFP.aspx
[4] http://www.medical.theconferencewebsite.com/

real-world objects. Obviously, this strategy is domain independent, but require more human intervention to get more accurate results. Third, synonyms can be programmatically generated. For example, we can generate rules that connect the integer and textual representation of numbers such as *36th* and *Thirty-Sixth*.

2.2 Query Rewriting

We now describe how to rewrite queries when given a set of synonym rules $\mathcal{R}$. The following is an example that illustrate the procedure of expanding one string to another according to synonym rules.

Example 4. Considering the query $\mathcal{Q}=$"$k_1\ k_2\ k_3$" and synonym rule set $\mathcal{R} = \{r_1 :$"k_3"$\rightarrow$"k_4"$, r_2 :$"$k_2\ k_3$"$\rightarrow$"$k_2\ k_5$"$, r_3 :$"$k_1\ k_2$"$\rightarrow$"$k_6\ k_7$"$\}$. This means "k_3", "$k_2\ k_3$" and "$k_1\ k_2$" can be replaced by "k_4", "$k_2\ k_5$" and "$k_6\ k_7$" respectively. Consequently, we can get extended strings as follows:

- "$k_1\ k_2\ \underline{k_3}$" $\xrightarrow{r_1}$ "$k_1\ k_2\ \underline{k_4}$"
- "$\underline{k_1\ k_2}\ k_4$" $\xrightarrow{r_3}$ "$\underline{k_6\ k_7}\ k_4$"
- "$k_1\ \underline{k_2\ k_3}$" $\xrightarrow{r_2}$ "$k_1\ \underline{k_2\ k_5}$"
- "$\underline{k_1\ k_2}\ k_3$" $\xrightarrow{r_3}$ "$\underline{k_6\ k_7}\ k_3$"

Finally, $\mathcal{Q}$ will be converted to $\{$"$k_1\ k_2\ k_3$", "$k_1\ k_2\ k_4$", "$k_6\ k_7\ k_4$", "$k_1\ k_2\ k_5$", "$k_6\ k_7\ k_3$"$\}$, where underline denotes matching transformations.

We can see from example 4 that "$k_1\ k_2\ k_5$" can not be transformed to "$k_6\ k_7\ k_5$" although "$k_1\ k_2$" can be replaced by "$k_6\ k_7$"(r_3). Here we only allow 1-level transformation which means a gram which is generated as a result of substitution is prohibited participating in a subsequent transformation. It's easily to see there might be the case that the string transformation would never stop if we don't restrict the transformation level. Even we restrict the transformation level to 1, it is still NP-Complete to determine whether one string could be transformed to another string by applying rules.

3   Keyword Search with Synonym Rules

In this section, we review existing related proposals about keyword search and SLCA. Then we will describe our methods based on these notations.

3.1   Notation

We model XML docuemnt as a tree using conventional labeled ordered tree model. Nodes in the tree corresponds to an XML element and is labeled with a tag denoted by $\lambda(n)$. Given two nodes u and v, $u \prec v$ denotes u is ancestor of v and $u \preceq v$ denotes $u \prec v$ or $u = v$. We assign to each node a numerical id $pre(v)$ that is compatible with preorder numbering, in the sense that if a node v_1 precedes a node v_2 in the preorder left-to-right depth-first traversal of the tree then $pre(v_1) < pre(v_2)$. The usual $<$ relationship is also compatible with Dewey numbers [18]. For example, $0.1.0.0.0 < 0.1.1.1$.

We begin by formally introducing the concepts of *Lowest Common Ancestor(LCA)* and *Smallest Lowest Common Ancestor(SLCA)*.

Definition 1 (LCA). *Given m nodes $n_1, n_2, \ldots n_m$, u is called LCA of nodes $n_1, n_2, \ldots n_m$ iff u is ancestor of each node n_i for $1 \leq i \leq m$ and $\nexists$ node v, $u \prec v$ that v is also ancestor of each node n_i. This can be denoted as $u = lca(n_1, n_2, \ldots n_m)$.*

Given sets of nodes $S_1, S_2, \ldots S_m$, the *LCA* of sets $S_1, S_2, \ldots S_m$ is the set of LCA for each combination of nodes in S_1 through S_m which can be denoted as the flowing expression:

$$lca(S_1, S_2, \ldots S_m) = \{u | u \in lca(n_1, n_2, \ldots n_m) | n_1 \in S_1, n_2 \in S_2, \ldots n_m \in S_m\}$$

Definition 2 (SLCA). *Given a set of nodes $S_1, S_2, \ldots S_m$, u is called SLCA of $S_1, S_2, \ldots S_m$, iff $u \in lca(S_1, S_2, \ldots S_m)$ and $\forall v \in lca(S_1, S_2, \ldots S_m)$, $u \nprec v$. This can be denoted as $slca(S_1, S_2, \ldots S_m) = \{u | u \in lca(S_1, S_2, \ldots S_m) \wedge v \in lca(S_1, S_2, \ldots S_m), u \nprec v\}$.*

Given a set of nodes $\mathcal{S}$, the basic idea of SLCA is that it will return the closest ancestor node which doesn't contain any descendent node that is also the ancestor of each node in $\mathcal{S}$. Given two nodes v_1 and v_2 and their Dewey number dw_1 and dw_2, $lca(v_1, v_2)$ is the node with Dewey number that is the *longest common common prefix* of dw_1 and dw_2. For example, the LCA of nodes "database(0.1.0.0.0)" and "Jenny(0.1.1.1.0)" is the node in "proceedings(0.1.1)" in Figure 1. It's easily to get that $slca(S_1, S_2, \ldots S_m) = RemoveAncestor(lca(S_1, S_2, \ldots S_m))$.

Given a query $\mathcal{Q}$ which contains a list of m keywords $k_1, k_2, \ldots k_m$(for ease of presentation, we don't make any distinguish between query($k_1, k_2, \ldots k_m$) and query("$k_1\ k_2 \ldots k_m$") with an input XML document $\mathcal{D}$), the answers of $slca(k_1, k_2, \ldots k_m)$ are the result nodes of $slca(S_1, S_2, \ldots S_m)$, where S_i denotes the sorted keyword list of k_i for $1 \leq i \leq m$, i.e., the list of nodes whose label directly contains k_i sorted by id.

3.2   SLCA Based Keyword Search with Synonym Rules

Given a query $\mathcal{Q} = \{k_1, k_2, \ldots k_m\}$, the Indexed Lookup Eager Algorithm(IL) [18] first get the sorted inverted list S_i for each k_i, and then sort the inverted list according to the size of S_i to make sure S_1 is the list with smallest size. IL algorithm compute their slca by $slca(slca(S_1, \ldots S_{m-1}), S_m)$.

When computing any $slca(S_i, S_j)$ for $1 \leq i < j \leq m$ and $i = j - 1$, the algorithm efficiently removing the ancestor nodes according to the following two lemmas.

Lemma 1. *For any two nodes v_i, v_j and s set S, if $pre(v_i) < pre(v_j)$ and $pre(slca(v_i, S)) > pre(slca(v_j, S))$ then $slca(v_i, S) \prec slca(v_j, S)$*

Lemma 2. *Given any two nodes v_i,v_j and s set S such that $pre(v_i)$ ¡ $pre(v_j)$ and $pre(slca(v_i, S)) < pre(slca(v_j, S))$, if $slca(v_i, S)$ is not an ancestor of $slca(v_j, S)$, then for any v such that $pre(v) > pre(v_j)$, $slca(v, S) \not\prec slca(v_j, S)$*

Given a set of synonym rules $\mathcal{R}$ and a query $\mathcal{Q}$, we denoted the slca as slca($\mathcal{Q}$,$\mathcal{R}$). When introducing the concept of synonym rules, we should first get all the possible strings that could be produced after applying rules. Then we can compute the query result for each generated string. we can get the final answer by removing the ancestors of these query results. This can be formally defined by the following expression where transform($\mathcal{Q}, \mathcal{R}$) denotes all the strings generated by applying rules.

$$slca(\mathcal{Q}, \mathcal{R}) = \{removeAncestor(slca(\mathcal{Q}_1), \ ... \ \mathcal{Q}_i) \ ... \ slca(\mathcal{Q}_k)),$$
$$\mathcal{Q}_i \in transform(\mathcal{Q}, \mathcal{R}), 1 \le i \le k\}$$

Next, we will give a deeply analysis for each possible transformation. We first give the sorted inverted list for the keywords used later as shown in Table 1.

Table 1. Part of sorted inverted keyword list in Figure 1

Jennie	S_1=[0.1.2.1.0, 0.1.3.1.0, 0.2.0.0, 0.3.1.1.0]
database	S_2=[0.1.0.0, 0.1.3.0.0, 0.3.0.0]
data	S_3=[0.1.1.0.0, 0.3.1.0.0, 0.3.2.1.0]
base	S_4=[0.1.1.0.0, 0.1.2.0.0, 0.1.2.2.0, 0.3.1.0.0, 0.3.2.0.0]
datum	S_5=[0.1.2.0.0]
Jenny	S_6=[0.0.0, 0.1.1.1.0]

Condition 1, Split. Given a query $\mathcal{Q}$ ="database Jennie" in Figure 1 and a set of synonym rules $\mathcal{R}$ = "database"→"data base". After transformation, the query will be expanded to $\mathcal{Q}_1$ = "database Jennie" and $\mathcal{Q}_2$ ="data base Jennie". Therefore, we should first compute slca($\mathcal{Q}_1$) and slca($\mathcal{Q}_2$) and then remove the ancestor nodes to get the final result. As the keyword list for "data"(S_3), "base"(S_4), "database"(S_2) and "Jennie"(S_1) are listed in Table 1. For slca($\mathcal{Q}_2$), we should first compute slca(S_3,S_4) and then compute their slca with S_1. The procedure can be expressed as slca($\mathcal{Q}$,$\mathcal{R}$)=ra(slca(S_2,S_1),slca(S_3,S_4,S_1)) where ra() denotes remove ancestor operation. As we know "database" and "data base" represents the same object, then the expression can be denoted as slca($\mathcal{Q}, \mathcal{R}$) = slca(ra(slca(S_3, S_4), S_2), S_1) which could be modeled as a tree as shown in Figure 3(a). The optimize technique here we used is to execute *ra* operation as soon as possible due to slca is much more time-consuming compared with ra. Some early pruned nodes(by ra) would have no chance to participate in the next slca operation. Then, we can get the final result by slca($\mathcal{Q}, \mathcal{R}$) = slca(ra([0.1.1.0.0, 0.3.1.0.0, 0.3.2],[0.1.0.0, 0.1.3.0.0, 0.3.0.0]),S_1)=slca([0.1.1.0.0, 0.3.1.0.0, 0.3.2,0.1.0.0, 0.1.3.0.0, 0.3.0.0],[0.1.2.1.0, 0.1.3.1.0, 0.2.0.0.0, 0.3.1.1.0, 0.3.2.1.0])=[0.1.3, 0.3.1, 0.3.2].

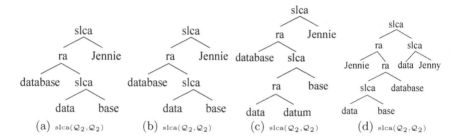

Fig. 2. Possible Transformation

Condition 2, Merge. Considering a query Q = "data base Jennie" in Figure 1 and a set of synonym rules R = "data base"→"database". we will get new strings Q_1 = "data base Jennie" and Q_2 ="database Jennie" after transformation. We can see the procedure is very similar to Condition 1. slca(Q, R) = RemoveAncestor(Q_1, Q_2) = [0.1.3, 0.3.1, 0.3.2]. The corresponding tree model is shown in Figure 3(b).

Condition 3, Mix. For query Q = "data base Jennie" in Figure 1 and a set of synonym rules R = "data base" → "database", "data" → "datum". The new generated queries will be Q_1 = "data base Jennie", Q_2 = "database Jennie" and Q_3 = "datum base Jennie". According to the xml tree model as shown in Figure 3(c), we can easily get the expression of the procedure slca(Q,R) = slca(Q_1, Q_2, Q_3). Thus the result will be slca(Q,R) = [0.1.2, 0.1.3, 0.3.1, 0.3.2].

Condition 4, Overlap. Given a qeury Q = "data base Jennie" in Figure 1 and a set of synonym rules R = "data base" → "database", "base Jennie" → "Jenny". The new queries will be Q_1 = "data base Jennie", Q_1 = "database Jennie" and Q_1 = "data Jenny". The result will be [0.1.1, 0.1.3, 0.3.1, 0.3.2](Figure 3(d)).

We have listed all the possible conditions we might meet when introducing the synonym rule semantics. **Split** means the one substring are expanded to more strings according to the size. Similarly, **Merge** refers to the inverse process. While, **Mix** and **Overlap** refers to the condition that Split and Merge come up with the same substring. For example, "data base" could be expanded to "data base", "database" and "datum base" according to R = {"*database*" → "*database*", "*data*" → "*datum*"}. However, the overhead of computing slca based on synonym rules is much too expensive. The cost will be in exponential scale to the size of useful synonym rules.

4 Optimizations

In this section, we will give some optimization techniques to speed up the procedure of computing slcas when introducing synonym rules.

Given a string s and a set of synonym rules R, the goal is to find all possible matching transformations. The observation is that if *lhs* is a substring of s, then *lhs* is a prefix of some suffix of *s*. We can construct a trie over all the distinct *lhs* in T and then process every string in *lhs* of $\mathcal{R}$. After that, for a given string, use each of its suffixes to look up the trie, i.e. we need to traversal each input string from rear to head in word level. For each substring we then need to scan from head to rear to check whether there is a matching for the sub of this substring. The details are straightforward and we defer them to the full version of the paper.

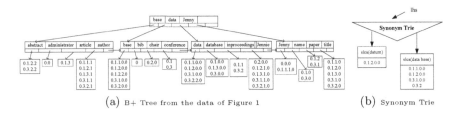

(a) B+ Tree from the data of Figure 1   (b) Synonym Trie

Fig. 3. Index Structure

Traditional strategies first get all sorted inverted list for each keyword occurs in the XML Document and then maintain a B^+ structure to speed up the look up option as shown in Figure 3(a). Differently, apart from the B^+ structure we also store the slca of the right-side of each rule. i.e., given a rule "lhs" → "rhs", we pre-compute slca("rhs") and then store the result into the trie which is called *synonym trie* as shown in Figure3(b). Note that, we only store slca of string with non-empty result. The benefit is that when getting a transformation, we also get the slca of "rhs".

4.1   Transformation Matching Based IL Algorithm

Next we will explain how to use this synonym trie to speed up our query procedure. Here we combine transformation matching operation and IL algorithm together denoted as *Transformation Matching based Indexed Lookup Eager Algorithm(TM-IL)*. The algorithm is based on transformation matching operations. As shown in Algorithm 1, ♯6 - 10 denotes the IL Algorithm, while ♯11 - 16 denotes the matching transformations. when there is a matching, we need to query the slca result of this matching substring(♯11, TMIL) and then pass the query result to next iteration(♯15, TMIL). However, if the query result is null, this might happen due to i) no matching, ii) no keywords in the xml document for the query, and iii) null slca for the matching keywords, the algorithm will goes into next loop. Finally, if we reach to the head of input string *s*, the algorithm will call computeSlca(canList) to compute and return slca of candidate inverted lists(♯20, TMIL).

For example, we consider again the query used in Section 3.2, Condition 4 applied on the data of Figure 1. "base Jennie" will be firstly detected(♯7, TMIL), then we will get slca("Jenny")=S_1 as shown in Figure 1 which will be added to the candidate list canList and passed to next iteration(♯10, TMIL). For the next iteration, s="data", canList={S_6} and rList=empty, it will get inverted list S_3 for "data" at ♯7, and compute slca(S_3,S_6) at ♯8, then pass the result to next

---

Algorithm 1. TM-IL Algorithm

Procedure TMILCall(s)

1: result = empty; //receive final results
2: rList = empty; //store intermediate keyword lists
3: canList = empty; //store candidate keyword lists
4: **return** result.addAll(TMIL(s,canList,rList)); //combine the result and remove ancesotrs

Procedure TMIL(s,canList,rList)

1: **if** s!=null **then**
2:    **for** $(i = |s| - 1; i \geq 0; i - -)$ **do**
3:       $s_1 = s.substring(i, |s|)$;
4:       **for** $(j = 1; j \geq |s_1|; j + +)$ **do**
5:          $s_2 = s_1.substring(0, j)$;
6:          **if** (j==1) **then**
7:             queryResult=queryBPTree(s_2);
8:             canList = slca(queryResult,canList)
9:             rList.addAll(TMIL(s.substring(0,i),canList,rList))
10:            **end if**
11:            qeuryResult=qeurySynonymTrie(s_2); //query slca(s_2) from synonym trie.
12:            **if** (qeuryResult != null) **then**
13:               declare *canList'* and add *queryResult* to this list.
14:               canList'.add(IL(s_1.substring(j,$|s_1|$))); //IL denotes Index Lookup Eager Algorithm
15:               rList.addAll(TMIL(s.substring(0,i),canList',rList))
16:            **end if**
17:      **end for**
18:   **end for**
19: **else**
20:    **return** computeSlca(canList);
21: **end if**

Procedure computeSlca(canList)

1: for each list in canList, compute their slca by IL();

---

iteration($\sharp$9). The result for "data Jennie" will be add to result list *rList* at $\sharp$20 for the next iteration. After all the return operations, the final result will be combined together($\sharp$4, TMILCall).

5   Conclusion and Future Work

We propose in this paper a SLCA based keyword search with synonym rules over xml documents. We formally defines the semantics of synonym rules and analyze the possible matching transformations. In addition, we also apply this method to effectively and efficiently compute slcas over XML Documents. To the best of our knowledge, this is the first work that takes into account of synonym rules for xml keyword search. Our theoretical analysis shows that our method is orthogonal to other xml keyword search techniques.

There are several avenues for future work. For instance, we can use synonyms to improve the quality of ranking based xml keyword search, such as XSearch [6], XRank [7] and so on.

Acknowledgement. This research is partially supported by NSF China (No. 61170011) and 863 National High-tech Research Plan of China (No. 2012AA011001).

References

1. Arasu, A., Chaudhuri, S., Kaushik, R.: Transformation-based framework for record matching. In: ICDE, pp. 40–49 (2008)
2. Arasu, A., Ganti, V., Kaushik, R.: Efficient exact set-similarity joins. In: VLDB, pp. 918–929 (2006)
3. Bao, Z., Ling, T.W., Chen, B., Lu, J.: Effective xml keyword search with relevance oriented ranking. In: ICDE, pp. 517–528 (2009)
4. Bilenko, M., Mooney, R.J.: Adaptive duplicate detection using learnable string similarity measures. In: KDD, pp. 39–48 (2003)
5. Chaudhuri, S., Ganti, V., Kaushik, R.: A primitive operator for similarity joins in data cleaning. In: ICDE, p. 5 (2006)
6. Cohen, S., Mamou, J., Kanza, Y., Sagiv, Y.: Xsearch: A semantic search engine for xml. In: VLDB, pp. 45–56 (2003)
7. Guo, L., Shao, F., Botev, C., Shanmugasundaram, J.: Xrank: Ranked keyword search over xml documents. In: SIGMOD Conference, pp. 16–27 (2003)
8. Kondrak, G.: N-Gram Similarity and Distance. In: Consens, M.P., Navarro, G. (eds.) SPIRE 2005. LNCS, vol. 3772, pp. 115–126. Springer, Heidelberg (2005)
9. Koudas, N., Sarawagi, S., Srivastava, D.: Record linkage: similarity measures and algorithms. In: SIGMOD Conference, pp. 802–803 (2006)
10. Li, G., Feng, J., Wang, J., Zhou, L.: Effective keyword search for valuable lcas over xml documents. In: CIKM, pp. 31–40 (2007)
11. Miller, D.R.H., Leek, T., Schwartz, R.M.: A hidden markov model information retrieval system. In: SIGIR, pp. 214–221 (1999)
12. Murray, G.C., Teevan, J.: Query log analysis: social and technological challenges. SIGIR Forum 41(2), 112–120 (2007)
13. Salton, G., Buckley, C.: Term-weighting approaches in automatic text retrieval. Inf. Process. Manage. 24(5), 513–523 (1988)
14. Sarawagi, S., Kirpal, A.: Efficient set joins on similarity predicates. In: SIGMOD Conference, pp. 743–754 (2004)
15. Tsuruoka, Y., McNaught, J., Tsujii, J.-i., Ananiadou, S.: Learning string similarity measures for gene/protein name dictionary look-up using logistic regression. Bioinformatics 23(20), 2768–2774 (2007)
16. Wikipedia, http://en.wikipedia.org/
17. Winkler, W.E.: The state of record linkage and current research problems. Technical report, Statistical Research Division, U.S. Census Bureau (1999)
18. Xu, Y., Papakonstantinou, Y.: Efficient keyword search for smallest lcas in xml databases. In: SIGMOD Conference, pp. 537–538 (2005)
19. Xu, Y., Papakonstantinou, Y.: Efficient lca based keyword search in xml data. In: CIKM, pp. 1007–1010 (2007)

XML Concurrency Control Protocols: A Survey[*]

Weifeng Shan[1,2], Husheng Liao[1], and Xueyuan Jin[1]

[1] Colleage of Computer Sciences, Beijing University of Technology, Pingleyuan 100#,
Chaoyang District, Beijing 100024, China
[2] Department of Disaster Information Engieering, Institue of Disaster Prevention,
College street, Yanjiao Develpment Zone, Sanhe 065201, China
Shanweifeng@cidp.edu.cn, {Liaohs,Jinxueyun}@bjut.edu.cn

Abstract. As XML has become a standard format for data exchange and data representation, ensuring the correctness of concurrent update on XML data becomes a critical issue. This paper presents an overview of some of the most important XML concurrency control protocols so far, such as locking-based, timestamp-based and optimistic XML concurrency control protocols. Summary and comparison are given for each protocol from point of implementation principle, optimistic algorithm, multiversion technology and serializability aspects. Some current researching trends in XML concurrency control protocols are discussed.

Keywords: XML Concurrency Control Protocol, Locking-based Protocol, Timestamp-based Protocol, Optimistic Concurrency Control Protocol, XML Database Transactions.

1 Introduction

In recent years, Extensible Markup Language (XML) has become the standard format for data representation and data exchange on the Internet and attracted many researchers to study. Existing works on XML data have been concentrated on query optimization, storage, and indexing, etc. As the content of XML documents may change over time, updating XML becomes a crucial issue in distributed applications because many clients may read or/and write the same XML document concurrently.

As is known to all, there are some side effects while multi users concurrently read/write the same shared data [1]. In order to solve these problems, transaction technology is firstly introduced in the relational database area. As an import component of transaction, concurrency control protocol is responsible for the isolation property of the transaction while multi programs accessing the same database concurrently.

Many concurrency control protocols for relational and object-oriented database systems have been proposed. They cannot provide high concurrency for XML

---

[*] This research is supported by Beijing Natural Science Foundation (4122011).
This research is supported by Beijing Natural Science Foundation (4082003).

Z. Bao et al. (Eds.): WAIM 2012 Workshops, LNCS 7419, pp. 299–308, 2012.

because of the semi-structure and hierarchical properties of XML and special access interfaces such as DOM and XPath. In the past few years, various XML concurrency control protocols have been proposed, and they can be classified in different dimensions. According to the concurrency control mechanism used, they can be divided into two classes: locking-based and non-locking. We can also classify them into single-version and multiversion protocols. According to assumed access standard API to the XML documents, there are three classes: DOM-based, XPath-based and mixed protocols which are appropriate for both DOM and XPath standards.

There are essentially three ways to store XML documents: file system, relational or object-oriented database systems, and Native XML database systems. As the lack of synchronization mechanism of storing in a file, and low degree of concurrency for XML of storing in relational or object-oriented database systems, Native XML database systems are regarded as a best alternative to store XML because they can provide efficient synchronization mechanism and high degree of concurrency [2-5]. Ref. [6] compared the performance of several DOM-based XML concurrency control protocols, but it did not involve other protocols. Ref. [7] introduced an evaluation framework for XML concurrency control protocols and tested the performance of OO2PL, taDOM, PL-SAT and XLP. However, little attention has been paid on summarization of the advantages and disadvantages of different protocols. Ref. [2] only introduced protocols based on 2PL. Some new progresses have been made in the study of XML concurrency control protocols in recent years. To summarize the state-of-the-art technologies on XML concurrency control protocols, we analyze and compare each protocol presented so far from implementation principle, optimistic algorithm, multiversion technology and serializability aspects, and discuss some developing trends of XML concurrency control protocols.

This paper is organized as follows. Section 2 shows the protocols based on locking, including node locking and path index locking protocols. None-locking based protocols, including timestamp-based and optimistic, are discussed in Section 3 and Section 4 compares all protocols presented in this paper and summarizes the advantages and disadvantages of each type protocol. Section 5 concludes the paper.

2   Locking-Based XML Concurrency Control Protocols

In traditional transaction processing research areas, two-phase locking (2PL) is a concurrency control schema that guarantees the serializability [1]. The protocol utilizes locks to block other transactions from accessing the same data. There are two basic types of locks in 2PL protocols: shared (S) and exclusive (X) locks.

Many locking-based XML concurrency control protocols are tailored and transplanted from 2PL protocols. To enhance the high degree of concurrency, these protocols may utilize more lock types. On the base of locking different objects, these protocols can be classified into node locking and path index locking protocols.

2.1   XML Concurrency Control Protocols Based on Node Locking

XML nodes, sub trees or edges between two nodes are locked while a read and/or write operation wants to access them in the node locking protocols. They may be classified into three categories: DOM-based, XPath-based and mixed protocols which may be used for both DOM and XPath standards.

DOM-Based XML Concurrency Control Protocols

These protocols provide navigation and update operations for DOM API. Ref. [5, 8] introduced four concurrency control protocols based on 2PL: Doc2PL, Node2PL, NO2PL and OO2PL. Doc2PL is the most simplified protocol, which locks the whole XML document. Obviously, it is impossible that many users concurrently update the same XML document. In Node2PL, a shared lock is required for the node while a transaction is about to read it, and an exclusive lock is acquired for its parent node while a transaction wants to update it. The NO2PL protocol acquires locks for all nodes whose pointer is traversed or modified. OO2PL is different from Node2PL and NO2PL that lock nodes, it locks pointers. While a transaction wants to read or update one node, it must check the compatibility with the locks on all four pointers of the node. Serializability is ensured in four protocols even in the presence of ID jumps.

Although *2PL(Doc2PL, Node2PL, NO2PL and OO2PL) protocols are proposed as universal XML concurrency control protocols for DOM API, they have low degree of concurrency for ignoring the semi-structured characteristic of XML data[9].

Multi-granularity locking (MGL) protocol which applies to hierarchies of objects is used in the relational world. To avoid lock conflicts when objects set different levels are locked, intention locks are introduced except lock modes S and X. Although a MGL protocol can also be applied to XML document trees, it is too strict in most cases, because both S and X mode on a node would always lock the whole subtree below. This restriction does not suit XML where transactions may concurrently update different subtrees of the current node. Hence, MGL does not provide high degree of concurrency [10]. Ref. [7, 11-14] introduced a serial of taDOM* (taDOM2, taDOM2+, taDOM3 and taDOM3+) protocols which refined the MGL ideas and provided tailored lock modes for high concurrency in XML trees. They introduced a far richer set of locking modes. There are 8 lock modes in taDOM2, 12 modes in taDOM2+, and more than 20 different lock modes in taDOM3+. These protocols ensure the serializability of transaction and provide higher performance than *2PL protocols.

Ref. [15] confirmed that an updated XML should be a valid XML which is restricted by DTD. Repetitive parts which are updated with actions such as inserting/deleting could be identified based on a given DTD, nodes corresponding to symbols enclosed with '*' or '+'. List locking protocol was proposed which locks on the list of repetitive children nodes for handling concurrent insertion/deletion of subtrees and it allows update actions on descendents during inserting/deleting subtrees. It uses less number of lock modes for synchronizing structure update actions compared to the other protocols.

Ref. [16] exploited the DOM operations' semantics to increase concurrency. By analyzing DOM operations, it proposed a new semantic-based concurrency control protocol (SCD). Although the performance of it outperformed both taDOM* and OO2PL protocols on throughput and response time, it did not guarantee the serializability of transaction.

Except MGL protocols, Tree-based Locking (TL) is another scheme used to deal with concurrent operation on the shared tree objects. Ref. [17] proposed a dynamic tree-based locking protocol to support the update of XML tree structure, and it also guaranteed the serializability of transaction and was dead-lock free. Ref. [18] proposed a new XML concurrency control protocol, called Multi Lightweight Multi Granularity Locking (LWMGL), based on TL and MGL. To achieve high concurrency, LWMGL locked at the level of precise elements by using light weight locks of granule size. Phantom problems were resolved and the numbers of pseudo-conflicts were reduced in LWMGL.

XPath-Based XML Concurrency Control Protocols
There are two aspects of XPath to achieve high concurrency. Firstly, only those result nodes whose node type is specified in the last location step are actually accessed. Secondly, each axis of a location step specifies the direction to navigate in the tree of XPath mode.

Ref. [19] proposed two XPath-based protocols: Path Lock Propagation (PL-PRO), which sat very few locks but required more work for judging conflicts, and Path Lock Satisfiability (PL-SAT), which used a multitude of read locks but made checking for conflicts easy. Both PL-PRO and PL-SAT are serializable, however, they only support two axes of XPath: '/' and '//'.

Ref. [20, 21] presented another protocol, called XPath Locking Protocol (XLP), which defined five lock modes: P-, R-, W-, I- and D- for navigate, read, write, insert and delete actions, respectively. Although this protocol has better performance than *2PL protocols, it only supports a sub set of XPath and possibly results in dead-lock. Based on XLP, Ref. [22] proposed XML Path Locking by Child Consideration (XPLC), which allowed two transactions to concurrently update below a node. By using AC and IRC lock modes, it allows a transaction to read a subset of a subtree and at the same time permits another transaction to insert a node with different type under the root of that subtree.

Ref. [23] introduced a novel XML concurrency control protocol which used multiversion technology and supported full-fledged XPath expressions. It effectively resolves phantom problems by using logical locking approach. As it is necessary for every active write transactions to create a copy of nodes, it will consume more memory than those do not use multiversion technology. Ref. [24] presented a similar protocol, namely MPX.

Similar with Ref. [16], Ref. [25] also considered the operations' semantics to increase the degree of concurrency. Two write operations can be considered in conflict whereas they are not, if we do not take into account the semantics of the operations, but only the path expression. Therefore, a Semantics-based Fine Granular Concurrency Control protocol for XML Data (SeCCX) was proposed, which locked the objects according to the semantics of user's operation and sat different restrictions on different objects. SeCCX protocol has good performance and conflict serializability property.

Different from the previous XPath-based protocols, XML Region Lock (XR-Lock) protocol was proposed in Ref. [26]. All nodes in a XML tree are encoded with Relative

Region Coordinate (RRC) [27] solution and the lock objects in XR-Lock are nodes in a relative region. It ensures the serializability property of transaction. Similarly, Ref. [28] introduced a protocol based on Local Path Locking. Based on node coding of XML nodes, it uses the scan function to judge the locking status of the target node ancestor and descendant, and locks the target node or its parent node according to the operation type to reduce the frequent request for lock to improve the concurrency of transaction.

XML Concurrency Control Protocols Supporting Both XPath and DOM

These protocols support both XPath and DOM standards, even SAX standard. Simple XML Concurrency Control Protocol (SXCCP+) was proposed in Ref. [29], where each accession to an XML document are translated into a set of primitive and indivisible equivalent operations, such as $C(n)$ operation for test of child node existence, $R(n)$ operation for node content access, $U(n)$ operation for node modification, $D(n)$ operation for node deletion, and $I(n', n, l)$ operation for insertion of n' node into n node at lst position of children of n node. Only these primitive operations are taken into account in SXCCP+, so it is autonomous from specific XML document access method.

2.2   XML Concurrency Control Protocols Based on Path Index Locking

Unlike protocols based on node locking, these protocols guarantee the isolation of transactions by locking nodes in path index, not nodes in XML document tree [9]. The advantages of these protocols are they need fewer locks than the node locking protocols because the size of nodes in path index is smaller than that of nodes in XML document tree. However, they also have some disadvantages, e.g., they may fail to detect all pseudo-conflicts, and conflict detection is NP-complete while paths contain predictions, and it is difficult to deal with IDREF problems, etc.

Ref. [30] proposed DGLOCK protocol which is based on DataGuides. Although it reduces the number of locks, it does not prevent phantoms and does not guarantee for serializability, and it also does not support descendant axis.

Ref. [31, 32] presented an XPath-based DataGuide Locking protocol (XDGL) based on DGLOCK, which ensured strict serializability and combined predicate and logical locks to provide protection from phantom appearance. It supports a subset of XPath express, including child axis and descendant axis.

In Ref. [33], Snapshot based Concurrency Control Protocol (SXDGL) was introduced. As it uses multiversion technology, SXDGL eliminates data contention between read-only and update transactions. Moreover, SXDGL also takes into account the hierarchical structure and semantics of XML data model determining conflicts between concurrent XML operations.

3   None-Locking XML Concurrency Control Protocols

Besides locking-based protocols, there are many protocols based on none-locking, such as timestamp-based and optimistic XML concurrency control protocols.

3.1   Timestamp-Based XML Concurrency Control Protocols

In timestamp-based XML concurrency control protocols, the commit orders of transactions are determined by their timestamps. Ref. [34] proposed two protocols based on timestamp: XML Timestamp Ordering (XTO) and XML dynamic Commit Ordering (XCO). The XCO protocol had lower abortion rate than the XTO protocol. It is achieved by delaying the assignment of timestamp of transactions.

A multiversion timestamps concurrency control protocol for XML data (XStamps) was shown in Ref. [35]. The XStamps protocol was designed based on Multiversion Timestamps Ordering Protocol (MTOP). It divides transactions into read transactions and write transactions which contain at least one update operation. Read transactions operate on the read version of data while write transactions operate on the write version. Read transactions do not conflict with write transactions.

3.2   Optimistic XML Concurrency Control Protocols

Optimistic protocol is a concurrency control method that assumes that multiple transactions can execute concurrently without affecting each other, therefore transactions can proceed without locking the data resources that they affect. While a transaction is about to commit, it will verify that no other transaction has modified its data. If validation fails, the committing transaction will roll back and restart.

Ref. [36] proposed two protocols OptiX and SnaX. A transaction follows three phases: a working phase, a validation phase and an update phase. In the working phase, nodes that are read and written in the transaction T_i are stored in the set $RS(T_i)$ and $WS(T_i)$ in OptiX, respectively. While SnaX provides snapshot isolation avoiding keeping track of reads. Once a transaction wishes to commit, it goes into the validation phase. No two transactions can be concurrently in validation phase. In OptiX, a transaction T_i passes validation if for all concurrent transaction T_j that already validated, $WS(T_j) \cap RS(T_i) = \emptyset$. While in SnaX, a transaction T_i is validated if for all concurrent transaction T_j that already validated, $WS(T_j) \cap WS(T_i) = \emptyset$. If T_i passes validation, it goes into the update phase. The changes carried out by T_i are updated into the database. Otherwise, T_i must abort and all its temporary space is freed. Both OptiX and SnaX do not guarantee for the serializability and phantom problem is not solved.

On the basis of OptiX and SnaX, Ref. [37] proposed a novel optimistic path-based concurrency control over XML documents for XPath-based queries and updates. It aims to reduce the validation phase duration by detecting conflicts via analyzing XPath expressions. The advantage of this approach is that conflict detection is not dependent on the size of the database, nor on the amount of modified fragments. However the drawback of this approach is that not every conflict can be detected from only XPath expressions comparison. The paper also gives a solution to deal with the problems of '*' sign and '//' axis of an XPath expression.

Table 1. Comparison of XML Concurrency Control Protocols

XML Concurrency Control Protocols	Mechanism	Multiversioned	Optimistic/ Pessimistic	Serializability
*2PL [5]	Node locking	No	Pessimistic	Yes
taDOM*[11]	Node locking	No	Pessimistic	Yes
List Locking[15]	Node locking	No	Pessimistic	Yes
SCD[16]	Node locking	No	Pessimistic	No
LWMGL[18]	Node locking	No	Pessimistic	Yes
PL-SAT[19]	Node locking	No	Pessimistic	Yes
PL-PRO[19]	Node locking	No	Pessimistic	Yes
XLP[20]	Node locking	No	Pessimistic	Yes
XPLC[22]	Node locking	No	Pessimistic	Yes
Ref. [23]	Node locking	Yes	Pessimistic	Yes
MPX[24]	Node locking	Yes	Optimistic	No
XR-Lock[26]	Node locking	No	Pessimistic	Yes
Ref. [28]	Node locking	No	Pessimistic	Yes
SeCCX[25]	Node locking	No	Pessimistic	Yes
SXCCP+[29]	Node locking	No	Pessimistic	Yes
DGLOCK[30]	Path index locking	No	Pessimistic	No
XDGL[31, 32]	Path index locking	No	Pessimistic	Yes
SXDGL[33]	Path index locking	Yes	Pessimistic	Yes
XTO[34]	Timestamp	No	Pessimistic	Yes
XCO[34]	Timestamp	No	Pessimistic	Yes
XStamps[35]	Timestamp	Yes	Pessimistic	Yes
OptiX[36]	Optimistic	Yes	Optimistic	No
SnaX[36]	Optimistic	Yes	Optimistic	No
Ref. [37]	Optimistic	Yes	Optimistic	No

4   Summarization for XML Concurrency Control Protocols

As is shown in Tab. 1, Most of XML concurrency control protocols are locking-based protocols, such as *2PL, taDOM*, XPL, and XDGL, etc. Only few of protocols are optimistic, like SnaX and OptiX. Most of protocols may ensure the serializability. Only several protocols use multiversion technology, e.g. MPX, SXDGL, SnaX, OptiX and XStamps.

Tab. 2 summarizes advantages and disadvantages for each type protocol. In node locking protocols, different protocols have different lock modes and lock granular, the numbers of locks increase with the size of XML document tree. This not only consumes more storage, but also increases the difficulty of management of locks. While in protocols based on path index locking, the numbers of locks is independent for the real XML tree, they need fewer locks and have smaller lock management costs than node locking protocols.

Table 2. Comparison of different categories of XML Concurrency Control Protocols

Concurrency Protocols	Advantages	Disadvantages	Usage Scenario
Node locking protocols	Universal, mature, and has better performance for most purposes	The number of locks increases with the size of XML document. Difficult to deal with deadlock, livelock, phantom, IDREF problem and pseudo-conflicts	Used for most purposes
Path index locking protocols	Less locks than node locking protocols, easy to manage locks	It's difficult to deal with deadlock, livelock, phantom, IDREF problem, pseudo-conflicts and prediction.	Auxiliary solution
Timestamp-based protocols	No deadlocks, none-blocking	Starvation is possible	Low conflict environments
Optimistic protocols	No deadlocks, none-blocking	Cost of abortion of transaction is high	Low conflict environments

Both timestamp-based and optimistic protocols are lock free, so they are deadlock free and none-blocking. Timestamp-based protocols need additional storage to store timestamps for every data item, likewise, optimistic protocols also need temporary memory to store the repetition of every data item. They are suit for low conflict environment. When conflicts are rare, transactions can complete without waiting and lead to higher throughput than other methods.

To sum up, there are several trends in the research of XML concurrency control protocols. (1) XPath should be studied deeply for its location steps contain rich information which can be used to enhance the degree of concurrency. (2) Semantic of XML operations may reduce pseudo-conflicts. (3) A DTD can be used to effectively detect conflicts between two transactions. (4) Multiversion is important technology to improve the performance, although it consumes more storage than other schemes. (5) Each protocol has its own advantage and disadvantages, so there is a trend to compose multi protocols to enhance the performance of transactions.

5   Conclusions

This paper introduces and summarizes various kinds of XML concurrency control protocols, e.g. locking-based, timestamp-based, optimistic, DOM-based, XPath-based protocols. We also compare more than twenty protocols from implementation mechanism, multiversion, optimistic or pessimistic control and conflict serializability. Finally, we discussed the trend in researching of XML concurrency control protocols.

References

1. Qwikum, F., Vossen, G.: Transactional Information Systems: Theory, Algorithms, Practice of Concurrency Control and Recovery. Morgan Kaufmann, San Francisco (2002)
2. Wang, J.W., Li, H.N.: Research Development of XML database. Journal of Nanhua University (Science and Technology) 20(03), 42–46 (2006)
3. Fiebig, T., Helmer, S., Kanne, C.-C., Moerkotte, G., Neumann, J., Schiele, R., Westmann, T.: Natix: A Technology Overview. In: Chaudhri, A.B., Jeckle, M., Rahm, E., Unland, R. (eds.) NODe-WS 2002. LNCS, vol. 2593, pp. 12–33. Springer, Heidelberg (2003)
4. Fomichev, A., Grinev, M., Kuznetsov, S.: Sedna: A Native XML DBMS. In: Wiedermann, J., Tel, G., Pokorný, J., Bieliková, M., Štuller, J. (eds.) SOFSEM 2006. LNCS, vol. 3831, pp. 272–281. Springer, Heidelberg (2006)
5. Helmer, S., Kanne, C.C., Moerkotte, G.: Evaluating lock-based protocols for cooperation on XML documents. Sigmod Record 33(1), 58–63 (2004)
6. Jankiewicz, K., Siekierska, A., Siekierski, M.: Quantity comparison of concurrency control methods for XML database systems based on DOM API. Int. J. Web Eng. Technol. 4(4), 534–554 (2008)
7. Haustein, M., Harder, T., Luttenberger, K.: Contest of XML lock protocols. In: Proceedings of the 32nd International Conference on Very Large Data Bases, pp. 1069–1080. VLDB Endowment (2006)
8. Helmer, S., et al.: Lock-based protocols for cooperation on XML documents. In: 14th International Workshop on Database and Expert Systems Applications, pp. 230–234 (2003)
9. Wang, Y.: Study on XML Transaction Model and Concurrency control. Zhejiang University, Hangzhou (2006)
10. Bächle, S., Härder, T.: The Real Performance Drivers behind XML Lock Protocols. In: Bhowmick, S.S., Küng, J., Wagner, R. (eds.) DEXA 2009. LNCS, vol. 5690, pp. 38–52. Springer, Heidelberg (2009)
11. Haustein, M.P., Härder, T.: Optimizing lock protocols for native XML processing. Data & Knowledge Engineering 65(1), 147–173 (2008)
12. Haustein, M.P., Härder, T.: Adjustable Transaction Isolation in XML Database Management Systems. In: Bellahsène, Z., Milo, T., Rys, M., Suciu, D., Unland, R. (eds.) XSym 2004. LNCS, vol. 3186, pp. 173–188. Springer, Heidelberg (2004)
13. Haustein, M.P., Härder, T.: taDOM: A Tailored Synchronization Concept with Tunable Lock Granularity for the DOM API. In: Kalinichenko, L.A., Manthey, R., Thalheim, B., Wloka, U. (eds.) ADBIS 2003. LNCS, vol. 2798, pp. 88–102. Springer, Heidelberg (2003)
14. Bächle, S., Härder, T., Haustein, M.P.: Implementing and Optimizing Fine-Granular Lock Management for XML Document Trees. In: Zhou, X., Yokota, H., Deng, K., Liu, Q. (eds.) DASFAA 2009. LNCS, vol. 5463, pp. 631–645. Springer, Heidelberg (2009)
15. Lee, E.: Multi-granularity locks for XML repetitive. In: Fourth Annual ACIS International Conference on Computer and Information Science, pp. 222–227 (2005)
16. Jea, K.F., Chang, T.P., Chen, S.Y.: A Semantic-Based Protocol for Concurrency Control in DOM Database Systems. Journal of Information Science and Engineering 25(5), 1617–1639 (2009)
17. Pang, Y.M., Tan, Z.J., Wang, W.: Concurrent locking protocols for XML. Journal of Computer Research and Development 41(07), 1232–1239 (2004)
18. Choi, Y., Moon, S.: Lightweight multigranularity locking for transaction management in XML database systems. Journal of Systems and Software 78(1), 37–46 (2005)
19. Dekeyser, S., Hidders, J.: Path locks for XML document collaboration. In: Proceedings of the Third International Conference on Web Information Systems Engineering, pp. 105–114 (2002)

20. Jea, K.F., Chen, S.Y., Wang, S.H.: Concurrency control in XML document databases: XPath locking protocol. In: Ninth International Conference on Parallel and Distributed Systems, pp. 551–556 (2002)
21. Jea, K.F., Chen, S.Y.: A high concurrency XPath-based locking protocol for XML databases. Information and Software Technology 48(8), 708–716 (2006)
22. Izadi, K., Asadi, F., Haghjoo, M.S.: XPLC: a novel protocol for concurrency control in XML databases. In: 2007 IEEE/ACS International Conference on Computer Systems and Applications, pp. 450–453 (2007)
23. Choi, E.H., Kanai, T.: XPath-based Concurrency Control for XML Data. In: Proceedings of the 14th Data Engineering Workshop, pp. 302–313 (2003)
24. Wang, Y., Chen, G., Dong, J.-x.: MPX: A Multiversion Concurrency Control Protocol for XML Documents. In: Fan, W., Wu, Z., Yang, J. (eds.) WAIM 2005. LNCS, vol. 3739, pp. 578–588. Springer, Heidelberg (2005)
25. Rong, C., Lu, W., Zhang, X., Liu, Z., Du, X.: SeCCX: Semantics-Based Fine Granular Concurrency Control for XML Data. In: Shen, H.T., Pei, J., Özsu, M.T., Zou, L., Lu, J., Ling, T.-W., Yu, G., Zhuang, Y., Shao, J. (eds.) WAIM 2010. LNCS, vol. 6185, pp. 146–155. Springer, Heidelberg (2010)
26. Zhang, W.S., Liu, D.X., Sun, W.: XR-lock: Locking XML Data for Efficient Concurrency Control. In: Proceedings of the 5th World Congress on Intelligent Control and Automation, pp. 3921–3925 (2004)
27. Kha, D.D., et al.: An XML indexing structure with relative region coordinate. In: Proceedings of 17th International Conference on Data Engineering, pp. 313–320 (2001)
28. Kang, H., et al.: XML Concurrency Control Protocol Based on Local Path Locking. Computer Engineering 36(21), 7–10 (2010)
29. Jankiewicz, K.: SXCCP+: Simple XML Concurrency Control Protocol for XML Database Systems. Control and Cybernetics 38(1), 215–237 (2009)
30. Grabs, T., Bohm, K., Schek, H.J.: XMLTM: Efficient transaction management for XML documents. In: Proceedings of the Eleventh International Conference on Information and Knowledge Management (CIKM 2002), pp. 142–152 (2002)
31. Pleshachkov, P., Chardin, P., Kuznetsov, S.: A DataGuide-Based Concurrency Control Protocol for Cooperation on XML Data. In: Eder, J., Haav, H.-M., Kalja, A., Penjam, J. (eds.) ADBIS 2005. LNCS, vol. 3631, pp. 268–282. Springer, Heidelberg (2005)
32. Pleshachkov, P., Chardin, P., Kuznetsov, S.: XDGL: XPath-based concurrency control protocol for XML data. In: Database: Enterprise, Skills and Innovation: 22nd British National Conference on Databases, pp. 145–154 (2005)
33. Pleshachkov, P., Kuznetcov, S.: SXDGL: Snapshot Based Concurrency Control Protocol for XML Data. In: Barbosa, D., Bonifati, A., Bellahsène, Z., Hunt, E., Unland, R. (eds.) XSym 2007. LNCS, vol. 4704, pp. 122–136. Springer, Heidelberg (2007)
34. Helmer, S., Kanne, C.-C., Moerkotte, G.: Timestamp-Based Protocols for Synchronizing Access on XML Documents. In: Galindo, F., Takizawa, M., Traunmüller, R. (eds.) DEXA 2004. LNCS, vol. 3180, pp. 591–600. Springer, Heidelberg (2004)
35. Khin-Myo, W., et al.: XStamps: a multiversion timestamps concurrency control protocol for XML data. In: Proceedings of the 2003 Joint Conference of the Fourth International Conference on Information, Communications and Signal Processing and Fourth Pacific-Rim Conference on Multimedia, pp. 1650–1654 (2003)
36. Sardar, Z., Kemme, B.: Don't be a pessimist: Use snapshot based concurrency control for XML. In: 22nd International Conference on Data Engineering (ICDE 2006), pp. 130–133 (2006)
37. Berrabah, D., et al.: Optimistic path-based concurrency control over XML documents. In: 5th International Conference on Soft Computing As Transdisciplinary Science and Technology, CSTST 2008, pp. 390–396 (2008)

Using Conceptual Scaling for Indexing XML Native Databases

Dhekra Ayadi, Olfa Arfaoui, and Minyar Sassi-Hidri

National Engineering School of Tunis,
BP. 37 Le Belvédère 1002 Tunis, Tunisia
{dhekra.ayadi,olfa.arfaoui}@yahoo.fr, minyar.sassi@enit.rnu.tn

Abstract. The essential purpose of indexing techniques is to find methods that ensure faster access to a given well-defined data and thus avoid a sequential scan of document. Two approaches are used for indexing XML documents: indexing based on the values and structural one. However, indexing XML documents for research purposes can be a complex task especially when we consider content and structure. The aim is to provide a combined structure while assuring hierarchical levels of data content and structure representation. In this paper, we propose to use conceptual scaling-based Formal Concept Analysis for indexing both content and structure.

1   Introduction

The considerable development that has known Internet in recent years, particularly with the advent of the World Wide Web, has led to an exponential growth of the number of users of this type of resources. In order to facilitate the exchange and the data standardization, the W3C has introduced an XML (extensible Markup Language) [1], [2], which is a simple and flexible text format derived from SGML (Standard Generalized Markup language).

XML is a semi-structured data, which separates the contents of documents instructions presentations. These characteristics suggest an explosion of XML documents.

In this context, the issue of information retrieval (IR) in XML corpus becomes crucial. Indeed, according to databases domain, XML is a new approach for modeling information. Therefore, the development of systems for storing and efficient querying of XML documents requires the development of new indexing techniques. Recall that the purpose of indexing techniques is to accelerate access to specific data, other than a sequential scan of the entire data set.

Several studies have been adopted to try and improve storage and indexing techniques [3], [4], [5]; those from the databases community and those from recent research in the field of XML databases. All these methods aim to develop indexing systems corresponding to the specific needs of XML.

We focus in this work on methods of storage and indexing corresponding to the specific needs of XML documents.

Z. Bao et al. (Eds.): WAIM 2012 Workshops, LNCS 7419, pp. 309–318, 2012.

From the perspective of IR, the access to such documents raises new problems related to both structural and content information as well as the large volume of XML documents. Taking into account the structural dimension is expected to better meet the expectations of different users. It updates the reflection on the granularity problem of returned information.

Hence the importance of developing a new powerful technique for indexing XML documents, knowing that a good indexing technique has a direct influence on the optimization of the querying process of this type of document in terms of search time and query processing.

However, it is important to maintain the hierarchical structure of the document elements and the order of indexed items in order to avoid recalculating it each time and avoiding the accessing of the document in a sequential manner to determine a relationship between the elements.

The use of Formal Concept Analysis (FCA) [6] for indexing XML documents is interesting to resolve the indexing problems while putting into consideration the following objectives:

- The extraction of keys research, namely the most representative words in the documents and the structural information they contain.
- For semi-structured documents, the structural dimension is added to the content, and the following questions arise:
    - What do we index in the document structure?
    - How to connect this structure to the document content?
    - Depending on what dimension (items level, documents and collections) the indexing terms should be weighted?

An indexing scheme for XML documents should cover the following aspects:

- Allow reconstruction of the XML document decomposed in storage structures.
- Allow processing of the path expressions on the XML structure
- Accelerate navigation in XML documents.
- Allow processing of vague and precise predicates on the content of XML documents.
- Allow search by keywords.

The rest of the paper is organized as follows: section 2 presents the new approach to indexing XML documents. Section 3 concludes the paper and presents future work.

2   Indexing XML Document Based on FCA

The aim of our approach is the application of the FCA principles on XML documents in order to index them and subsequently facilitate the querying process of such documents. We adopted as the storage system of XML documents, the free native XML database eXist [7].

Our indexing approach is mainly composed of three steps:

- Step 1: XML tree traversal: is to traverse the XML tree and extract the textual data in the form of a set E.
- Step 2: Conceptual classification: is to build the concept lattice associated to each parent nodes generated following an ascending traversal of the document.
- Step 3: Conceptual scaling: The lattice structures obtained are combined into a single structure called fuzzy nested lattice.

Fig. 1 shows the overview of our indexing approach.

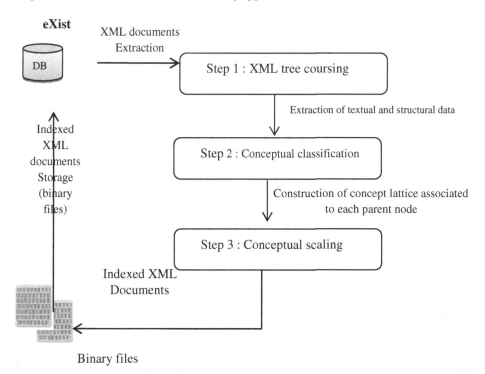

Fig. 1. Overview of our indexing approach

2.1   XML Tree Traversal

This step is to browse the tree and extract the XML data as a separate set of textual data called E.

Fig. 2 shows an example of XML document. The data that we may have in our example are extracted from leaves and nodes as follow: beginner, CSS 2, Daniel Glazman, Eyrolles, Training ... XML, Michael *J.* YOUNG, Microsoft Press, intermediate, eng, Training ... ASP.Net, Richard Clark.

Let take $E = \{D1, D2, ..., D11\}$ the data set representing the leaf nodes of the XML tree.

```
<bib>
<book level="beginner">
<title>CSS 2</title>
<author>Daniel GLAZMAN</author>
<publisher>Eyrolles</publisher>
</book>
<book niveau="beginner">
<title>Training ... XML</title>
<author>Michael J. YOUNG</author>
<author> Daniel GLAZMAN </author>
<publisher>Microsoft Press</publisher>
</book>
<book niveau="intermediate" lang= "eng" >
<title>Training ... ASP.Net</title>
<author>Richard CLARK</author>
<publisher>Microsoft Press</publisher>
</book>
</bib>
```

Fig. 2. Example of XML document

Next, the tree coursing is made with an ascending manner; its aim is to extract the structural data of parent nodes. According to our example, the first book is a parent node and it has as structural data $S = \{\text{level, title, author, publisher }\}$. Fig. 3 gives the steps of data extraction from an XML document.

Procedure 1 - Data Extraction

INPUTS : XML document
OUTPUT : A set of objects and attributes
BEGIN
Step 1: XML Tree traversal and data extraction in a set E.
Step 2: Ascending traversal and extraction of structural data from parent nodes.
END.

Fig. 3. Data extraction procedure

So we have extracted all objects and attributes in order to construct, in the second step, the simple concept lattice while basing on theoretical foundations of the FCA.

2.2   Conceptual Classification

This step consists in the building of the concept lattice associated to each parent nodes generated following an ascending traversal of the XML tree of step 1. Conceptual classification procedure is given in Fig. 4.

Procedure 2 - Conceptual classification

INPUTS : A set of objects and attributes
OUTPUTS : Simple Galois lattice
BEGIN
REPEAT
Step 1: For each parent node we construct the lattice whose context consists of all E words as context objects and all the leaf nodes as context attributes.
Step 2: For parent nodes which have the same parent we construct the lattice whose context consists of the objects set and the attributes set presenting respectively the set of parent nodes and the set of leaf nodes.
Until get the root of the XML tree.
End.

Fig. 4. Conceptual Scaling procedure

In this step, we relied on the FCA theory to construct simple Galois lattice.

The following tables show some examples of formal context of our example "bib.xml" with Boolean values.

The first node on the left noted "<book> [0]" is the first node to be processed. For this node, we construct the corresponding context which consists of the set of words E (obtained during the first stage) and the set of the child nodes in our example the nodes are: <level>, <title>, <author >, <publisher> representing respectively the set of objects and attributes of this context.

Table 1 defines the context node "<book> [0]", the lines represent the set of objects and the columns represent the set of attributes.

Table 1. Binary context of the node <book> [0]

R	D1	D2	D3	D4	D5	D6	D7	D8	D9	D10	D11
					<book> [0]						
<level>	1	0	0	0	0	0	0	0	0	0	0
<lang>	0	0	0	0	0	0	0	0	0	0	0
<title>	0	1	0	0	0	0	0	0	0	0	0
<author> [0]	0	0	1	0	0	0	0	0	0	0	0
< author > [1]	0	0	0	0	0	0	0	0	0	0	0
<publisher>	0	0	0	1	0	0	0	0	0	0	0

Similarly for the second and third node denoted "<book> [1]" and <book> [2].

Tables 2 and 3 define the contexts nodes "<book> [1]" and "<book> [2]". The lines of each one of these tables represent the set of objects and the columns represent the set of attributes.

Table 2. Binary context of node <book> [1]

	<book> [1]										
R	D1	D2	D3	D4	D5	D6	D7	D8	D9	D10	D11
<level>	1	0	0	0	0	0	0	0	0	0	0
<lang>	0	0	0	0	0	0	0	0	0	0	0
<title>	0	0	0	0	1	0	0	0	0	0	0
<author> [0]	0	0	0	0	0	1	0	0	0	0	0
< author > [1]	0	0	1	0	0	0	0	0	0	0	0
<publisher>	0	0	0	0	0	0	1	0	0	0	0

Table 3. Binary context of the node<book> [2]

	<book> [2]										
P	D1	D2	D3	D4	D5	D6	D7	D8	D9	D10	D11
<level>	0	0	0	0	0	0	0	1		0	0
<lang>	0	0	0	0	0	0	0	0	1	0	0
<title>	0	0	0	0	0	0	0	0	0	1	0
<author> [0]	0	0	0	0	0	0	0	0	0	0	1
< author > [1]	0	0	0	0	0	0	0	0	0	0	0
<publisher>	0	0	0	0	0	0	1	0	0	0	0

The nodes presented above have the same parent <bib>. So all the child nodes of these nodes become leaves.

Therefore in the context node <bib>, the lines represent the set of objects and the columns represent the set of attributes which are respectively the set of parent nodes (<book> [0] , <book> [1] , <book> [2]) and all the leaf nodes (<level>, <title>, <lang>, <author>, <author>, <Publisher>). Table 4 illustrates this context.

Table 4. Binary context of the root node< bib>

	< bib >		
R	<book>[0]	<book>[1]	<book>[2]>
<level>	1	1	1
<lang>	0	0	1
<title>	1	1	1
<author> [0]	1	1	1
< author > [1]	0	1	0
<publisher>	1	1	1

The interest of a concept lattice is to organize information about groups of objects with common properties. Taking the example above of Table 4, $(\{< book > [0], < book > [1], < book > [2]\}, \{< level >, < title >\})$ and $(\{< book > [0], < book > [1], < book > [2]\}, \{< publisher >\})$ are both concepts.

The second concept means that objects <book> [0], <book> [1] and <book> [2] have in common the attribute <publisher>.

Some algorithms are limited to obtaining concepts without building the lattice [8], [9] while others include the order identification between concepts and thus allow the construction of the entire lattice [6], [10].

The construction of the lattice itself offers the advantage of a very orderly classification of the concepts. It is then easy to select sets of objects or attributes that have a stronger influence than others. This influence is often characterized by the support of the concept.

We represent the lattice of concepts related to each attribute. Figures 5, 6, 7 and 8 below show concept lattices defined in Tables 3, 4, 5 and 6.

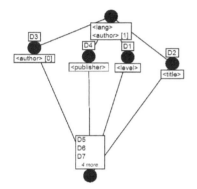

Fig. 5. Concept Lattice of book node [0]

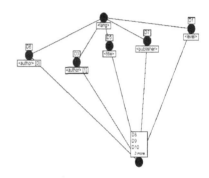

Fig. 6. Concept Lattice of book node [1]

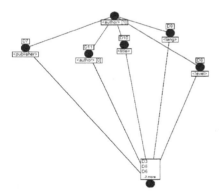

Fig. 7. Concept Lattice of book node [2]

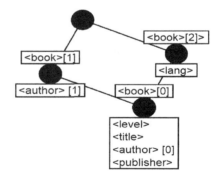

Fig. 8. Concept lattice of the root node <bib>

Several algorithms have been proposed for the construction of concept lattices mentioned above, construct mesh without considering any specific context. Their complexity is exponential, and several techniques have been developed to reduce computation time [11].

2.3   Conceptual Scaling

The indexing XML documents for research purposes can be complex especially when it needs to index text documents putting into consideration the content and structure. This leads us to say that the indexing process must also address the structure and content of the document.

We recall that the concept lattice is constructed from a binary context. So, if we want to index an XML document, indexing will necessarily relate to the content and the structure. This is insufficient because the binary contexts are not satisfactory to index multi valued attributes. Hence, the use of conceptual scaling_based nested lattice which aims to extend the impact of the FCA [12].

This lattice is constructed basing on two or more simple lattices. Next, we consider each node of this lattice and we insert the other lattice. The other nodes of other lattices that perform the attribute of the first lattice are highlighted.

The principle of nesting is summarized in Fig. 9.

Procedure 3 - Imbrication

INPUTS: Simples Galois Lattices.

OUTPUTS: Nested Galois lattices.

BEGIN

Combine the structures of simple Galois lattice obtained in a single structure called fuzzy nested lattice.

END.

Fig. 9. Imbrication procedure

An example is given in Table 5.

Table 5. Binary context of the node <book> [1] and the root node < bib >

R	D1	D2	D3	D4	D5	D6	D7	D8	D9	D10	D11	<book>[0]	<book>[1]	<book>[2]
				<book> [1]									< bib >	
<level>	1	0	0	0	0	0	0	0	0	0	0	1	1	1
<lang>	0	0	0	0	0	0	0	0	0	0	0	0	0	1
<title>	0	0	0	0	1	0	0	0	0	0	0	1	1	1
<author>[0]	0	0	0	0	0	1	0	0	0	0	0	1	1	1
<author>[1]	0	0	1	0	0	0	0	0	0	0	0	0	1	0
<publisher>	0	0	0	0	0	0	1	0	0	0	0	1	1	1

Take the example the two lattices corresponding to <book> [1] and <library>. Consider the lattice <bib> spread its nodes to insert the lattice <book> [1]. The result is shown in Figures 10 and 11.

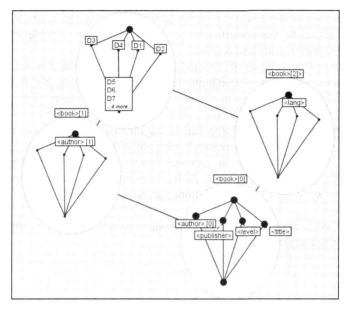

Fig. 10. Conceptual Scaling of attributes <book> [1] et < bib >

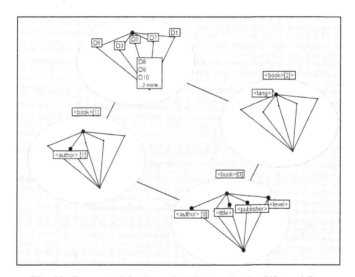

Fig. 11. Conceptual Scaling of attributes <book> [1] et < bib >

3   Conclusion

The main goal of this work is to define an approach based on the FCA for indexing XML native databases. This technology is increasingly used for different reasons. This is a rapidly expanding area due, in part, to changes of using XML in the exchange, storage and sharing of computerized data, on the other hand, to the complexity related to the construction of querying systems capable of satisfying queries combining structure and content.

It has been used as a means of storing data in a structured format in order to exploit them through the execution of search queries. In this context, this new technology has proven its ability to solve many problems while providing a neutral representation of data between different architectures.

Many approaches have been proposed in the literature and a number of solutions have been found, although many research ways are still open.

For this, we have proposed an approach based on conceptual scaling-based on FCA. We will be able to achieve an XML indexing system based this hierarchical structure. This system will be used to improve the flexible querying on such kind of document.

As future work, we propose to compare our approach with other approaches and applying it on a querying process.

References

1. Bradely, N.: The XML Companion, Paperback, Subsequent Edition,
   http://www.ebay.com/ctg/Xml-Companion-Neil-Bradley-
   2001-Paperback-Subsequent-Edition-/2193386#
2. Extensible Markup Language (XML) Recommendation, http://www.w3.org/TR/2000
3. Luk, R.W., Leong, H., Dillon, T.S., Shan, A.T., Croft, W.B., Allan, J.: A survey in indexing and searching XML documents. Journal of the American Society for Information Science and Technology 53(3), 415–435 (2002)
4. Weigel, F., Meuss, H., Bry, F., Schulz, K.U.: Content-Aware DataGuides: Interleaving IR and DB Indexing Techniques for Efficient Retrieval of Textual XML Data. In: McDonald, S., Tait, J.I. (eds.) ECIR 2004. LNCS, vol. 2997, pp. 378–393. Springer, Heidelberg (2004)
5. Li, Q., Moon, B.: Indexing and querying XML data for regular path expressions. In: Proceedings of the 27th VLDB Conference, Roma, Italy (2001)
6. Ganter, B., Wille, R.: Formal concept analysis, Mathematical foundations. Springer, Berlin (1999)
7. Meier, W.: eXist: An Open Source Native XML Database (2002)
8. Chein, M.: Algorithme de recherche des sous-matrices premières d'une matrice. Bull. Math. Soc.Sci. Math. R.S. Roumanie 13, 21–25 (1969)
9. Norris, E.M.: An algorithm for computing the maximal rectangles in a binary relation. Revue Roumaine de Mathématiques Pures et Appliqueés 23(2), 243–250 (1978)
10. Bordat, J.P.: Calcul pratique du treillis de Galois d'une correspondance. Math. Sci. Hum. 96, 31–47 (1986)
11. Choi, V.: Faster Algorithms for constructing Concept (Galois) Lattice. In: SIAM Conference on Discrete Mathematics (2006)
12. Ganter, B., Wille, R.: Conceptual Scaling. In: Roberts, F. (ed.) Applications of Combinatorics and Graph Theory to the Biological and Social Sciences, pp. 139–167 (1989)

Indexing Compressed XML Documents

Ahmed Jedidi, Olfa Arfaoui, and Minyar Sassi-Hidri

National Engineering School of Tunis,
BP. 37 Le Belvédère 1002 Tunis, Tunisia
jedidi.ahmed@gmail.com, olfa.arfaoui@yahoo.fr,
minyar.sassi@enit.rnu.tn

Abstract. XML data compression process seems to be inevitable to solve some problems related to the evolutionary growth of such data. Therefore, the indexing of compressed XML data, meanwhile, remains an important process and needs improvement and development in order to exploit the compressed data for querying and information retrieval. This work consists in studying and analyzing some suitable compressors to improve the indexing compressed XML documents process in order to query them later. We propose a new indexing process which leads in compressed XML data by re-indexing compressed XML data under XMill compressor.

1   Introduction

eXtensible Markup Language (XML) [1] is an extensible textual format of document description defined by World Wide Web Consortium (W3C) [2]. This is actually a subset of SGML [3]. This is a simple and universal language for describing a wide variety of data. Its open specifications and the use of languages for the definition of the document types facilitate the exchange of information and its long-term preservation.

However, this observation is shared by both users and industrial need to make the language more compact. Depending on the context, the challenge is to optimize the storage space or to accelerate the exchange of Web services that circulate over the Internet. Two large compression solutions are needed: one at the infrastructure level, the other at the application one.

Compressed XML format is a new approach for modeling information. Developing systems for storing and efficient querying of compressed XML documents also requires the development of new indexing techniques. Recall that the purpose of indexing techniques is to accelerate access to specific data.

We focus in this work on XML data compression techniques and their performance in indexing process and information retrieval (IR) in XML documents.

Several studies have been adopted this idea and have improved it. They are based on the method of queried compressed documents [4]. However, with the existence of large volumes of XML documents, these solutions are insufficient to ensure an efficient compression and to adopt an effective indexing technique.

The aim of this work is to try to solve this problem while putting into consideration the following objectives:

Z. Bao et al. (Eds.): WAIM 2012 Workshops, LNCS 7419, pp. 319–328, 2012.
© Springer-Verlag Berlin Heidelberg 2012

- The choice of an efficient XML data compression method.
- The re-indexing of compressed documents.
- The study of the effectiveness of this method on compressed documents.

The rest of this paper is organized as follows: section 2 presents a comparative study of XML data compression techniques. Section 3 presents our approach for re-indexing compressed XML data. Section 4 concludes the paper and highlights future work.

2   Comparative Study of XML Data Compression Techniques

XML is based on the separation between data and presentation. It uses tags to delimit pieces of data.

To handle and process such data, XML documents can be represented as trees whose leaves contain textual data. This tree has several types of nodes which are: elements, attributes and values.

During recent years, a large number of XML compressors were been proposed in the literature. These compressors can be classified into three main characteristics which are: (1) The consideration of the XML documents structure, (2) the ability to support queries, (3) the operating speed of the compressor.

In the first characteristic, the compressors can be divided into two main groups.

In the first group, text file compressors, XML data can be stored as text files, so the first logical approach to compress this data is to use traditional tools for compressing text documents. This group of compressors is called "blind –XML" since it traits XML documents exactly by the same way as text documents by applying traditional compression of text. We can give as an example GZip [5] and BZip2 [6] compressors.

The second group of compressor is designed so that to be specific to the XML format, by taking into account the structure of the XML document to achieve a compression ratio better than that achieved by text compressors. These compressors can be further classified according to their dependence or not to the XML schema.

XML Schema dependent compressors need to have access to the XML schema of the document to start the compression process such as XAUST [7].

In XML Schema independent compressors, the existence of the XML diagram is not necessary to start the process of compression such as SCMPPM [8].

For the second feature, we distinguish those that allow the querying of those which do not allow.

The first compressors allow the execution of queries on compressed XML documents. However, the main objective of this type is to avoid decompressing of the document during query execution. By default, all queried XML compressors are not really perfect such as XGrind [9].

Other XML compressors do not allow the execution of queries on compressed XML data. The main objective of these compressors is to achieve the highest compression ratio. By default, the text file compressors belong to the group of non-queried compressors such as XAUST [7] and SCMPPM [8] compressors.

The third feature indicates whether the operating speeds of the compressor is so online or off line.

In what follows we present a comparative study between different types of compressors which was listed above.

To achieve this comparison, it is necessary to highlight some theoretical indicators needed to make the comparison between different compressors. These indicators are: (1) the compression ratio, (2) the fraction between the original file and the compressed file, and (3) the rate of memory usage.

2.1   About Compression Rate

So far, works presented in the literature have proposed two different expressions to define the compression ratio of a compressed XML document.

The first indicator is the compression rate [10] noted by $CR1$, it expresses the number of bits needed to represent a data byte of the original file. This indicator indicates that the lower value is that corresponding to the most efficient compressor. The second indicator is the compression ratio, noted by $CR2$, shows the fraction of memory space between the original file and compressed files. The value which is relatively higher corresponds to the most performing compressors.

The main difference between the two compression ratio is that CR1is proportional to the size of the memory space used to store the compressed document, while CR2shows only the percentage reduction of the compressed file size in relation to the original file size.

$$CR1 = \frac{sizeof(compressed\ file) * 8}{sizeof(original\ file)}\ bit\ /byte \tag{1}$$

$$CR2 = \left(1 - \frac{sizeof(compressed\ file) * 8}{sizeof(original\ file)}\right) * 100\% \tag{2}$$

We present a comparison between several XML compressors through experimental results based on the indicators described above. Table 1 highlights a comparison between Gzip, which is a text compressor, and XMill [11] which is a specific compressor for XML files. This comparison is made with experiments on the six most popular data sources that cover a wide range of XML data.

Table 1. Comparison between Gzip and XMill

XML Files	Document size before compression (KB)	Size of the compressed document (KB)		CR1 (bits/byte)		CR2 (%)	
		Gzip	XMill	Gzip	XMill	Gzip	XMill
Weblog	32722	1156	726	0,282	0,177	96,5	97,8
SwissProt	21254	2889	1739	1,088	0,654	86,4	91,8
DBLP	40902	7418	6149	1,451	1,203	81,9	85
TPPC-H	32295	2912	1514	0,721	0,375	91	95,3
Xmark	103636	13856	8313	1,07	0,642	86,6	92
hakespeare	7882	2152	1986	2,184	2,016	72,7	74,8

Through this table, we illustrate the difference between the two compression ratios (CR1and CR2) obtained following the execution of two compressors Gzip and XMill on different data sources.

We see that the percentage of the original file and compressed file for Gzip is slightly less than XMill. It can be said that the performance of Gzip and XMill is the same for both compressors. However, by examining the values of actual size of compressed documents generated by XMill, we can suggest that is smaller than that generated by Gzip.

Since the previous analysis was essentially based on a comparison between a text file compressor and a specific XML compressor (respectively Gzip and XMill), it becomes necessary to carry out this comparative study on specific XML data compressors, this through the comparison of their compression ratio.

The experimental comparison results of different compression ratio of compressors namely XMLPPM [10], XMill [11], XMLZIP [10] and XGrind [9] are represented in Fig. 1.

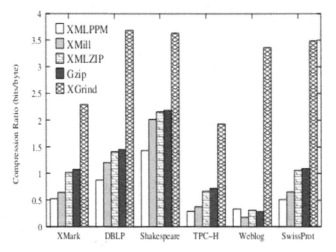

Fig. 1. Comparison of the compression ratio between XMILPPM, XMill, XMLZIP and XGrind

A shown in Fig.1, the compression ratio for the three compressors XMLPPM, XMill and XMLZip are systematically higher than those obtained by the text file compressors namely Gzip , we also note that the compression ratio of XGrind is worse than Gzip, XMLZip, XMill and XMLPPM.

The compression ratio of Gzip and XMLZip are very similar for all test files, however they are higher than those of XMill and XMLPPM. Based on this comparison we concluded that the compressor XMLPPM achieves the best compression ratio among all compressors used in this comparative study.

2.2   About Compression and Decompression Time

In the compression process, the time factor remains very important. For this reason, we focus on this factor. We will study it by examining Fig. 2 and Fig.3.

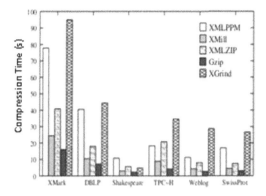

Fig. 2. Compression times given by XMLPPM, XMill, XMLZip, Gzip and XGrind

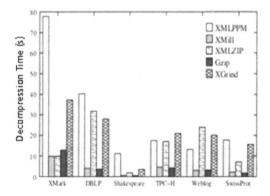

Fig. 3. Decompression times given by XMLPPM, XMill, XMLZip, Gzip and XGrind

As shown in Fig. 2 and Fig. 3, the compression and decompression times of XMLPPM, XMill, XMLZip, Gzip, and XGrind on the six data sets, except XMill, are longer, in the compression and decompression than Gzip.

The compression Time of XMill is slightly longer than that of Gzip. This is perhaps due to its pre-compression phase. Indeed, XMill separates the data structure, before applying Gzip to compress data containers.

In addition, the pre-compression phase causes a waste of time for the compression process because XMill must also reconstruct the original XML structure, after the decompression of data containers.

Therefore, this surcharge of additional decompression slightly increases the total decompressions. However, when a file is compressed by XMill its size becomes much smaller than the size of a file compressed by Gzip.

2.3   About Memory Consumption

The memory consumption is another factor to measure the performance of the chosen compression process. Fig. 4 shows a difference in the use of the memory between the various compressors on the same data sets.

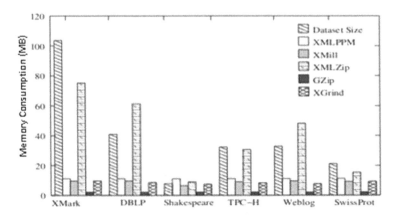

Fig. 4. Memory consumption

We notice that among all compressors, Gzip uses less memory whatever the input documents. XMill and XGrind use approximately 8 MB of memory on all data sets, while XMLPPM uses about 3 MB of memory more than XMill. This latter uses a main memory whose size is 8 MB (by default) and when the memory consumption reaches this size, XMill write all buffered data on the disk and then start the compression process. XGrind uses the Huffhomme coding [9] to compress data. Thus, it requires only small size to maintain the status of Huffhomme model.

We conclude our comparative studies by the following consequence: XMill compressor is the most performing one.

3   Indexing Compressed XML Data

In this section, we describe our approach for indexing compressed XML documents.

The aim of our work is to find an algorithm that solves the problem of adapting an indexing algorithm to compressed XML data while keeping the performance offered by the XMill compressors.

As we have presented in the section 2, the structure of an XML document compressed by XMill is done in several steps: the first contains the document structure and the other contain data. Our work consists of separating the structure from the data during the compression process.

3.1   Overview of the System

Fig. 5 shows the overview of our approach.

As shown in Fig. 5, our approach follows two main steps: pre-processing and post-processing. In this work, we limit ourselves to the pre-processing step.

The purpose of our approach is to index the data compressed by XMill so that the user can query data.

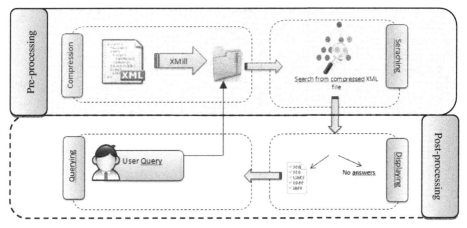

Fig. 5. Overview of the approach steps

3.2   Re-Indexing and Compression Step

Based on the Dietz schema numbering system [12], our idea is to involve us in the construction phase of the container structure, this is as follows: instead of linking each node of the XML tree symbols and numerical identifiers we associate to each node a quadruple obtained by applying the Dietz numbering system and the tag identifier (obtained by applying the XMill algorithm), that is to say, in the traversal of the XML file and during the preparation of different containers, instead of associating symbol (defined by the algorithm XMill), to data dictionary of, we allocate it the values of quadruplet.

Let us take the example of the XML file in Fig. 6. This figure describes a paper presented through its title and its date of publication and the conference with which it is associated:

```
<paper><entryyear="2003"><journal>
  <title>Secret Sharing</title>
  </journal></entry><entry
year="2003"><conference><title>XML  Water
Mark</title></conference></entry></paper>
```

Fig. 6. Example of an XML File

To simplify the explanation in what follows, we propose the following symbols which constitute a quadruplet allowing identifying every node of the XML tree:

- PH = traversal of the tree from top to bottom and from left to right.
- PB = traversal of the tree from bottom to top and from left to right.
- Niv= the depth of the tree.
- Idi = the node identifier.

A node is identified by the quadruplet *(PH, PB, Niv, Idi)*.

The new format of the XML document in Fig. 6 is illustrated in Fig. 7.

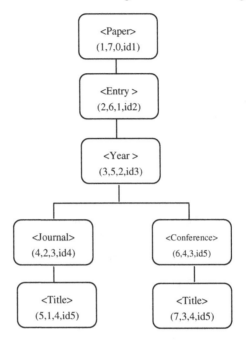

Fig. 7. The theoretical approach applied to an XML document

For example the tag "Title" has the quadrupled (5, 1, 4, id5). The number 5 shows the position of the XML tree traversal from top to down. 1 shows the position of the XML tree traversal from bottom to top. The number 4 defines the depth of this node (the level) in the XML file and *id5* corresponds to the node identifier obtained following the application of the XMill algorithm. This is shown in Fig.8.

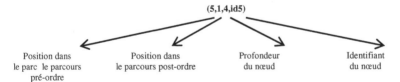

Fig. 8. Example of proposed numbering plan

The purpose of re-indexing and compression procedure is to improve the indexing phase by the introduction of the Dietz numbering scheme in the indexing algorithm of the B + tree. Thus, this procedure can improve the compression process through the assignment of a quadruplet *(PH, PB, Niv, Idi)* at each node of the XML tree. These integers (the quadruplet) will be recorded later in the container structure. For each node in the XML tree repeating (See Fig. 9).

Procedure 1 - Procedure for re-indexing and compression

INPUT: The tree corresponding to the input XML file.

OUTPUT : The new XML tree where each node is identified by the quadruplet *(PH, PB, Niv, idi)*.

BEGIN

> **Step 1:** XML tree traversal by adopting the PH traversal.
>
> **Step 2:** determination of the integer obtained following the PH traversal.
>
> **Step 3:** XML tree traversal by adopting the PB traversal.
>
> **Step 4:** Determination of the integer obtained following the PB traversal.
>
> **Step 5:** Determination of the integer Niv corresponding to the node depth in the XML tree.
>
> **Step 6:** Application of the XMill compression procedure on the tree nodes to retrieve the node identifier (Idi).
>
> **Step 7:** Recording of the quadruplet (PH, PB, Niv, Idi).

END.

Fig. 9. Re-Indexing and compression procedure

4   Conclusion

XML provides flexibility in publishing and exchanging data on the Web. However, the language is by nature verbose and thus XML documents have a large size by comparison with other specifications containing the same data content. This problem becomes an obstacle in front of the XML applications, because it significantly increases the cost of data storing, processing and exchange. In this fact, it becomes necessary to resolve these problems, so the idea to compress these data becomes unavoidable.

Indeed, the compression of XML data has proven its effectiveness in solving many problems which have emerged following the significant growth of large repositories of XML data. This phase was completed by the appearance of a set of compression tools and techniques that each one has its own characteristics.

Therefore, the new XML data format which we call compressed XML is a new approach for modeling information. In this context, the issue of information retrieval in compressed XML corpus becomes crucial. This certainly requires the review of techniques for indexing compressed XML documents because a good indexing technique directly affects the query process.

Our re-indexing and compression process is an attempt that aims to improve the indexing process at the compression step of the XMill algorithm. We have modified and improved the indexing step which was based on the B+ tree algorithm via the implementation of the Dietz numbering plan.

As future work, we propose to test our approach on others large XML data sets and taking into account the flexibility in the querying process.

References

1. Baldonado, M., Chang, C.-C.K., Gravano, L., Paepcke, A.: The Stanford Digital Library Metadata Architecture. Int. J. Digit. Libr. 1, 108–121 (1997)
2. World Wide Web Consortium, XQuery 1.0: An XML Query Language, W3C Working Draft (2004)
3. Goldfarb, C.: The SGML Handbook. Oxford University Press, Oxford (1990)
4. Cheney, J.: Compressing XML with Multiplexed Hierarchical PPM Models. In: Proceedings of the Data Compression Conference, Washington, DC, USA (2001)
5. Gailly, J.-L.: Gzip, version 1.2.4., http://www.gzip.org/
6. Seward, J.: bzip2, version 0.9.5d., http://sources.redhat.com/bzip2/
7. Subramanian, H., Shankar, P.: Compressing XML Documents Using Recursive Finite State Automata. In: Farré, J., Litovsky, I., Schmitz, S. (eds.) CIAA 2005. LNCS, vol. 3845, pp. 282–293. Springer, Heidelberg (2006)
8. Adiego, J., De la Fuente, P., Navarro, G.: Merging prediction by partial matching with structural contexts model. In: Proceedings of the 2004 IEEE Data Compression Conference, p. 522 (2004)
9. Tolani, P.M., Haritsa, J.R.: XGRIND: A Query-friendly XML Compressor. In: IEEE Proceedings of the 18th International Conference on Data Engineering (2002)
10. Cheney, J.: Compressing XML with Multiplexed Hierarchical PPM Models. In: Data Compression Conference, pp. 163–172 (2001)
11. Hartmut, L., Dan, S.: XMill: An Efficient Compressor for XML Data. In: SIGMOD Conference, pp. 153–164 (2000)
12. Dietz, P., Sleator, D.: Two Algorithms for Maintaining Order in a List. In: 19th Annual ACM Symposium on Theory of Computing, pp. 365–372. ACM Press (1987)

Path-Based XML Stream Compression with XPath Query Support[*]

Bingyi Qian, Hongzhi Wang, Jianzhong Li, and Hong Gao

P.O. Box 318, Harbin Institute of Technology
Rukata.qian@gmail.com {wangzh,lijzh,honggao}@hit.edu.cn

Abstract. XML stream has redundancy that results in waste of processing time and bandwidth in transmission from sender to receiver. It is significant to compress XML stream for smaller bandwidth and more efficient query processing. This paper presents a compression for XML streams technology. This compressing method divides XML stream into structure and context, and then encodes them respectively. In the meantime, it keeps homomorphism between compressed XML stream and original one. It divides XML stream into paths and encodes them into path encodings that supports XML stream to be compressed during the arrival of XML stream, which obviously increases the efficiency of XPath query process on XML stream. The XPath query execution on compressed XML stream is also presented. Experimental results demonstrate the effectiveness and efficiency of the methods proposed in this paper.

Keywords: XML stream, compression, XPath query, path encoding.

1   Introduction

XML has been gradually becoming a significant standard of data expression and conversion in web and other information systems. It is widely used in air traffic control [2], data exchange [7], web-based services [8], content management [9], etc. In all formats of XML data, XML stream is broadly utilized in many applications, including subscription and publishing of network information, E-mail monitoring. Hence, the effective management of XML stream is in great demand. XML stream is a kind of real time unlimited stream that arrives at local host in chronological order, and can only be accessed in chronological order. However, XML stream has a common flaw of all kinds of XML data. That is data redundancy, which causes too much occupation in bandwidth and longer processing time. Therefore, efficient XML stream compression strategies are in necessary.

In general, there are two types of XML compression technology. One of them separates structure and context and encodes them respectively in the first place, then

---

[*] This paper was partially supported by NGFR 973 grant 2012CB316200 and NSFC grant 61003046, 6111113089. Doctoral Fund of Ministry of Education of China (No. 20102302120054). The Fundamental Research Funds for the Central Universities (No. HIT.NSRIF.2013064).

Z. Bao et al. (Eds.): WAIM 2012 Workshops, LNCS 7419, pp. 329–339, 2012.

assembles two parts. This kind of compression technology keeps homomorphism between original XML and compressed one. Two representative methods of this type is XGRIND[11] and XPRESS[12], respectively. These methods usually have high compression ratio, while they pay a high storage space for intermediate results in order to support efficient XPath Query. However, compression on XML stream requires low consumption of memory in query processing. Hence, such type of compression technologies is not suitable for XML stream.

The other type of compression technology stores and compresses context and structure respectively. Take XMILL [13] for example. XMILL has high compression ratio, but it has to decompress the whole compressed document to support queries, which obviously wastes too much time in query processing. So it is not considered as compression technology for XML streams.

Compression technology on XML stream requires not only high compression ratio, but also efficient query processing and transmission-continuation support. Unfortunately, current methods are not satisfactory. Thus we design a novel XML stream compression method for it.

For the former requirement, we separate structure and context in XML stream and encode them respectively according to [11] and [12]. As to the latter, with the consideration of the feature of XML streams that it is unable to be compressed according to the common parts of different fragment of XML document, we compress each path separately, where path means a sequence of tags from root to leaf. Another basic principal of our method is that the order of leaves in XML stream is absolutely same as the chronological order of XML stream arrival. That is to say, if an XML stream is divided into paths and encoded respectively, it can support transmission-continuation and efficient query processing, for the XML stream being partly decompressed during query.

The basic idea of our compression method is to encode structure and context of XML stream respectively. The structure of XML stream can be divided into paths, which are encoded to intervals. Considering that the elements sharing the same incoming path have the same data type. Thus the values of them are compressed with the same compressor.

The contributions of this paper include:

• This paper presents a XML compression that separates tags and context and compresses them respectively, at the same time keeps the homomorphism between original XML streams and compressed one. A decent-sized XML stream can be processed in main memory.
• The encoding method utilizes structure information but not schema one. The latter one is out of requirement.
• The Context of XML stream is separated from original one and encoded according to data type and range of value.
• Demonstrate the efficient query processing technologies on compressed XML stream.

The remaining part of this paper is organized as follows. Compression method for XML stream is presented in section 2. Section 3 presents an XPath query technology on compressed XML streams. The result and analysis of experiments is in section 4.

2   Path-Encoding XML Compression

In this section, a path-encoding compression method for XML streams is presented. At first, the architecture of the compressor is provided. Then the compression technology of structure is discussed in 2.2.

2.1   Architecture of Path-Encoding Compression

The architecture of path-encoding compression is discussed in this section. Fig.1 presents the entire architecture of the method. Path-encoding compression is proceeding during the sending of XML stream, and if the process of sending is interrupted unexpectedly, it can resend the stream from the interrupted position in the XML stream. During the compression, the schema information is updated synchronously between sender and receiver with a tag-encoding table. Before sender starts sending a path to receiver, it firstly matches tags in the path with tag-encoding table, if there are tags that don't exist in the table, add these new tags into table and re-calculates the tag encodings. When the table is updated on the sender, send the updated one to receiver and replace the old one on the receiver. After several times of updating, the whole tag-encoding table is established in the receiver. Structure and context of XML stream is separated and encoded respectively.

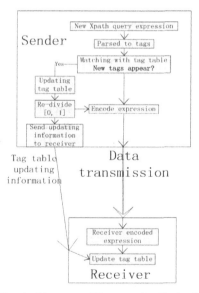

Fig. 1. The Architecture of Compression Technology

In this technology of compression, XML stream is converted to a pair of ($[x, y]$, v) after compression, inspired by the thoughts in [12]. The format of $[x, y]$ is a float-point interval that encodes paths in the XML stream. Paths means a sequence of tags from a root tag to a leaf tag. And v presents the compressed value of leaf tag in the corresponding path. In this way, a complex XML stream is encoded into a sequence of tuples of float-point intervals and values, which achieves high compression ratio.

Algorithm 1 encodes path into float-point interval. Before encoding the path, we should encode tags in the XML stream firstly. Supposed there are n tags in XML stream. We divide [0. 1] into n equal intervals, each interval is the encoding of a tag. Then algorithm 1 calculates the left value and right value as follows.

Algorithm 1. Calculate left value and right value of interval

Input: path $a_1/a_2/.../a_n$

Output: the corresponding path-encoding

PathEncoding()
1 $left = 0$
2 $right = 1$
3 **for** $i = 1$ **to** k **do**
4      $left = left + l_{a_i}/(right - left)$
5      $right = right + r_{a_i}/(right - left)$

The algorithm has the following properties:

1.   The whole XML tree can be restored from [left, right].
2.   In regard to $[l_1,r_1]$ and $[l_2,r_2]$, if $l_1 \leq l_2 \leq r_1 \leq r_2$, the path formatted by $[l_1,r_1]$ is the prefix of the path formatted by $[l_2,r_2]$.

For example, the number of tags in example 1 is five, so [0, 1] is divided into five intervals, [0.0, 0.2], [0.2, 0.4], [0.4, 0.6], [0.6, 0.8] and [0.8, 1.0]. In order to encode path $a/b/d$, in the first loop, $left$ is 0.0 and $right$ is 1.2. In the second loop, $left$ is 0.16 and $right$ is 1.6. The third loop calculates the interval as [0.4, 2.3].

We define trackback-encoding that encodes return paths. Return path is used to solve the ambiguity problem in generating path encodings. For example, supposed there are two paths in a tree, $a/b/d$ and $a/b/e$. The structure of the tree has two possibilities shown in fig. 2.

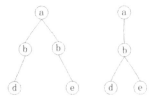

Fig. 2. Two possibilities

Without the trackback-encoding, the encodings of these two XML fragments are the same. In order to distinguish these two kinds of fragments, we insert trackback-encoding to the stream. The format of trackback-encoding is $[n, 0]$ and the corresponding value is null. Supposed return path is between path A and path B, the length of public ancestors of path A and path *B* is *x* and the length of path *A* is *y*. Difference between the two is *n*. The format of *v* depends on the type and value range of value in order to get a higher compression ratio.

Example 1. This example presents a simple fragment of XML stream that uses this technology of compression to encode structure part. The original XML stream is as follows.

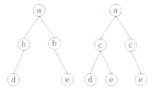

Fig. 3. Simple version of XML stream

There are five kinds of tags in XML stream. So the [0, 1] is divided into five segments. The corresponding table is shown below.

Table 1. Tag-encoding mapping table

Tags	Encodings
A	[0.0, 0.2]
B	[0.2, 0.4]
C	[0.4, 0.6]
D	[0.6, 0.8]
E	[0.8, 1.0]

Obviously there are five paths in XML stream, the encodings of each path calculated by algorithm is showed in the following table.

Table 2. Path-encoding mapping table

Paths	Encodings
a/b/d	[0.4, 2.3]
Return path	[1.0, 0.0]
a/b/e	[0.7, 2.8]
a/c/d	[1.0, 3.0]
a/c/d	[1.0, 3.0]
a/c/e	[1.5, 3.4]
Return path	[1.0. 0.0]
a/c/e	[1.5, 3.4]

Therefore, the compressed XML stream structure is a sequence of float-point intervals. The sequence is [0.4, 2.3], [1.0, 0.0], [0.7, 2.8], [1.0, 3.0], [1.5, 3.4], [1.0, 0.0], [1.5, 3.4].

2.2   Compression Algorithms for Structures

XML stream is transmitted from sender to receiver as a series of XML fragments. Firstly, a table of tags is established on sender and receiver respectively and when the table on sender changes, it sends updating information to receiver to synchronous update the table on receiver. When a path is to be sent to the sender, the tag table will update its data if tags in a path are not recorded in the tag table and then the [0, 1] interval is re-divided into n intervals, where n is the number of tags in current tag table. Otherwise the query processing will go on to encode the path in Algorithm 2.

Algorithm 2. Path encoding

Input: The root nodes of tag trees and HashMap of tag encodings.

Output: the sequence of coding and corresponding path-encoding

PATH-ENCODING(*TreeRoot*, *TagCodesMap*, *PathCodesList*)
1   **if** *TreeRoot* == **null do**
2   **return**
3   *left* ← 0.0, *right* ← 1.0
4   **if** *TreeRoot.mParent* != **null do**
5   *left* ←*TreeRoot->Parent->Left*
6   *right* ←*TreeRoot->Parent->Right*
7   *TagCode*←*TagCodesMap->*Get(*TreeRoot->Name*)
8   *left* ← *left* + *TagCode->Left*/(*right* - *left*)
9   *right* ← *right* + *TagCode->Right*/(*right* - *left*)
10 *TreeRoot->*SetInterval(*left*, *right*)
11 **if** *TreeRoot->Childs* == **null do**
12   *PathCodesList->*Add(*left*,*right*)
13   **return**
14 *count* ←*TreeRoot->Childs->*Size()
15 **for** *i* = 0 **to** *count* **do**
16   PATH-ENCODING(*TreeRoot->Childs*[*i*],*TagCodesMap*, *PathCodesList*)
17   **if** *i* == *count* - 1 **do**
18   *length* ← DATE-BACK(*TreeRoot->Childs*[*i*])
19   **if** *length* > 1 **do**
20   *PathCodesList->***add**("", *length* - 1, 0)

Algorithm 2 is used to encode paths of XML stream. It is based on recursion. Firstly, the algorithm judge the value of root node, determining whether it is null or not in lines 1-2.If it is null, then return. Otherwise the algorithm continues. The path encodings are initialized in lines 3-6, if current node is root node, then they are 0.0 and 1.0. Otherwise, the values are decided by the parent node of current node. After that, the interval of current node is achieved in line 7. Then the upper and lower bound of the interval is calculated in lines 8-9. If current node is a leaf node, then current interval is returned in lines 10-13. Otherwise the following nodes in the path are encoded recursively in lines 14-16. Particularly, if there is a return path, the length of it will be calculated firstly. Afterwards, when the length of returned path is larger than 1, [*return-path-length, 0*] is inserted into compressed structure. In example 2.1, each path is encoded to encodings that are showed in table 2.

3   XPath Query on Compressed XML Stream

In this section, the technology of XPath query on path-encoding compressed XML stream is presented.

The definition of XPath from World Wide Web Consortium (W3C) is as follows [14]. XPath, the XML Path Language, is a query language for selecting nodes from an XML document. XPath uses path expressions to select nodes or node-sets in an XML document or stream. These path expressions look very much like the expressions that used in a traditional computer file system.

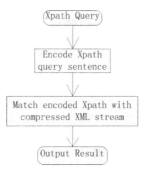

Fig. 4. Query Process Steps

Since in section 2, XML streams have been compressed into a sequences of pairs with format ([*x, y*], *v*). For the convenience of query processing, XPath query expressions with slashes are converted to the format of a sequence of pairs too. Ones with double slashes also can be parsed in this technology. Double slash means that all of tags that has same name with the tag after double slash are needed. Because slash defined in this paper has the same meaning with double one, we do not distinguish them in the remaining part of this paper.

Therefore, an encoding method is designed to convert XPath query expressions in Section 3.1. Then query processing is presented in Section 3.2. The whole steps of query processing on compressed XML stream are summarized in Fig. 3.

3.1   Encoding of XPath Query Expression

At the time the compressor processing receives an XPath query expression, it firstly converts the expression to a series of tuples of ([*x, y*], *v*), and then matches the encoded path with encoded XML stream.

This section introduces the procedure of encoding XPath query expression. Firstly, match each tag in XPath query expression to corresponding tag encoding using tag-mapping established in Algorithm 2. Especially, / and // are treated in the same way in

Algorithm 2. That is, if there isn't a tag after / or //, the algorithm will find all the children tags of the tag before it and then encode these paths with different leaf tags into corresponding encodings. Or the algorithm just encodes the path from root to leaf. Then use Algorithm 1 to encode the XPath query expression.

Example 2. Two XPath query expressions are provided as follows.

XPath query expression 1: *a/c/d*
XPath query expression 2: *a/c//*

According to tag-encoding mapping table, encodings of tags in expression 1 are [0.0, 0.2], [0.4, 0.6] and [0.6, 0.8], and [0.0, 0.2], [0.4, 0.6] and [0.8, 1.0] in expression 2. Based on algorithm 4, the final formats of the expressions can be calculated. Take expression 1 as example, in the first loop, the left value of the interval is 0.0 and the right one is 1.0. By Such Analogy the final result of expression 1 is [1.0, 3.0]. As for expression 2, we should find all these tags in the child nodes of *a/c*. Fig. 3 shows that there are three paths that match the query expression, a/c/d, *a/c/e* and *a/c/e*. Considering that there is a return path between the two *a/c/e*, the compressed expression 2 is [1.0, 3.0], [1.5, 3.4], [1.0, 0.0], [1.5, 3.4].

3.2   Path Matching

The basic idea of path matching in this method is that it is unnecessary to decompress any compressed XML streams during query processing, in order to accelerate the process. The encoded XPath query expressions just have to be matched with encoded XML streams because they have same format by scanning the whole compressed XML streams. The pseudo-code of this algorithm is shown as follows.

Algorithm 3. Inquirement of XPath query expression.

Input: XPath query expression.

Output: The required context of XPath query.

QUERY(*Path, TagCodeMap, PathCodeList, ValueArray*)
1 *tags* ← **Split**(*Path*)
2 *tagCode* ← **null**
3 *left* ← 0.0, *right* ← 1.0
4 **for** *tag* **in** *tags* **do**
5   **if** "" == *tag* **do**
6   **continue**
7   *tagCode* ← *TagCodeMap*->Get(*tag*)
8   *left* ← *left* + *tagCode*->*Left* / (*right* - *left*)
9   *right* ← *right* + *tagCode*->*Right* / (*right* - *left*)
10 **for** *pathCode* **in** *PathCodeList* **do**
11   **if** *path*->*left* <= *left* &&
12   *left* <= *path*->*rigth* && *path*->*right* <= *right* **do**
13   *ValueArray*->Add(*path*->*Value*)
14 **return** *ValueArray*

In this algorithm, firstly the expression is split into tags in line 1. Then encode the expression in lines 2-9. After the expression has converted to path encoding, it matches same float-point interval with equal left value and equal right value in compressed XML stream in lines 10-14. With the encoded expression1 in example 3.1, we can match path *a/c/d* in original XML stream in example 2.1. As for expression 2, it has converted to a sequence of intervals, [0.5, 2.6], [1, 0], [0.5, 2.6]. This sequence is matched with the compressed XML stream, and the final results of expression 2 are *a/c/e* and *a/c/e*.

4   Experiment

In order to validate the effectiveness of this compression technology, some experiments were accomplished using real-life data sets and benchmark.

4.1   Experimental Environment

The experiments are performed on a PC with Core 2 T5870 and 2G main memory. Operation system is Microsoft windows XP and the platform is Eclipse Java SDK 1.6.

Data Set. We generate different size of XML stream to validate the feasibility and compression ratio of the compression technology. The whole data comes from XMark. The results of experiment are as follows.

Queries. In order to present XPath query processing on different kinds of data set, we design five special queries to test them respectively.

- *PATH1 /BOOKSTORE/COMPUTER//COMPUTERSAPPLICATION*
- *PATH2/BOOKSTORE/COMPUTER/PROGRAMMING/*
- *PATH3 /BOOKSTORE/LITERATURE/SHARESPEAR/MACBETH/PRICE*
- *PATH4 /BOOKSTORE/COMPUTER*
- *PATH5 /BOOKSTORE/*

4.2   Experimental Results

The efficiency of our compression method is evaluated by the compression ratio of structure. The results are in fig. 5 and fig. 6. From fig.5, it is observed that the compression ratio of structure of XML stream is growing up with the increase of XML stream. It is because that larger data quantity results in larger redundancy. From Fig. 6, the query time increases linearly with the growing of size of XML stream. It demonstrates that our method has good scalability for query processing.

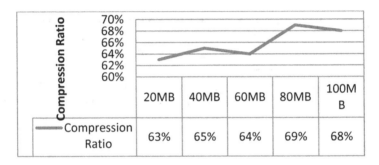

Fig. 5. The compression ratio of five kinds of size XML streams

Compression Ratio	20MB	40MB	60MB	80MB	100MB
Compression Ratio	63%	65%	64%	69%	68%

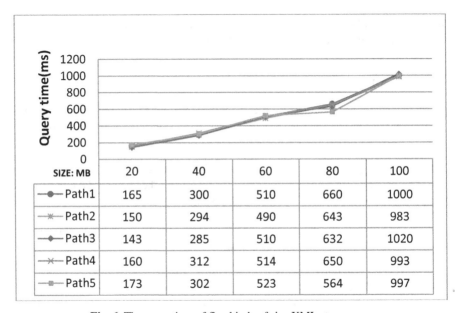

Fig. 6. Thequery time of five kinds of size XML streams

SIZE: MB	20	40	60	80	100
Path1	165	300	510	660	1000
Path2	150	294	490	643	983
Path3	143	285	510	632	1020
Path4	160	312	514	650	993
Path5	173	302	523	564	997

5   Conclusion

In this paper, to solve the redundancy of XML stream, this paper presents efficient compression technology with XPath query support. This method contains two steps: the first step is to divide structure and context of XML stream; the second is to encode the structure in path, which ensures it needs no decompression in the procedure of query. Consequently, complex queries oriented structure can be accomplished in main memory. The experiment employs real-life XML stream to validate this compression, which can be widely used in various XML databases and Web databases with XML support.

References

1. Liu, J., Xia, K.-J.: Application of data display according to demand in web. In: Proceedings of Computer Engineering and Design (2008)
2. Snoeren, A.C., Conley, K., Gifford, D.K.: Mesh based content routing using XML. In: Symposium on Operating Systems Principles (2001)
3. Min, J.-K., Park, M.-J., Chung, C.-W.: Xpress: A Queriable Compression for XML Data. In: SIGMOD Conference, pp. 122–133 (2003)
4. Liefke, H., Suciu, D.: XMill: an Efficient Compressor for XML Data. In: Proc of ACM SIGMOD (2000)
5. Buneman, P., Grohe, M., Koch, C.: Path Queries on Compressed XML. In: Proc of VLDB (2003)
6. Wang, H., Li, J., Luo, J., et al.: XCpaqs: Compression of XML Document with XPath Query Support. In: Proc of ITCC (2003)
7. Carlisle, D., Ion, P.: Mathematical Markup Language (MathML) Version 2.0 (second Edition). W3C Recommendation (October 21, 2003),
   http://www.w3.org/TR/2003/REC-MathML2-20031021/
8. Gudgin, M., Hadley, M., Mendelsohn, N., et al.: SOAP Version 1.2 Specification (2007),
   http://www.w3.org/TR/soap12
9. Clark, J.: Xsl Transformations (xslt) Version 1.0 (2005),
   http://uddi.xml.org/uddi-org
10. Wang, H., Li, J., Luo, J., et al.: Twig Query Processing in Path-based Compressed XML Data. In: Proc. of NDBC (2004)
11. Tolani, P.M., Haritsa, J.R.: XGRIND: A Query-friendly XML Compressor. In: Proc. of the 18th International Conference on Data Engineering (2002)
12. Min, J., Park, M.-J., Chung, C.: XPRESS: A Queriable Compression for XML Data. In: Proc. of ACM SIGMOD (2003)
13. Liefke, H., Suciu, D.: XMILL: an Efficient Compressor for XML Data. In: Proc of ACM SIGMOD (2000)
14. Berglund, A., Boag, S., Chamberlin, D., et al.: XML Path Language (XPath) 2.0, 2nd edn. (2010), http://www.w3.org/TR/2010/REC-xpath20-20101214/

Uncertain XML Functional Dependencies Based on Tree Tuple Models

Teng Lv[1], Weimin He[2], and Ping Yan[3],[*]

[1] Teaching and Research Section of Computer,
Army Officer Academy, Hefei 230031, China
[2] Department of Computing and New Media Technologies,
University of Wisconsin-Stevens Point, Stevens Point, Wisconsin 54481, USA
[3] School of Science, Anhui Agricultural University, Hefei 230036, China
{Lt0410,want2fly2002}@163.com, whe@uwsp.edu

Abstract. With the increase of uncertain data in many new applications, such as sensor network, data integration, web extraction, etc, uncertainty both in relational databases and XML datasets has attracted more and more research interests in recent years. As functional dependencies are critical and necessary to schema design in relational databases and XML datasets, it is also significant to study the functional dependencies and their applications in uncertain XML datasets. This paper proposed three new kinds of functional dependencies based on tree tuple mode for uncertain XML datasets. We also give a set of sound and complete inference rules and two applications, such as to test whether an uncertain XML dataset satisfies a given functional dependency and to find a closed set for a given uncertain XML dataset.

Keywords: uncertain XML, functional dependency, inference rule, closed set.

1 Introduction

XML (Extensible Markup Language) has become the de facto standard of data exchange and is widely used in many fields. With the increase in applications such as data integration, web extraction, sensor networks, etc, XML datasets may be obtained from heterogeneous data sources and are not always deterministic. In such cases, XML datasets may contain uncertain data for the same attribute or element due to different data sources, information extraction, approximate query, and data measurement. Although functional dependencies of uncertain XML datasets are much more complicated than the counterparts of traditional relational databases and XML datasets, it is possible and necessary to study them in uncertain XML datasets as shown in the paper. Functional dependencies of uncertain XML datasets are also critical to schema design of uncertain XML datasets, such as lossless decomposition, normal forms, query optimization, etc.

Related Work. Although there has been a lot of significant work in functional dependencies for relational databases and XML datasets, none of them can be directly

---

[*] Corresponding author.

Z. Bao et al. (Eds.): WAIM 2012 Workshops, LNCS 7419, pp. 340–349, 2012.

applied to uncertain XML datasets. We analyze the related work in the following three aspects:

(1) **For traditional relational databases**, functional dependencies are thoroughly studied for several decades[1,2]. Ref.[3] proposed a concept of functional dependencies in relational databases which can deal with slight variations of data values. Ref.[4] proposed the conditional functional dependencies to detect and correct data inconsistency. It is obviously that the techniques of traditional relational databases can not be directly applied to XML due to the significant difference in structure between XML documents and relational databases.

(2) **For traditional XML datasets**, functional dependencies are also thoroughly studied for some years. There are two major approaches to define functional dependencies in XML research community, i.e. path-based approach and sub-tree/sub-graph-based approach. In path-based approach[5-11], XML datasets are represented by a tree structure, and some paths of the tree with their values are used in defining XML functional dependencies. In Sub-tree/Sub-graph-based approach[12,13], functional dependencies of XML datasets are defined by sub-graph or sub-tree in XML datasets. A sub-graph or a sub-tree is a set of paths of XML datasets. As an improvement over Sub-tree/Sub-graph-based approach functional dependencies, Refs.[14,15] deal with XML functional dependencies with some constraint condition such that there exists a sub-tree is equal. The above XML functional dependencies can not deal with uncertainty in XML datasets, which is the research topic of the paper. Ref.[16] proposes an approach to discover a set of minimal XML Conditional Functional Dependencies (XCFDs) from a given XML instance to improve data consistency. The XCFDs extends XML Functional Dependencies (XFDs) by incorporating conditions into XFD specifications. It is easy to see that all the functional dependencies defined above can not deal with uncertainty in XML datasets.

(3) **For uncertain relational databases**, Ref.[17] proposes the probabilistic functional dependency for probabilistic relational databases which associated with a likelihood of the traditional functional dependency is satisfied. Ref.[18] proposes some kinds of functional dependencies for probabilistic relational databases, such as Probabilistic Approximate Functional Dependencies (pAFD), Conditional Probabilistic Functional Dependencies (CpFD), and Conditional Probabilistic Approximate Functional Dependencies (CpAFD), which combine approximate, conditional, and approximate/conditional characteristics into traditional functional dependencies to defined corresponding functional dependencies for probabilistic relational databases. Ref.[19] proposes horizontal functional dependencies and vertical functional dependencies for uncertain relational databases, which extends the traditional relational functional dependencies into the uncertain relational databases. Although these work of uncertain relational databases are meaningful and sinificant, they can not directly applied in uncertain XML datasets, as XML are more complicated in structure then relational databases.

Contributions. In this paper, we will study the functional dependencies of uncertain XML datasets based on tree tuple model. As a natural extension, a sound and complete set of inference rules is also studied in the paper. Two applications of the proposed functional dependencies and its inference rules are also given in the paper. The main contributions of the paper are detailed as followings:

(1) In this paper, we give three types of functional dependencies of uncertain XML dataset based on XML tree tuple model: Full Functional Dependencies (FFDs), Tuple-level Functional Dependencies (TFDs), and In-tuple-level Functional Dependencies (IFDs). We analyze the relationships among these tree types of functional dependencies.

(2) A sound and complete inference rules are given for the three types of uncertain XML functional dependencies.

(3) Two applications of the sound and compete inference rules are given. The first application is to find whether a given uncertain XML dataset satisfies a given uncertain XML functional dependency (FFD, TFD, or IFD). The second application is to find a closed set of a given uncertain XML dataset.

Organizations. The rest of the paper is organized as following: Three types of functional dependencies, including FFDs, TFDs, and IFDs, of uncertain XML datasets are given in Section 2. Then a sound and complete inference rules and two applications are given in Section 3. Finally Section 4 concludes the paper and points out the future directions of the paper.

2   Functional Dependencies of Uncertain XML Dataset

We first give some preliminary definitions such as DTD (Document Type Definition), XML tree, tree tuple, etc:

Definition 1 (DTD). A DTD[20] is defined to be $D=(E, A, P, R, r)$, where (1) E is a finite set of element types; (2) A is a finite set of attributes; (3) P is a mapping from E to element type definitions. For each $\tau \in E$, $P(\tau)$ is a regular expression α defined as $\alpha ::= S \mid \varepsilon \mid \tau' \mid \alpha \mid \alpha, \alpha \mid \alpha^*$, where S denotes string types, ε is the empty sequence, $\tau' \in E$, "|", " , " and " $*$ " denote nion, concatenation and Kleene closure respectively; (4) R is a mapping from E to the power set of A: $P(A)$; (5) $r \in E$ is called the element type of the root.

A path p in $D=(E, A, P, R, r)$ is defined to be $p = \omega_1 \cdots \omega_n$, where (1) $\omega_1 \in r$; (2) $\omega_i \in P(\omega_{i-1})$, $i \in [2, n-1]$; (3) $\omega_n \in P(\omega_{n-1})$ if $\omega_n \in E$ and $P(\omega_n) \neq \Phi$, or $\omega_n = S$ if $\omega_n \in E$, and $P(\omega_n) = \Phi$, or $\omega_n \in R(\omega_{n-1})$ if $\omega_n \in A$. Let $paths(D) = \{p \mid p \in D\}$, and $last(p) = \omega_n$ denotes the last symbol of path p.

Definition 2 (XML Tree). Let $D=(E, A, P, R, r)$. An XML tree T conforming to D (denoted by $T \models D$) is defined to be $T=(V, lab, ele, att, val, root)$, where (1) V is a finite set of nodes; (2) lab is a mapping from V to $E \cup A$; (3) ele is a partial function from V to V^* such that for any $v \in V$, $ele(v) = [v_1, \ldots, v_n]$ if $lab(v_1) \cdot \cdots \cdot lab(v_n)$ is defined in $P(lab(v))$; (4) att is a partial function from V to A such that for any $v \in V$, $att(v) = R(lab(v))$ if $lab(v) \in E$ and $R(lab(v))$ is defined in D; (5) val is a partial function from V to S such that for any $v \in V$, $val(v)$ is defined if $P(lab(v)) = S$ or $lab(v) \in A$; (6) $lab(root) = r$ is called the root of T.

Given a DTD D and an XML tree $T\models D$, a path p in T is defined to be $p = v_1 \cdot \cdots \cdot v_n$, where (1) $v_1 \in root$; (2) $v_i \in ele(v_{i-1}), i \in [2, n-1]$; (3) $v_n \in ele(v_{n-1})$ if $lab(v_n) \in E$, or $v_n \in att(v_{n-1})$ if $lab(v_n) \in A$, or $v_n = S$ if $P(lab(v_{n-1})) = S$. Let $paths(T) = \{p \mid p \in T\}$.

An uncertain XML tree $T_{uncertain}$ is an XML tree T with some distributional node type[21], such as IND and MUX. A node v of type IND specifies for each child w, the probability of choosing w. This probability is independent of the other choices of children. If the type of v is MUX, then choices of different children are mutually exclusive. That is, v chooses at most one of its children, and it specifies the probability of choosing each child (so the sum of these probabilities is at most 1. We use IND_SET and MUX-SET to denote the set of IND types and MUX types, respectively. So, an uncertain XML tree $T_{uncertain}$ is defined as $T_{uncertain} = (V, lab, ele, att, val, root, IND\_SET, MUX\_SET)$. An uncertain XML tree conforming to a DTD D is similar to the Definition 2 just omitting the distributional node types (IND and MUX) in the definition. In this paper, we only consider the IND type for clarity, so $T_{uncertain} = (V, lab, ele, att, val, root, IND\_SET)$. For MUX type, the definitions and

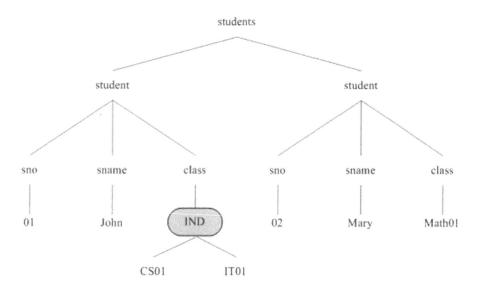

Fig. 1. An uncertain XML tree T_1

methods are similar to those of the paper. Fig.1 is an uncertain XML tree, in which the *class* can be "CS01" or "IT01" for *student* "John" independently. We omit the probability of each value in the paper for clarity.

Definition 3 (Tree Tuple). Given a DTD $D=(E, A, P, R, r)$ and an uncertain XML tree $T_{uncertain}=(V, lab, ele, att, val, root, IND\_SET)$ conforming to D, a tree tuple t for a node v in $T_{uncertain}$ is a tree rooted on node v with all of its decedent nodes.

In Fig.1, there are two student tree tuples: the first student tree tuple is the tree rooted on the first student node (its *sno* is "01"), and the second student tree tuple is the tree rooted on the second student node (its *sno* is "02").

Now, we give tree types of uncertain XML functional dependencies:

Definition 4 (Full FD). Given a DTD D, a Full Function Dependency (FFD) over D has the form $S_1 \rightarrow S_2$, where $S_1, S_2 \subseteq paths(D)$. For an uncertain XML tree $T \models D$, if T satisfies FFD $S_1 \rightarrow S_2$ (denoted by $T \models S_1 \rightarrow S_2$), then it implies that for any two tuples t_1 and t_2 in T, if $t_1(S_1) = t_2(S_1)$ then $t_1(S_2) = t_2(S_2)$.

For example, there exists a FFD in Fig.1:

FD$_1$: students.student.sno$\rightarrow$students.student.sname, which implies that for any two student tree tuples, the student number (*sno*) can uniquely determine the student name (*sname*).

Definition 5 (Tuple-level FD). Given a DTD D and an uncertain XML tree $T \models D$, a Tuple-level Function Dependency (TFD) has the form $\{S_h, [S_1 \rightarrow S_2]\}$, where $S_h, S_h.S_1, S_h.S_2 \subseteq paths(D)$. If T satisfies TFD $\{S_h, [S_1 \rightarrow S_2]\}$ (denoted by $T \models \{S_h, [S_1 \rightarrow S_2]\}$), then it implies that for each tree tuple rooted on node $last(S_h)$, $t(S_h.S_1)$ can uniquely determines $t(S_h.S_2)$.

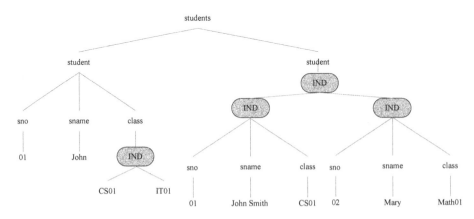

Fig. 2. An uncertain XML tree T_2

For example, there exists a TFD in Fig.2:

FD$_2$: {students.student, [sno→sname]}, which implies that for each student tree tuple (the first tree tuple is the tree rooted on the first student node, and the second tree tuple is the tree rooted on the second student node) alone, the student number (*sno*) can uniquely determine the student name (*sname*).

Definition 6 (In-tuple-level FD). Given a DTD D and an uncertain XML tree $T \models D$, an In-tuple-level Function Dependency (IFD) has the form $\{S_h.ID = n, [S_1 \rightarrow S_2]\}$, where $S_h, S_h.S_1, S_h.S_2 \subseteq paths(D)$. If T satisfies IFD $\{S_h.ID = n, [S_1 \rightarrow S_2]\}$ (denoted by $T \models \{S_h.ID = n, [S_1 \rightarrow S_2]\}$), then it implies that for the tree tuple rooted on node $last(S_h)$ with $ID = n$, $t(S_h.S_1)$ can uniquely determines $t(S_h.S_2)$.

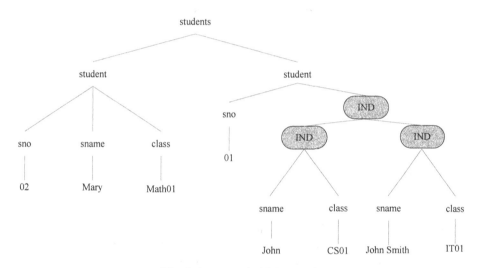

Fig. 3. An uncertain XML tree T$_3$

In the above definition, we assign each tree tuple with an ID number n to distinguish from each other. For example, in Fig.3, the first student's ID is 1, and the second student's ID is 2. There exists an IFD such that

FD$_3$: {Students.student.ID=1, [sno→sname]}, which implies that for the first student tuple whose ID is 1, the student number (*sno*=02) can uniquely determine the student name (*sname*=Mary).

By Definitions 4-6, the relationships among FFDs, TFDs, and IFDs are given as the following theorem:

Theorem 1. (1) Given a DTD D and an uncertain XML tree $T \models D$, if T satisfies FFDs $S_1 \rightarrow S_2$, then T also satisfies the corresponding TFD $\{S_h, [S_{10} \rightarrow S_{20}]\}$ and IFD $\{S_h.ID = n, [S_{10} \rightarrow S_{20}]\}$, where $S_h.S_{10} = S_1$,

$S_h.S_{20} = S_2$, and $S_h.ID = 1,2,\cdots,n$ (n is the number of the tree tupels rooted on $last(S_h)$). (2) Given a DTD D and an uncertain XML tree $T \models D$, if T satisfies TFDs $\{S_h,[S_1 \rightarrow S_2]\}$, then T also satisfies the corresponding IFDs $\{S_h.ID = n,[S_1 \rightarrow S_2]\}$, where $S_h.ID = 1,2,\cdots,n$ (n is the number of the tree tupels rooted on $last(S_h)$).

Proof. From Definitions 4 and 5, we can see that TFDs are satisfied by every tree tuples separately and FFDs are satisfied by all tree tuples together, so TFDs are special cases for FFDs, so FFDs imply the corresponding TFDs. From Definition 6, we can see that IFDs are satisfied by some specific tree tuples, so IFDs are also special cases for FFDs and TFDs, so FFDs and TFDs imply the corresponding IFDs.

3   A Set of Sound and Complete Inference Rules of Uncertain XML FDs and Its Applications

In this section, we will give a set of sound and complete inference rules of uncertain XML FDs. The definitions of soundness and completeness of a set of inference rules of uncertain XML FDs are similar to those in traditional relational databases.

Theorem 2. The following set of inference rules for uncertain XML dataset are sound and complete with respect to FFDs:

(1) Reflexivity: If $S_2 \subseteq S_1$, then $S_1 \rightarrow S_2$.

(2) Transitivity: If $S_1 \rightarrow S_2$ and $S_2 \rightarrow S_3$, then $S_1 \rightarrow S_3$.

(3) Augmentation: If $S_1 \rightarrow S_2$ and $S_3 \supseteq S_4$, then $S_1 \bigcup S_3 \rightarrow S_2 \bigcup S_4$.

Proof. The soundness of the set of inference rules of FFD is easy to get by FFD definition. We explain the completeness as following: for an uncertain XML dataset D, we can construct a new certain XML dataset D_1 by extracting all the alternative tuples of D. It is easy to see that a FFD $S_1 \rightarrow S_2$ holds in D if and only if it also holds in D_1. Hence if any FFD can be derived by a set of inference rules, the corresponding FFD can be also derived by the corresponding set of inference rules of Theorem 2.

For TFDs and IFDs, it is easy to see that they also have the same set of sound and complete inference rules according to Theorem 1. So we do not give such a set of sound and complete inference rules here for TFDs and IFDs.

Applications. The above set of inference rules is very useful in some applications:

(1) One useful application is **to test whether an uncertain XML dataset D satisfies an FD f**. Specifically in our paper, for an FFD $S_1 \rightarrow S_2$, we test whether all the tuples w.r.t. S_1 and S_2 satisfy $S_1 \rightarrow S_2$ according to Definition 4; for a TFD $\{S_h,[S_1 \rightarrow S_2]\}$, we test whether each tuple w.r.t. S_h, S_1, and S_2 satisfies

$\{S_h,[S_1 \rightarrow S_2]\}$ separately according to Definition 5; for an IFD $\{S_h.ID = n,[S_1 \rightarrow S_2]\}$, we test whether the tree tuple rooted on node $last(S_h)$ with $ID = n$ satisfies $\{S_h.ID = n,[S_1 \rightarrow S_2]\}$ according to Definition 6.

Suppose the number of tree tuples to be tested is n. The above test procedure can be solved by $O(n)$ as the number of tree tuples to be tested is no more than n.

(2) Another useful application is **to find a closed set for a given uncertain XML dataset D**. Similar to the definition of relational databases, a closed set S_C is a set of paths in D, i.e. $S_C \subseteq paths(D)$, if for any path $S \notin S_C$, D does not satisfies any FFD $S_C \rightarrow S$, TFD $\{S_h,[S_1 \rightarrow S_2]\}$, or IFD $\{S_h.ID = n,[S_1 \rightarrow S_2]\}$, where $S_h.S_1 \in S_C$ and $S_h.S_2 = S$. Intuitively, no path outside of S_C can be functionally determined (including FFDs, TFDs, and IFDs) by S_C. To obtain the closed set S_C for a given uncertain XML dataset D, we give the following method:

To find a closed set for a given uncertain XML dataset D:

(i) Initialize a base set $S_B = \Phi$. For every pair of tree tuples t_1 and t_2 in D, find the set of all paths S such that $t_1(S) = t_2(S)$, and let $S_B = S_B \bigcup S = \{s_1, s_2, \ldots, s_n\}$.

(ii) Initialize the closed set $S_C = \{s_1\}$. For $\forall s_j \in S_C$, let $S_C = S_C \bigcup \{s_j \cap b_i\}$, where $i = 2,3,\cdots,n$.

The intuitive meaning of the above method is following: (i) all pairs of tree tuples are compared on the common path set S. If they are equal on S, then S is included in the base set S_B for a closed set S_C; (ii) the intersections of all the paths in base set S_B are used to generate the final closed set S_C.

Suppose the number of tree tuples to be compared is n. In the worst case, each tree tuple should be compared with the other $(n-1)$ tree tuples. So in the worst case, we must compare $n(n-1)$ times in step (i). As step (ii) can always be finished in constant time, the complexity of the above method can be solved in $O(n^2)$.

4   Conclusions and Future Work

This paper studies the functional dependencies of uncertain XML datasets, which extends the notions of functional dependencies of uncertain relational databases and traditional functional dependencies of XML datasets based on tree-tuple model. Three kinds of functional dependencies such as FFDs, TFDs, and IFDs are given to capture three kinds of data dependencies of uncertain XML data sets. A set of sound and complete inference rules for uncertain XML datasets are also given. As two applications, we also study the problem such as how to test whether an uncertain

XML dataset satisfies a given XML functional dependencies and how to generate a closed set for a given uncertain XML dataset.

An interesting work in the future is to study functional dependencies of uncertain XML datasets with probability values. In this paper, we do not consider the probability values of uncertain XML datasets. When probability values are added in the distributional nodes of uncertain XML datasets, functional dependencies may be different from that of the paper. Another interesting work in the future is to study the lossless decomposition of uncertain XML datasets. Functional dependencies provide a useful guide way to decompose an uncertain XML dataset losslessly, so it is a natural and interesting extension of uncertain XML functional dependencies proposed in the paper.

Acknowledgments. The work is supported by Natural Science Foundation of Anhui Province (No.1208085MF110) and Science Research Plan Project of Anhui Agricultural University (No.YJ201012).

References

1. Abiteboul, S., Hull, R., Vianu, V.: Foundations of Databases. Addison-Wesley, Boston (1995)
2. Ullman, J.D.: Principles of Database and Knowledge-Base Systems, vol. 1. Computer Science Press, New York (1988)
3. Koudas, N., Saha, A., Srivastava, D., Venkatasubramanian, S.: Metric functional dependencies. In: Proc. of the 25th International Conference on Data Engineering, pp. 1275–1278. IEEE Computer Society Press, New York (2009)
4. Bohanno, P., Fan, W., Geerts, F., Jia, X., Kementsietsidis, A.: Conditional functional dependencies for data cleaning. In: Proc. of the 23rd International Conference on Data Engineering, pp. 746–755. IEEE Computer Society Press, New York (2007)
5. Janosi-Rancz, K.T., Varga, V., Nagy, T.: Detecting XML Functional Dependencies through Formal Concept Analysis. In: Catania, B., Ivanović, M., Thalheim, B. (eds.) ADBIS 2010. LNCS, vol. 6295, pp. 595–598. Springer, Heidelberg (2010)
6. Li Lee, M., Ling, T.-W., Low, W.L.: Designing Functional Dependencies for XML. In: Jensen, C.S., Jeffery, K., Pokorný, J., Šaltenis, S., Bertino, E., Böhm, K., Jarke, M. (eds.) EDBT 2002. LNCS, vol. 2287, pp. 124–141. Springer, Heidelberg (2002)
7. Liu, J., Vincent, M.W., Liu, C.: Functional Dependencies, from Relational to XML. In: Broy, M., Zamulin, A.V. (eds.) PSI 2003. LNCS, vol. 2890, pp. 531–538. Springer, Heidelberg (2004)
8. Liu, J., Vincent, M., Liu, C.: Local XML functional dependencies. In: Proc. of 5th ACM CIKM International Workshop on Web Information and Data Management, pp. 23–28. ACM, New York (2003)
9. Vincent, M.W., Liu, J.: Functional Dependencies for XML. In: Zhou, X., Zhang, Y., Orlowska, M.E. (eds.) APWeb 2003. LNCS, vol. 2642, pp. 22–34. Springer, Heidelberg (2003)
10. Vincent, M., Liu, J., Liu, C.: Strong functional dependencies and their application to normal forms in XML. TODS 29, 445–462 (2004)
11. Yan, P., Lv, T.: Functional dependencies in XML documents. In: Proc. of the 8th Asia Pacific Web Conference Workshop, pp. 29–37. Springer, Heidelberg (2006)

12. Hartmann, S., Link, S.: More Functional Dependencies for XML. In: Kalinichenko, L.A., Manthey, R., Thalheim, B., Wloka, U. (eds.) ADBIS 2003. LNCS, vol. 2798, pp. 355–369. Springer, Heidelberg (2003)
13. Hartmann, S., Link, S., Trinh, T.: Solving the Implication Problem for XML Functional Dependencies with Properties. In: Dawar, A., de Queiroz, R. (eds.) WoLLIC 2010. LNCS, vol. 6188, pp. 161–175. Springer, Heidelberg (2010)
14. Lv, T., Yan, P.: Removing XML data redundancies by constraint-tree-based functional dependencies. In: Proc. of ISECS International Colloquium on Computing, Communication, Control, and Management, pp. 595–599. IEEE Computer Society, Washington, DC (2008)
15. Lv, T., Yan, P.: XML normal forms based on constraint-tree-based functional dependencies. In: Proc. of Joint 9th Asia-Pacific Web Conference and 8th International Conference on Web-Age Information Management Workshops, pp. 348–357. Springer, Heidelberg (2007)
16. Vo, L.T.H., Cao, J., Rahayu, W.: Discovering conditional functional dependencies in XML data. In: Proc. of the 22nd Australasian Database Conference, pp. 143–152. Australian Computer Society, Sydney (2010)
17. Wang, D.Z., Dong, L., Sarma, A.D., Franklin, M.J., Halevy, A.Y.: Functional dependency generation and applications in pay-as-you-go data integration systems. In: Proc. of the 12th International Workshop on the Web and Databases,
http://www.cs.berkeley.edu/~daisyw/webdb09.pdf
18. De, S., Kambhampati, S.: Defining and mining functional dependencies in probabilistic databases, http://arxiv.org/pdf/1005.4714v2
19. Sarma, A.D., Ullman, J., Widom, J.: Schema design for uncertain databases. In: Proc. of Alberto Mendelzon Workshop on Foundations of Data Management,
http://ilpubs.stanford.edu:8090/820/
20. Fan, W., Libkin, L.: On XML integrity constraints in the presence of DTDs. JACM 49, 368–406 (2002)
21. Kimelfeld, B., Sagiv, Y.: Modelling and querying probabilistic XML data. ACM SIGMOD Record 37, 69–77 (2008)

XML Document Classification Using Closed Frequent Subtree

Songlin Wang, Yihong Hong, and Jianwu Yang[*]

Institute of Computer Sci. & Tech., Peking University,
Beijing 100871, China
{wang_sl05,hongyihong,yangjw}@pku.edu.cn

Abstract. An efficient classification approach for XML documents is introduced in this paper, which lies in combining the content with the structure of XML documents to compute the similarity between the categories and documents. It is based on the Support Vector Machine (SVM) algorithm and the Structured Link Vector Model (SLVM) which used closed frequent subtrees as the structural units. The document tree pruning strategy was applied to improve the classification system while the link information between the documents was considered to get better classification results. We did experiments on the INEX XML mining data sets combining these techniques, and the results showed that our approach performs better than any other competitor's approach on XML classification.

Keywords: XML Document, Text Classification, Frequent Subtree.

1 Introduction

XML is the W3C recommended markup language for semi-structured data. Its structural flexibility makes it an attractive choice for representing data in application domains. With the rapid growth of XML documents, this arises many issues concerning the effective management of XML documents.

How to capture both content and structural information from XML documents is one of the key research areas of the XML document mining. As the XML documents can be represented as a tree, some researchers used edit distance between the document trees to measure the similarity between the documents. Besides, frequent items (path/tree/tags) have been applied in some classification methods. Sometimes frequent items cannot convey the structural information exactly especially when there are a large number of nodes in the XML documents. In this paper, we mine the closed frequent subtrees as structural units, which can not only maintain the structural and content characteristics shared by the XML documents with the same class effectively, but also be more efficient than the frequent subtrees in the data mining.

In this paper, the Structured Link Vector Model(SLVM)[1] extended from the conventional vector space model (VSM) is introduced to represent the XML

---

[*] Corresponding author.

Z. Bao et al. (Eds.): WAIM 2012 Workshops, LNCS 7419, pp. 350–359, 2012.
© Springer-Verlag Berlin Heidelberg 2012

documents, and the closed frequent subtrees was used as structural units. Moreover, the links between the documents has been considered, which can be essential for improving the performance of the categorization. We have used the collection of INEX (the Initiative for the Evaluation of XML Retrieval) for all of the experiments and our method performs well for XML document classification.

This paper is organized as follows. Some related work will be described in section 2. In section 3, we show the architecture of our classification method using closed frequent subtree, and also discuss the SLVM for document representation. Frequent subtree will be introduced in section 4. In section 5, we show some proposed approaches for the classification. In section 6, we show the results of our experiment, and the paper ends with a discussion of future research and conclusion in section 7.

2   Related Work

Although there are various approaches to proceed the XML document classification, they are mainly divided into two streams. One is structure-based which only considered the structural information of XML documents[2, 3] and the other one considered both the content and structural information of XML documents[4, 5].

A classifier called XRules[2] is developed, which is structure oriented and aims at discovering a set of structural rules to define the individual classes. Basically, these rules that reflect the regular structural patterns of each class are learned during the training phase. J. De Knijf etc. described a classification method for XML data based on frequent attribute trees[3]. They used frequent patterns as binary features in a decision tree algorithm, and showed that the combination of emerging attribute tree patterns and standard classification methods is promising to tackle the classification of XML documents.

A classification approach combined the structure information with the content information is proposed to classify XML documents, which relies on a generative Bayesian classifier[4]. Here, the generative model assumes that there are two types of belief, structural and textual, which will be combined to get the single evidence about a document's assignment to classes. Costa G. and Ortale R. used a particular rule-learning technique for the induction of interpretable classification models, called XCCS. They separated the individual classes of XML documents by looking at the presence within the XML documents themselves of certain features, which provide information on their content and structure[5]. A novel complete framework for XML document classification was presented in [6]. A XML document is represented as a higher order logic term where both its content and structure are captured. A decision-tree learning algorithm was later driven by precision/recall breakeven point (PRDT) for the XML classification problem.

In this paper, we present a classification approach based on the Structured Link Vector Model (SLVM) used closed frequent subtree as structural units.

3   Using Closed Frequent Subtrees for Classification

3.1   System Overview

The architecture of our classification system is presented in Fig.1. The structure of XML document (document tree) can be easily gotten from the dataset. As some XML documents are complex, some pruning strategies have been employed. After some preprocessing, we used closed frequent subtree mining algorithms to find the closed frequent subtrees.

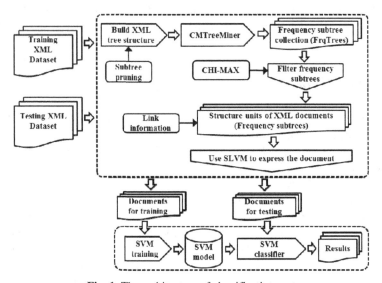

Fig. 1. The architecture of classification system

We applied the Structured Link Vector Model (SLVM) used closed frequent subtrees as structural units for the XML documents classification, which use the SVM algorithm in SVMTorch[7]. The content information and the structural information can be used as the features of the classification. In addition, the links between the documents were considered because links can provide much richer information to classifiers for classification.

3.2   Basic Representation Using Structured Link Vector Model

The Structured Link Vector Model (SLVM), which forms the basis of this paper, was originally proposed in[1] to represent XML documents. It was extended from the conventional vector space model (VSM) by incorporating document structures (represented as term-by-element matrices), referencing links (extracted based on IDREF attributes), as well as element similarity (represented as an element similarity matrix).

Vector Space Model (VSM) has been widely used to represent unstructured documents as document feature vectors which contain term occurrence statistics. This bag of terms approach assumes that the term occurrences are independent of each other.

Definition 3.1. Assume that there are n distinct terms in a given set of documents D. Let doc_x denote the x-th document and d_x denote the document feature vector represented by formula (1) and (2).

$$d_x = \left[d_{x(1)}, d_{x(2)} \cdots \cdots, d_{x(n)}\right]^T,$$ (1)

$$d_{x(i)} = TF(w_i, doc_x) IDF(w_i).$$ (2)

Where $TF(w_i, doc_x)$ is the frequency of the term w_i in doc_x ; $IDF(w_i) = log(|D|/DF(w_i))$ is the inverse document frequency of w_i to discount the importance of the frequently appearing terms; $|D|$ is the total number of the documents; $DF(w_i)$ is the number of documents containing the term w_i.

Applying VSM directly to represent XML documents is not desirable as the document syntactic structure tagged by their XML elements will be ignored.

The Structured Link Vector Model (SLVM) can be considered as an extended version of vector space model to represent the XML documents. Intuitively, SLVM represents an XML document as an array of VSMs. Each row of the array was specific to an XML element.

Definition 3.2. SLVM represents an XML document doc_x using a document feature matrix $\Delta_x \in R^{n \times m}$, given as formula (3).

$$\Delta_x = \left[\Delta_{x(1)}, \Delta_{x(2)} \cdots \cdots, \Delta_{x(m)}\right],$$ (3)

$$\Delta_{x(i)} = \left[\Delta_{x(i,1)}, \Delta_{x(i,2)} \cdots \cdots, \Delta_{x(i,n)}\right].$$ (4)

Where m is the number of distinct XML elements; $\Delta_{x(i)} \in R^n$ is the TF/IDF feature vector representing the i-th XML element e_i, given as $\Delta_{x(i,j)} = TF\left(w_j, doc_x \cdot e_i\right) * IDF(w_j)$ for all $j=1$ to n; $TF\left(w_j, doc_x \cdot e_i\right)$ is the frequency of the term w_j in the element e_i of doc_x.

The elements of documents are used as structural unit in [1]. Here closed frequent subtrees are used as structural unit.

Definition 3.3. The normalized document feature matrix is defined as formula (5).

$$\tilde{\Delta}_{x(i,j)} = \Delta_{x(i,j)} / \sum_k \Delta_{x(i,k)}.$$ (5)

Where the factor caused by the varying size of element content is discounted via normalization.

Similarity Measures. Similarity between two documents doc_x and doc_y is typically computed as the cosine value between their corresponding document feature vectors, given as formula (6).

$$sim\left(doc_x, doc_y\right) = \frac{d_x d_y}{||d_x|| ||d_y||} = \tilde{d}_x \tilde{d}_y^T = \sum_i^n \tilde{d}_{x(i)} \tilde{d}_{y(i)}.$$ (6)

Where n is the total number of terms and $\tilde{d}_x = d_x / ||d_x||$ denotes normalized d_x. So, the similarity measure can also be interpreted as the inner product of the normalized document feature vectors.

4   Frequent Subtree

We used the frequent tree to describe the structural information as it is an effective way to describe the structural information and widely adopted. In this section, we will introduce frequent subtree and use it as the structural units in the SLVM to proceed the XML documents classification.

4.1   Frequent Subtree

An XML document can be abstracted as a rooted ordered tree. The top element is the root of the document tree. The sub-elements of the elements are the children of the corresponding node in the document tree. And the order of siblings in the tree depends on the order of XML elements in XML document.

Given a tree dataset $TD = \{T_1, T_2, T_3, \dots, T_n\}$ with n number of trees, there exists a subtree T', $Support(T')$ is defined as the number of the trees in D where T' is a subtree. A subtree T' is frequent if its support is not less than a user-defined minimum support threshold. In other words, T' is a frequent subtree of the trees in TD such that $Support(T)/|TD| \geq \min\_supp$, where $\min\_supp$ is the support threshold and $|TD|$ is the number of trees in the tree dataset TD.

We define a frequent tree T to be maximal if none of the proper supertrees of T is frequent; and a frequent tree T is closed if none of the proper supertrees of T has the same support that T has.

As a part of Mining Closed Frequent Item sets, a lot of research work has been done in frequent subtree mining. Some significant results have been achieved, including FREQT, HybridTreeMiner, Treeminer, PCITMiner and so on.

In this paper, we utilize CMTreeminer[8] to mine closed frequent substrees from XML document collection. And the closed frequent substrees which act as structural units are utilized to extract the content information from the XML documents.

4.2   Classification Using Closed Frequent Subtree

We represent the document according to the structured link vector model (SLVM), and use the closed frequent subtrees as the structural units. Finally, each document is represented by a document feature matrix. SVM classier is used for classification. In training phase we finds the closed frequent subtrees that are most closely related to the category variable, then determine the category of the document based on the SVM classification model and the closed frequent subtrees gotten in the training phase.

5   Proposed Approaches

5.1   Tree Pruning

In the previous section, all frequent subtrees are discovered by traversing the enumeration tree in a graph database, but it is not necessary to grow the complete enumeration tree because some elements of the enumeration tree have no useful information to the class variable under certain conditions. Therefore the enumeration tree can be pruned efficiently. In this section, we introduce algorithms that prune unwanted branches of the search space.

An algorithm which aims at estimating the weights of the elements for a given class has been introduced in[9]. This weight is integrated afterwards in the influence function of the terms, as presented in Table 1.

Table 1. Position of the element e_i for each category

	C	$\bar{c}$	$C = c \cup \bar{c}$
e_i	r_i	$nr_i = n_i - r_i$	n_i
$\bar{e_i}$	$c - r_i$	$N - n_i - C + r_i$	$N - n_i$
E	c	$\bar{c} = N - c$	N

Where E is the set of elements in the dataset; C is the set of categories of the giving dataset, and c is one category in the set C; $\bar{c}$ contains all other categories except c; r_i is the number of times that element e_i appears in the documents belong to category c; nr_i is the number of times element that e_i appears in the documents that not belong to category c.

The weight $w_{e_i}(c)$ of an element e_i for a category c is defined by the formula (9) [9] .

$$w_{e_i}(c) = log \frac{r_i \times (\bar{c} - nr_i)}{nr_i \times (c - r_i)}. \tag{7}$$

The weight w_{e_i} of an element is the averaged using the training set, using the formula (10).

$$w_{e_i} = \frac{1}{|C|} \Sigma_{c \in C} w_{e_i}(c). \tag{8}$$

After sorting the weight of elements, we keep the top-N elements for the document.

5.2   Link Information

As the external structure of the collection, links can provide much richer information to classifiers for classification. In this section, we will show how those links add relevant information for the categorization of documents.

The link-class information of document d_i is defined by formula (11-12).

$$l(d_i) = \left[l_1(d_i), l_2(d_i) \dots \dots, l_{|c|}(d_i) \right], \tag{9}$$

$$l_k(d_i) = \frac{\sum_{j=1}^{|N_{d_i}^k|} l_{train,k}(d_j)}{\sum_{j=1}^{|N_{d_i}^k|} a_k(d_j)}, k = 1, \dots, |C|. \tag{10}$$

Where $l_{train,k}(d_i) = 1$ if document d_i belongs to class k in the training set; $l_{train,k}(d_i) = 0$ otherwise. $|C|$ is the number of categories; $a_k(d_j) = 1$ if any target documents that d_j links to belong to category k, $a_k(d_j) = 0$ otherwise. $|N_{d_i}^k|$ is the number of documents that d_i links to belong to class k.

For all XML documents in the corpus, $l(d)$ is defined as the link-class matrix $l(d) \in R^{n \times |C|}$, and n is the number of documents in the corpus. We used iterative method to optimize the weight of $l_k(d_i)$ used the formula (13).

$$l_k^t(d_i) = \frac{1-q}{n} + q \times \sum_{d_j \in M_k(d_i)} \frac{l_k^{t-1}(d_j)}{L(d_j)} \tag{11}$$

Where $M_k(d_i)$ is the set of documents belong to class k that link to d_i; $L(d_j)$ is the number of out links on document d_j; n is the total number of documents in the collection; q is a damping factor; t is the times of iteration. $l_k^0(d_i)$ is initialized using the training set.

Each document can be denoted to a link feature vector using the out-link class information. We combined link information and closed frequent subtrees described in formula (14) to improve the performance of documents classification while the result is not ideal when only used the link feature vector.

$$S(c|d) = \alpha \times S_l(c|d) + (1-\alpha) \times S_t(c|d). \tag{12}$$

Where $S(c|d)$ is the final similarity between the document d and the category c; $S_l(c|d)$ is the score between the document d and the category c get from the SVM classier using the link information only; and $S_t(c|d)$ is the score between the document d and the category c using the closed frequent subtrees information based on SLVM; parameter α sets the importance of $S_l(c|d)$.

6   Experiments and Discussion

6.1   Datasets

We used the training collection and testing collection of INEX 2008, 2009 and 2010 XML Mining Track for the experiments.

XML documents are defined by their logical structure and content. The current Wikipedia collection contains structure as document structure such as sections, titles and tables, semantic structure as entities mined by YAWN[10], and navigation structure as document to document links.

The track of INEX 2009 and 2010 are multi-label classification task that a document can belong to one or more categories. All of the categories are extracted from the Main Topic Classifications in the Wikipedia from the 1st and 2nd level.

Table 2. Datasets

	Training set	Testing set	Categories	Elements	Number of Links	Multilable classification
INEX2008	11,433	102,900	15	3763	636,187	N
INEX2009	11,028	43,861	39	5213	4,554,203	Y
INEX2010	23,976	120,649	36	10411	14,229,116	Y

6.2 Classification Results Comparison

We compared our method with the INEX of 2008, 2009 and 2010 XML mining track results. Tables 3, 4 and 5 show that the closed frequent subtree gives consistently better performance than other classifiers for the data sets of INEX.

Table 3. Compare the result with INEX2008

Team	Recall
SLVM-closedFrequentSubtree	**0.799**
SLVM- original	0.785
LaHC+	0.787
Vries+	0.785
boris+	0.738
kaptein+	0.698
romero+	0.681

+: Result Of INEX08;

Table 4. Compare the result with INEX 2009

Team	Micro F1	Macro F2	APR
SLVM-closedFrequentSubtree	0.609	0.571	**0.780**
SLVM-original	0.579	0.537	0.744
University of Wollongong+	0.512	0.479	0.680
University of Peking+	0.518	0.480	0.702
XEROX Research center+	0.600	0.571	0.678
University of Saint Etienne+	0.564	0.530	0.685
University of Granada+	0.262	0.253	0.729

+: Result Of INEX09;

Table 5. Compare the result with INEX 2010

Team	Micro F1	Macro F1
SLVM-closedFrequentSubtree	**0.591**	**0.535**
SLVM- original	0.538	0.491
the Queensland University of Technology +	0.536	0.473
University of Peking+	0.517	0.444

+: Result Of INEX10;

In Table 3, 4 and 5, we take the results of "SLVM-original" as the baseline which used XML elements as the SLVM structural units. The results of "SLVM-closedFrequentSubtree" which used the closed frequent subtrees as the structural units are better than "SLVM-original". The percentage gain 1.78%, 4.84% and 8.96% compared with "SLVM-original" on the three years dataset. In contrast with the data set of INEX2009 and INEX 2010, the document structure of collections in INEX2008 are simpler and have fewer elements on average, so fewer frequent trees could be constructed, and thus could not get enough structure information. On the other hand, the number of links in the corpus of INEX2008 is fewer too. Hence, the gain of results based on INEX2008 dataset is not as much as the ones on INEX2009 and 2010.

Compared with the best results of each year in the XML Mining Track of INEX, our method performs much better for XML document classification. In Table 6, Qut's submission used content only. They ignored the importance of the structural and link information. In their experiment, documents were represented in the bag of words vector space model, and SVM was applied to classify each document by treating each category[10]. Compared with their results, the *Micro F1* score of our method gain 10.3%. In Table 5, University of Granada got the best result in the INEX2009 Track[11]. In this track they proposed a new methodology for link-based document classification based on probabilistic classifiers and Bayesian networks, and provided a "graphical proof" which explained how the category of the linked documents tend to be similar to the category of the document itself. Even though they used the link and content information properly, the structure information of XML document cannot be used in their classification mode. Compared with their results, our results are much better. The *Micro F1* score gained 18.9% while the *APR* score increased14.7%.

In summary, classification using closed frequent subtree works better than the element-based approach and also outperforms other classifiers on the dataset of INEX. Classification on XML documents collection with richer structure information works more effective than on the document collection containing simple elements, because more frequent subtrees would be mined from the documents that have richer structure information.

7   Conclusions

In this paper, we considered both the content and structure of XML documents to classify XML documents. We mined the closed frequent subtrees for the class of attribute trees, and applied SLVM to represent the XML documents. According to the experiment results, we found the pruning strategies employed for mining closed frequent subtrees, the proper selection of the frequent subtrees and the use of the links between the documents can improve the performance of the categorization. Our experiments clearly show that our approach not only works better than the element-based approach but also outperforms other classifiers on the dataset of INEX2008, 2009 and 2010.

Acknowledgment. The work reported in this paper was supported by the National Natural Science Foundation of China Grant 60875033.

References

1. Yang, J., Chen, X.: A semi-structured document model for text mining. Journal of Computer Science and Technology 17, 603–610 (2002)
2. Zaki, M.J., Aggarwal, C.C.: XRules: an effective structural classifier for XML data. Presented at the Proceedings of the Ninth ACM SIGKDD International Conference on Knowledge Discovery and Data Mining, Washington, D.C. (2003)
3. De Knijf, J.: FAT-CAT: Frequent Attributes Tree Based Classification. In: Fuhr, N., Lalmas, M., Trotman, A. (eds.) INEX 2006. LNCS, vol. 4518, pp. 485–496. Springer, Heidelberg (2007)

4. Denoyer, L., Gallinari, P.: Bayesian network model for semi-structured document classification. Inf. Process. Manage. 40, 807–827 (2004)
5. Costa, G., Ortale, R., Ritacco, E.: Effective XML Classification Using Content and Structural Information via Rule Learning. In: 2011 23rd IEEE International Conference on Tools with Artificial Intelligence (ICTAI), pp. 102–109 (2011)
6. Wu, J.: A Framework for Learning Comprehensible Theories in XML Document Classification. IEEE Transactions on Knowledge and Data Engineering 24, 1–14 (2012)
7. Collobert, R., Bengio, S.: SVMTorch: support vector machines for large-scale regression problems. J. Mach. Learn. Res. 1, 143–160 (2001)
8. Yun, C., Yi, X., Yirong, Y., Muntz, R.R.: Mining closed and maximal frequent subtrees from databases of labeled rooted trees. IEEE Transactions on Knowledge and Data Engineering 17, 190–202 (2005)
9. Gery, M., Largeron, C., Thollard, F.: Integrating Structure in the Probabilistic Model for Information Retrieval. In: IEEE/WIC/ACM International Conference on Web Intelligence and Intelligent Agent Technology, WI-IAT 2008, pp. 763–769 (2008)
10. De Vries, C.M., Nayak, R., Kutty, S., Geva, S., Tagarelli, A.: Overview of the INEX 2010 XML Mining Track: Clustering and Classification of XML Documents. In: Geva, S., Kamps, J., Schenkel, R., Trotman, A. (eds.) INEX 2010. LNCS, vol. 6932, pp. 363–376. Springer, Heidelberg (2011)
11. de Campos, L.M., Fernández-Luna, J.M., Huete, J.F., Masegosa, A.R., Romero, A.E.: Link-Based Text Classification Using Bayesian Networks. In: Geva, S., Kamps, J., Trotman, A. (eds.) INEX 2009. LNCS, vol. 6203, pp. 397–406. Springer, Heidelberg (2010)

Author Index